D0805369

Psychology in the Service
of National Security

Psychology in the Service of National Security

EDITED BY
A. David Mangelsdorff

AMERICAN PSYCHOLOGICAL ASSOCIATION
WASHINGTON, DC

Copyright © 2006 by the American Psychological Association. All rights reserved. Except as permitted under the United States Copyright Act of 1976, no part of this publication may be reproduced or distributed in any form or by any means, including, but not limited to, the process of scanning and digitization, or stored in a database or retrieval system, without the prior written permission of the publisher.

Chapters 4 and 13 were authored or coauthored by an employee of the United States government as part of official duty and are considered to be in the public domain.

Published by
American Psychological Association
750 First Street, NE
Washington, DC 20002
www.apa.org

To order
APA Order Department
P.O. Box 92984
Washington, DC 20090-2984
Tel: (800) 374-2721; Direct: (202) 336-5510
Fax: (202) 336-5502; TDD/TTY: (202) 336-6123
Online: www.apa.org/books/
E-mail: order@apa.org

In the U.K., Europe, Africa, and the Middle East, copies may be ordered from
American Psychological Association
3 Henrietta Street
Covent Garden, London
WC2E 8LU England

Typeset in Goudy by Stephen McDougal, Mechanicsville, MD

Printer: Edwards Brothers, Ann Arbor, MI
Cover Designer: Berg Design, Albany, NY
Technical/Production Editor: Harriet Kaplan

The opinions and statements published are the responsibility of the authors, and such opinions and statements do not necessarily represent the policies of the American Psychological Association. Any views expressed in chapters 4 and 13 do not necessarily represent the view of the United States government, and the author's participation in the work is not meant to serve as an official endorsement.

Library of Congress Cataloging-in-Publication Data

Psychology in the service of national security / edited by A. David Mangelsdorff.
    p. cm.
    Includes bibliographical references and index.
    ISBN 1-59147-355-1
    1. Psychology, Military. 2. National security—United States. 3. United States—Defenses.
I. Mangelsdorff, A. David.

    U22.3.P795 2006
    355.001'9—dc22   2005030385

British Library Cataloguing-in-Publication Data
A CIP record is available from the British Library.

Printed in the United States of America
First Edition

R0408138372

CHICAGO PUBLIC LIBRARY

This work is dedicated to my parents, Maesie Rowland and Arthur F. Mangelsdorff, with sincerest gratitude for their instilling in me a love of reading, science, history, and curiosity. My legacy consists of the many students who have studied my musings and serve proudly. I wish them "fair winds and following seas"! It is also my honor to recognize and thank my colleagues for sharing their research, experience, and dedication, as well as for the numerous symposia, workshops, and discussions in which they participated, and for making this book possible.

CONTENTS

CONTRIBUTORS

L. Morgan Banks, U.S. Army Special Operations Command, Ft. Bragg, NC

Patrick H. DeLeon, U.S. Senate Staff, Washington, DC

Nicole M. Dudley, U.S. Army Research Institute for the Behavioral and Social Sciences, Arlington, VA, and Department of Psychology, George Mason University, Fairfax, VA

Paul A. Gade, U.S. Army Research Institute for the Behavioral and Social Sciences, Arlington, VA

Alan Gropman, National Defense University, Washington, DC

Jonathan D. Kaplan, U.S. Army Research Institute for the Behavioral and Social Sciences, Arlington, VA

Gerald P. Krueger, Wexford Group International, Vienna, VA

A. David Mangelsdorff, Army Medical Department Center and School, Fort Sam Houston, San Antonio, TX

Karl O. Moe, Uniformed Services University of the Health Sciences, Bethesda, MD

Loren Naidoo, Baruch College, City University of New York

Paul D. Nelson, American Psychological Association, Washington, DC

D. Stephen Nice, Naval Health Research Center, San Diego, CA

Robert S. Nichols, Rockville, MD

Peter F. Ramsberger, Human Resources Research Organization, Alexandria, VA

Morgan T. Sammons, National Naval Medical Center, Bethesda, MD

Mady Wechsler Segal, University of Maryland, College Park

Warren R. Street, Central Washington University, Ellensburg

Henry L. Taylor, Institute of Aviation, University of Illinois at Urbana–Champaign

Edmund Thomas, Naval Personnel Research and Development Center, San Diego, CA

Karl F. Van Orden, Naval Health Research Center, San Diego, CA

Martin Wiskoff, Northrop Grumman Mission Systems, Monterey, CA

Janice D. Yoder, University of Akron, Akron, OH

FOREWORD

PATRICK H. DeLEON

Over the years, it has been a genuine pleasure to witness the gradual maturation of professional psychology. Psychology has always been one of the most popular undergraduate majors at our universities, and its research and clinical findings are quickly absorbed by the popular media. Its recent growth as a profession is perhaps best reflected by the status of the American Psychological Association of Graduate Students (APAGS). Established in 1988 by the American Psychological Association (APA) Council of Representatives, APAGS today has a nonvoting seat on the APA Board of Directors and a membership of approximately 55,000 graduate students, which is comparable in number to the entire 1977 APA membership of 59,900. That was the year that President Jimmy Carter signed Executive Order No. 11,973 (President's Commission on Mental Health, 1977) and thereby established his landmark President's Commission on Mental Health. That was also the year that Missouri became the last state in the nation to license and certify practicing psychologists. The hallmark of any learned profession is the vitality of its future generations. Psychology is doing very well.

While serving as APA president as the 21st century arrived, I became increasingly aware of the extraordinary breadth of expertise that exists within our field and, equally importantly, of the extent to which colleagues within the public sector over the years truly have been on the cutting edge of addressing society's pressing priorities (DeLeon, 2002). A number of years prior to my tenure, former APA Presidents George Miller and Bill Bevan each issued inspirational calls for psychology to effectively focus its collective expertise on society's true needs (Bevan, 1980, 1982; G. A. Miller, 1969). These visionaries foresaw that the behavioral sciences (with their underlying data-based orientation) would ultimately be the key to effectively capitalizing on our nation's impressive resources and inherent strengths. Their clear directives have never been more relevant than they are today, whether one contemplates ensuring that our citizens receive high quality health care, addressing the escalating violence occurring within our schools or on our nation's highways, capitalizing on the unprecedented advances within the computer and communications fields, or developing an effective homeland security system that can be trusted and readily understood by those whom the government (and private industry) seeks to protect.

Over the past quarter century, I have been personally blessed with the opportunity to actively participate both within the APA governance and at the national public policy level. During this period, I have come to appreciate the critical importance of ensuring that the next generation of psychologists (and society's leaders at large) appreciate that the foundation for their aspirations has already been established by those who have gone before them—and, for those who desire to become personally involved in the public policy process, that they come to understand that the public sector (i.e., government) has a unique relationship to and responsibility for ensuring society's well-being. Within the health care arena, for example, under our constitutional form of government, health care providers who work within the state or federal system are not limited by the constraints that might be imposed on their private sector colleagues by local licensing boards or legislators, uninformed about the intricacies of nonphysician education. Instead, it is a fundamental responsibility of government to serve as a living laboratory for testing new and innovative models of health care delivery and the effective use of unique practitioner expertise. Lessons learned from these experiences allow educational institutions and researchers to develop new categories of health professionals and new treatment strategies based on our expanding knowledge base. As a result, the federal system has pioneered the use of nurse practitioners; nontraditional, nonphysician prescribers; telehealth care; and most recently, the use of individualized, integrated interdisciplinary medical records (Institute of Medicine, 1999). Federal health care delivery systems, such as those within the Veterans Administration and the Department of Defense, which possess the economic and staff resources that are simply not available within the private sector, have begun seriously investing in building the necessary infrastructure to systematically explore these yet uncharted clinical waters (Institute of Medicine, 2001).

Fortunately, the federal government's institutional commitment to fundamental change exists at the highest leadership level. On April 27, 2004, President Bush signed Executive Order No. 13,335 (Incentives for the Use of Health Information Technology, 2004), establishing the position of National Health Information Technology Coordinator in order to provide systematic leadership for the development and nationwide implementation of an interoperable health information technology infrastructure to improve the quality and efficiency of health care. That day the president stated,

> The way I like to kind of try to describe health care is, on the research side, we're the best. We're coming up with more innovative ways to save

lives and to treat patients. Except when you think about the provider's side, we're kind of still in the buggy era. . . . We're here to talk about how to make sure the government helps the health care industry become modern in order to enhance the quality of service, in order to reduce the cost of medicine, in order to make sure the patient, the consumer, is the center of the health care decision making process. . . . It's estimated that they spend $8,000 per worker on information technology in most industries in America, and $1,000 per worker in the health care industry. And there's a lot of talk about productivity gains in our society, and that's because companies and industries have properly used information technology. If properly used, it is an industry-changer for the good. It enables there to be a better cost structure and better quality care delivered, in this case in the health field. (Remarks in a Discussion on the Benefits of Health Care Information Technology, 2004, p. 699)

This is a conceptual theme that was to continue throughout that year's presidential debates and campaign—one consistently espoused by the candidates from both political parties.

The 21st century will, above all else, be an era of unprecedented change. Not only are the phenomenal advances occurring within the technological arena having a direct impact on almost every phase of daily life, within the educational community the various disciplines are increasingly engaging in interdisciplinary collaboration. For those individuals who are interested in having a direct impact on public policy, this evolution highlights the increasing importance of a broad-based, public health orientation in which the unique areas of expertise of fundamentally different disciplines combine to create a greater whole. The expertise of clinical psychology can and will merge with that of human factors analysis, resulting in a renewed occupational health psychology model. Nowhere is the intellectual and operational confluence of these two broad-based disciplines more evident than in the U.S. armed forces, as the range of ideas, subjects, and authors in *Psychology in the Service of National Security* readily shows.

Psychology has always been intrinsic to the mission of the military, whether in formulating selection criteria or making assessments for combat readiness. As described throughout this publication, there is a long and impressive history of psychological contributions to the underlying mission of each of the military services. The kinds of challenges facing military personnel—and their families—will undoubtedly contribute to the further seasoning of the profession of psychology, whether such challenges arise in special operations in remote and difficult terrain or in natural disasters in the United States. Building on their successes, individual colleagues have obtained well-deserved promotions to decision-making positions, resulting in psychological expertise becoming increasingly applied to a wider range of system problems. With change come exciting opportunities and challenges. The 21st century has definitely arrived. Aloha.

Psychology in the Service of National Security

INTRODUCTION

A. DAVID MANGELSDORFF

Homeland defense and national security are fundamental responsibilities of government. The meaning of *national security* has changed from threats of war and invasion by other nation states to a much broader definition that also includes community response to natural disasters, technological failures, domestic disturbances, and terrorism (domestic and transnational). Security needs of the United States shaped the evolution of this society, the roles and functions of its armed forces, the organization of national security, and the development of psychology.

Psychology's involvement in national security began during World War I with the assessment, selection, classification, and training of military personnel. During and after World War II, federal and military service laboratories were created and government programs established to fund psychological research, development, and educational programs at universities and with private contractors. The U.S. government is one of the largest employers of psychologists. Government support for national security continues with the Broad Area Announcement by the Department of Homeland Security (2004) to fund social and behavioral scientists to study terrorism.

The chapters in this volume examine how national security is organized, the national security needs psychologists have helped address, the cooperation between uniformed psychologists and behavioral scientists, lessons learned in military settings that have contributed to the maturation of the science and profession of psychology, and leadership by psychologists. It provides brief snapshots of the events and people who have addressed society's security needs, provided vision and leadership, and advanced the acceptance of psychology.

Government is one of the largest employers of psychologists; psychologists play a vital role in military readiness. The study of individual differences, personnel management, organizational behavior, personnel attitudes, and leadership had its origins in military settings. The maturation of psychology can be viewed in its interconnectedness with other disciplines and the ways it has addressed the security needs of society. The story of psychologists in the armed forces is the story of the evolution of the science and practice of psychology itself in addressing national security challenges.

This volume includes the perspectives of psychologists, social and behavioral scientists representing the uniformed services, research institutions, industry, and academia. Target audiences include readers interested in psychologists' roles in positioning the armed services on the leading edge in personnel assessment, human resource management, instruction, human factors design, racial integration, gender equality, stress management, and other fields.

Psychology includes many different fields of study that often overlap. This volume examines those related that are involved in military settings. For example, the introductory chapter (chap. 1) examines the maturation of psychology, occupational health psychology, the organization of national security, and the ways security needs shape the armed forces (e.g., personnel needs, preserving the force) and presents selected national security challenges. The book as a whole examines the contributions of psychology to promoting resilience and healthy behaviors and creating healthy workplaces in the armed services, which ultimately serves the nation's security needs. Successive sections offer chapters with perspectives on individual differences, personnel management, clinical and counseling psychology, and applied social psychology. There are overlaps among the chapters, just as there are overlaps in the needs addressed.

Acronyms are used throughout this volume. Military settings are the intersection of federal, academic, health, security, and organizational bureaucracies, each with its own unique (and overlapping) missions, operational environments, organizational culture, traditions, jargon, and acronyms. As a comparison, the bureaucracy of a much smaller entity, the American Psychological Association, includes an alphabet soup of committees and interest groups, including the Committee on Professional Practice and Standards (COPPS), the Committee for the Advancement of Professional Practice (CAPP), and the Committee on Structure and Function of Council (CSFC), to name a few. Acronyms are part of the uniqueness of bureaucratic cultures.

The overlaps in themes and presentations here reinforce the maturation of the profession of psychology within the military. Each branch of service has unique and varying missions, organizational cultures, and approaches. Historically, the services acted independently; more recent operations required joint service (with multiagency or multinational) participation, coordination, training, and execution. Conducting more missions with fewer resources requires additional planning, practice, and coordination.

The United States is in a global war on terrorism, and psychologists and policymakers are concerned with helping the public cope with threats to the homeland. The American Psychological Association's public education campaign "Resilience in a Time of War" addresses some of these themes. Concern for national security is heightened today; learning from the past is vital for educating practitioners and decision makers to deal with preparedness and threats. Psychologists and the American Psychological Association have been an integral part of the homeland defense efforts for all of the 20th century. Now everyone needs to be involved to address current security challenges. This book provides some of the historical background for what worked to meet past security challenges and suggests some goals for the future.

I

FROM MILITARY PSYCHOLOGY TO NATIONAL SECURITY PSYCHOLOGY

The security of the United States has been threatened by numerous crises, natural disasters, and enemy forces since the nation's beginning. The definition of *national security*, once understood primarily as the response to threats of war and invasion by other nation-states, has broadened to include community response to natural disasters, technological failures, domestic disturbances, and terrorism. Responsibility for the organization of national security has likewise evolved from being the sole domain of the armed forces to involving the shared efforts of multiple security organizations working together at local, state, and federal levels.

Security needs, demographic trends, and national and political interests have helped shape the evolution of society, the roles and functions of the armed forces, and the organization of national security as well as the development of the field of psychology and the roles of psychologists. The historical foundations of national security psychology are divided along four thematic lines: individual differences, personnel management, clinical and counseling psychology, and applied social psychology. Subsequent parts of this volume examine each of these themes in turn.

Service in the armed forces is demanding and stressful. The necessity of preserving the health, quality, and availability of service personnel has required the involvement of psychologists in work on realistic training, adjustment, prevention, occupational health, safety, and well-being, and the lessons learned in military settings are applicable to other security and service organizations concerned with homeland security and defense. Psychologists play a vital role in homeland defense and military readiness. The story of psychologists in the armed forces addressing national security challenges is the story of the evolution of the science and practice of psychology itself.

1

THE CHANGING FACE
OF NATIONAL SECURITY

A. DAVID MANGELSDORFF

The security of the United States has been threatened by numerous crises, natural disasters, and enemy forces since the nation's beginning in the late 1700s. Homeland defense and national security are fundamental responsibilities of government; the U.S. Constitution provides for the "common defense and general welfare" of the nation. National security has changed from threats of war and invasion by other nation-states to a much broader definition that also includes community response to natural disasters, technological failures, domestic disturbances, and domestic and transnational terrorism. Security needs of the United States shaped the evolution of its society, the roles and functions of its armed forces, the organization of national security, and the development of psychology. The maturation of psychology can be viewed in the context of its interconnectedness with other disciplines and how it has been a part of addressing of the security needs of society. Psychologists play a vital role in military readiness. The story of psychologists in the armed forces addressing national security challenges is the story of the evolution of the science and practice of psychology itself.

The views of the author are his own and do not purport to represent those of the Department of Defense, the Department of the Army, or Baylor University.

Psychology promotes the systematic study of behavior. As the field of psychology evolved, it gained scientific recognition and professional status through its research and publications. The cooperation and exchange of ideas among multidisciplinary problem-solving teams (including behavioral scientists in academia, private organizations, federal agencies, and the armed forces) permitted unique solutions to national security needs and often anticipated societal and professional trends in the United States during the 20th century. Selected principles developed by the military to meet wartime threats and national crises were often adapted for civilian or commercial uses, and vice versa. During the mid-1940s and after World War II, the federal government and associated national agencies concerned with security and defense interests employed more psychologists than at any other point in history. Leaders in psychology contributed significantly to wartime research efforts. The contributions of behavioral and social scientists as documented and published in the scientific and professional literature permitted further use of psychological concepts by the military.

ORGANIZATION OF NATIONAL SECURITY

The organizational structure of the armed services and components of national security evolved to fit foreign policy, political, and security challenges. These included reorganization of the armed services and political and legislative changes.

Armed Services

Aside from the military reorganization, the National Security Act established the National Security Council, a central place of coordination for national security policy in the executive branch, and the Central Intelligence Agency, the United States' first peacetime intelligence agency (U.S. Department of State, n.d.). The War Department was created in 1789 to defend the nation. As an organizational entity, the War Department lasted through World War II, when it became apparent that changes were needed. The National Security Act of 1947 created a major reorganization of the military establishment and foreign policy. The War Department and Navy Department merged into the Department of Defense under the Secretary of Defense and created separate branches for the Army, Navy, and Air Force. The 1947 act also established the Central Intelligence Agency. As part of the reorganization to the Department of Defense in 1947, the Army–Navy Medical Service Corps Act created corps recognizing administrative and scientific specializations (Ginn, 1978; Gray, 1997). The Total Force Policy was implemented in 1973, integrating the active, National Guard, and reserve forces into one force (Mangelsdorff, 1999b).

The Goldwater–Nichols Department of Defense Reorganization Act of 1986 (P.L. 99-433) requires an annual report to Congress detailing the U.S. national security strategy. The current national military strategy (NMS) is guided by the goals and objectives contained in the President's *National Security Strategy* (2002) for winning the war on terrorism. The NMS described "the ways and means to protect the United States, prevent conflict and surprise attack, and prevail against adversaries who threaten our homeland, deployed forces, allies and friends" (R. B. Myers, 2004, p. iii). The NMS priorities are (a) to win the war on terrorism, (b) to enhance the military branches' ability to fight as a joint force, and (c) to transform the armed forces. The transformation required technological, intellectual, and cultural adjustments. Military assistance to civilian authorities (as in a disaster) is a joint mission conducted in an interagency environment; the Department of Defense is authorized to assist by the Robert T. Stafford Disaster Relief and Emergency Assistance Act (P.L. 93-288; 1974) as amended by P.L. 106-390 in 2000.

Federal Emergency Management Agency

Hurricanes, storms, earthquakes, and fires are among the expected destructive forces assaulting the United States (National Oceanic and Atmospheric Administration, 2004). Hurricane Andrew (August 1992) affected 250,000 people, killed 15, and cost $26.5 billion in damages. Hurricane Floyd (September 1999) displaced more than 4 million people, killed 74, and inflicted damage totaling about $4.5 billion. In 2005, hurricanes Katrina, Rita, and Wilma inflicted significant damage as well. An organized program of disaster relief in the United States was incorporated into federal statutes in 1900; the American National Red Cross received congressional approval to provide disaster relief and planning. Until the 1970s, the federal approach to massive disasters was piecemeal. The first general Disaster Relief Act (P.L. 81-875; 1950) gave the President broad disaster assistance powers. In 1974, the Robert T. Stafford Disaster Relief and Emergency Assistance Act (P.L. 93-288) established the process of predisaster prevention and preparedness plans. In 1979, the Federal Emergency Management Agency (FEMA) was created, consolidating a variety of disaster and civil defense responsibilities. The Loma Prieta earthquake in 1989 and Hurricane Andrew in 1992 focused national attention on FEMA. In 2003, FEMA became part of the Department of Homeland Security (DHS).

In federally declared disasters, the Department of Defense can be mobilized to provide assistance. Special Forces medical personnel and the 44th Medical Brigade's Joint Task Force Andrew provided disaster relief and assistance in support of local civilian authorities in 1992 (Godbee & Odom, 1997; Holsenbeck, 1994). Terrorist attacks on New York and the Pentagon on September 11, 2001, required new considerations for handling casualties and responders (Castellano, 2003; Hammond & Brooks, 2001; Kendra & Wachtendorf, 2003; Simon & Teperman, 2001).

In the 1970s, ineffective multiagency responses to fighting fires in Southern California resulted in development of the Incident Command System (ICS) by the National Fire Academy; it is a good model for command, control, and coordination of resources and personnel (FEMA, 2004). Use of ICS is required for handling hazardous material incidents.

Department of Homeland Security

The Department of Homeland Security was formed under the Homeland Security Act of 2002. The DHS mission is to make the United States more secure by providing border and transportation security, protecting critical infrastructures (communications systems, power grids, transportation networks), monitoring chemical and biological threats, and assisting first responders (DHS, 2004). The National Disaster Medical System (NDMS) is a section within the Department of Homeland Security and FEMA responsible for managing and coordinating the federal medical response to major emergencies and federally declared disasters. NDMS serves as the lead federal agency for medical response under the National Response Plan, which forms the basis of how the federal government coordinates with state and local governments and the private sector during domestic incidents. NDMS works in partnership with the Departments of Health and Human Services, Defense, and Veterans Affairs.

Other Organizations With Readiness and Security Responsibilities

The National Academy of Sciences was established in 1863 (during the Civil War) to promote science and technology. The National Research Council was organized by the National Academy of Sciences in 1916 (during World War I) to further knowledge and to advise the federal government regarding scientific applications to support the war efforts.

The Centers for Disease Control and Prevention (CDC) is the lead federal agency for protecting the health and safety of citizens. Since 1946, the CDC has developed and applied disease prevention and control and environmental health, health promotion, and education activities designed to improve the health of Americans (CDC, 2004).

The National Institutes of Health (NIH) is the federal focal point for medical research in the United States (NIH, 2004). The mission of the National Institute of Mental Health (NIMH, 2004) is to reduce the burden of mental illness and behavioral disorders through research on mind, brain, and behavior (for a review of NIMH history, health policy, research, and training, see Pickren & Schneider, 2005). The Defense Science Board (2004) advises on changing roles and missions of science and medicine in relation to homeland protection. The board is appointed by the assistant secretary of defense as a standing committee reporting directly to him (Defense Science Board, n.d.).

EVOLVING NATIONAL SECURITY CHALLENGES

During the 20th century, national security needs evolved as the source of threats changed from other nation-states to terrorist organizations. Historically, the United States relied on the protection of its large ocean borders; on the availability of adequate time to mobilize its manufacturing industries and train sufficient military personnel; and on its ability to deploy overwhelming technology, equipment, and personnel resources to meet national needs. The last invasion by another country of the U.S. homeland was during the war of 1812. The illusion of security provided by these factors was shattered on September 11, 2001, with the assaults on New York and the Pentagon by Osama bin Laden and the al Qaeda terrorists' network. In response, the U.S. government began a global war on terrorism.

The national security challenges affecting the armed services involve people, preparation and training, performance and human factors, preservation, and threats. Securing and sustaining the armed services requires sufficient numbers of people who are adequately motivated and properly selected, classified, and assigned. The preparation and training of service members aims to ensure that they are capable of performing required tasks efficiently. Many human resource management concepts and practices began in military settings and were subsequently applied to other work settings, including personnel measurement and selection, classification of human abilities, personnel training, adjustment, leadership, human factors engineering, personnel attitudes, historical group debriefing, racial integration, integration of women, nontraditional roles, prohibition of discrimination, prevention and healthy behaviors, management of stress reactions, family support, organizational development, and demobilization.

How personnel adjust to military service and perform is affected by the number of recruits and trainers available, the adequacy of the education they are provided, the type and availability of equipment, the leadership, the situation, the current threats, the level of national support for the military, the political will to go to war (if necessary), and the options available. It is costly to compete in the marketplace to recruit, train, classify, and employ trained personnel and then have them leave the service. Retaining well-prepared personnel is important, particularly when resources are scarce. The medical departments of each service preserve the health of their forces. Preserving the force requires assessing personnel abilities, performance, attitudes, adjustment, leadership, training, and health. Finding the right mix of trained people and leaders, the responsibility of the secretary of defense and the joint chiefs of staff, is important. Force Health Protection, the military health care system, is emphasized during all phases of training; military personnel expect quality health care and support programs both for themselves and their families. Retention of personnel resources in an organization requires that people's careers have meaning. The nature of the threats (both external and domestic) must be considered by the secretary of defense, the military leaders (joint chiefs of staff), the military policymakers, and the politicians.

The roles of the armed services are complex and reflect a security continuum from wartime to operations other than combat (peacekeeping, humanitarian assistance, disaster assistance) to peacetime (homeland defense, civil disturbances). The developments and changes made in the military have often provided additional beneficial effects for the workplace and society.

Reasons that individuals join the armed services have varied since the establishment of the all-volunteer military in 1973; these reasons influence their commitment to remaining in the service. Realistic job previews by recruiters help recruits form appropriate expectations (Breaugh, 1983; McEvoy & Cascio, 1985; Meglino, Ravlin, & DeNisi, 2001; Phillips, 1998; Premack & Wanous, 1985; Wanous, 1992; Wanous, Poland, Premack, & Davis, 1992). The recruiting slogans tell recruits, "Be all you can be," or suggest that they will have opportunities for challenges or adventure. Other reasons recruits may join include the opportunity to serve the country, the promise of experience or job skills that may translate to future civilian occupations, avoidance of the draft (before 1973), or the availability of funds for educational opportunities (Servicemen's Readjustment Act of 1944 [GI Bill of Rights; P.L. 78-346; D. R. Segal, 1986]). Some view military service in terms of joining an institution temporarily; others view it as an occupation (Moskos, 1977, 1986; D. R. Segal, 1986) and become "professional soldiers" (Janowitz, 1960).

The American way of war is changing (Echevarria, 2004; Weigley, 1973). Detailed histories of operations, leadership, casualties, and lessons learned have been written following most military conflicts (Ashburn, 1929; Barnes, Woodward, Smart, Otis, & Huntington, 1870–1888; Coates, 1954–1976; Engelman & Joy, 1975; Ireland, 1921–1929; Joy, 1994; Link & Coleman, 1955; Neel, 1973; Pugh, 1957; Reister, 1954; Richards, 1910). Separate analyses focused on neuropsychiatry (Anderson, 1966; Bailey, Williams, Komora, Salmon, & Fenton, 1929; Deutsch, 1944; Glass & Bernucci, 1966, 1973; F. D. Jones, Sparacino, Wilcox, & Rothberg, 1994; F. D. Jones, Sparacino, Wilcox, Rothberg, & Stokes, 1995; discussion of recognition, treatment, and rehabilitation of combat stress reactions occurs in the section entitled Operational Stress Reactions, below). The military analyzes its performance to learn from its operational successes and failures. Internal and external assessments are required of organizations as well; the services emphasize quality improvement and review. The Quadrennial Defense Review is a periodic analysis to examine whether the armed services are organized appropriately in terms of objectives, defense posture, forces (personnel) available, and doctrine. Institutional histories have included those by and about the following organizations: the Office of Naval Research (2004), the Rand Corporation

(2004), the Naval Research Laboratory (2004), the Naval Health Research Laboratory (Frank, Luz, & Crooks, 1998), the Naval Health Research Center (Gunderson & Crooks, 1999), the Naval Personnel Research Center (E. D. Thomas, Yellen, & Polese, 1999), the Submarine Medical Research Laboratory (Weybrew, 1979), the Army Research Institute for the Behavioral and Social Sciences (Uhlaner, 1967b, 1968, 1977; Zeidner, 1986, 1987; Zeidner & Drucker, 1987), the Human Resources Research Organization (Crawford, 1984; Human Resources Research Organization, 2004; Ramsberger, 2001), Armstrong Laboratories, and the National Institute of Mental Health (2004; Pickren & Schneider, 2005). The histories provide invaluable information about the personnel, products, policies, decision making, organizational behavior, organizational structure, and politics of these institutions. As far back as 400 BC, Thucydides (military historian and Athenian general) wrote that one should study and learn from history. I hope readers will share that view.

HOW DEMOGRAPHICS SHAPE THE MILITARY

Military demographics must be viewed within the context of national demographics. The number of personnel available for service must be considered in the context of the nature of the security threats, the size and characteristics of the pool of potential personnel, the incentives provided for joining the armed services, and the role of the armed forces in society. The perceived strength of the United States has changed historically as a function of alliances and external threats. As the United States grew economically, greater numbers of immigrants came to settle, and U.S. commercial interests spread across the globe. Military strength has always represented national and political interests, so as the nature of security threats has changed, warfare has become more technological and lethal.

National Demographics

To appreciate the emergence of the United States as a nation, it is necessary to look at the national demographics beginning with the 1900 U.S. census of the 45 states and territories (U.S. Census Bureau, 2004). In 1900, the population reached 76.2 million (a major expansion from the 1790 census of 3.9 million citizens), 13.7% of whom were born outside the United States. Life expectancy was about 47 years; 39.6% lived in urban settings. Over 500,000 immigrants arrived in 1900, though one third later returned to their homelands. Workers labored 66-hour work weeks. Women stayed at home raising children; they had no vote and limited rights, education, and employment opportunities outside the home. The turn of the 20th century was a time of rapid industrialization, urbanization, and immigration (predominantly from Europe).

The U.S. population grew rapidly throughout the 20th century, reaching 151 million in 1950, 203 million in 1970, and 281 million in 2000. The mean age of the population has increased with each decade's census, the size of families has decreased, the educational level of the population has increased, and fewer people have served in the armed forces.

Military Demographics

The U.S. military stepped onto the international stage in the late 19th and early 20th centuries, protecting American interests abroad with the Spanish–American War of 1898, the Boxer Rebellion in China, and disturbances in the Philippines. In the Spanish–American War, 306,700 U.S. troops were involved (Directorate for Information Operations and Reports [DIOR], 2004).

Although World War I began in 1914, the United States delayed entry until 1917; in that year the U.S. Army had 190,000 soldiers (213,500 including National Guard and Reserves). When the United States entered the world conflict, it committed 4.7 million troops to the Allied side. Handling the huge volume of troops in 1917 created unique personnel and security needs and required new selection and classification procedures to obtain the large number of personnel. The nature of the warfare, weapons, and tactics used in Europe during World War I created other challenges for the armed services; preserving the fighting personnel demanded new strategies for managing adjustment to the military, operational stress reactions, sanitation, public health, medical logistics, decision making, and leadership. The threat was confined to one theater of operations, Europe.

Following World War I, the armed services reduced the number of personnel; they demobilized from 3,685,500 at the armistice in November 1918 to a force of 280,000 (National Defense Act of 1920), although the actual number was about 200,000 troops. In 1939, at the beginning of World War II, the size of the forces was about the same as before World War I (190,000–200,000 regular U.S. Army troops). The number of military personnel has expanded or contracted depending on national security needs and economic trends (DIOR, 2004). Downsizings occurred after World War II (Sparrow, 1951), after Vietnam in 1973, and after the fall of the Berlin wall in 1989. As of 2005, there were 1.4 million active duty personnel. Maintaining a large standing military force has not been popular with the American public, but large demobilizations and personnel policies to reduce the size of the military often have left it unprepared for conflicts.

The military competes in the occupational marketplace to meet its recruiting quotas; economic trends affect the ability of the armed forces to meet recruiting goals and to retain trained, skilled personnel. Force profiles during World War I show that enlisted personnel were in general male, unmarried, and of limited education. This personnel profile persisted through Vietnam in the 1970s, with the exception of World War II, when more women and married enlisted men served. With the all-volunteer force beginning in 1973, the military demographics have changed: The forces have had more women, more married enlisted men and women, and better educated personnel (DIOR, 2004).

For the Persian Gulf wars in 1990–1991 and 2003, the forces were older and better educated and were more likely to be married and reservists, and there were significantly more women. In 1990, the active force strength was 2.02 million, plus 1.15 million reserve or guard components. The armed services in 2004 consisted of 1.38 million active duty and 0.86 million reserve or guard component forces (DIOR, 2004). In 2005, about 350,000 U.S. military forces were deployed to nearly 130 nations performing a variety of operations, often with foreign military units (see http://www.globalsecurity.org).

PERSONNEL NEEDS

The following sections outline how the U.S. armed forces have met personnel needs since the beginning of the 20th century.

Selection, Classification, and Assignment

Meeting the personnel needs of the U.S. armed forces has been a continuous challenge. At the beginning of the 20th century, military recruiting stations had no systematic standards for assessing recruits. Franz (1912) developed a plan for the routine screening of psychiatric hospital patients, and questions similar to those used in the Binet–Simon test were used on a limited basis for screening recruits. In preparing to enter World War I in 1917, the most urgent military personnel issues involved recruitment, abilities assessment, personnel testing, selection, job classification, and placement. Psychologists pooled their knowledge and scientific methods to meet the military's needs in assessing and classifying large numbers of individuals on the basis of intellectual ability (Yerkes, 1917, 1918). The results included the written Army Alpha test for literate recruits, the pictorial Army Beta test for illiterate recruits, and individual examinations using some version of the Binet–Simon scale (Zeidner & Drucker, 1987). Under the direction of Robert Yerkes, president of the American Psychological Association, the profession of psychology mobilized to meet the critical security needs of 1917. Psychologists conducted personnel assessments, evaluated job assignments, developed training methods, rehabilitated disabled soldiers, and applied scientific techniques (Seidenfeld, 1966; Yerkes, 1918).

The Army assessments in preparation for World War I were the first intelligence survey of the U.S. population; more than 1.7 million inductees were assessed (Ginn, 1997; Uhlaner, 1977). Sanitary Corps psychology officers conducted military personnel studies on soldier attitudes, adjustment to army life, morale, propaganda, desertion, training, and political opinions during World War I and the demobilization period following the war (Keene, 1994). However, psychiatric screening of troops was not performed in any standardized format (Bailey et al., 1929; Glass, 1971).

During World War II, multiple forms of the Army General Classification Test were developed to assist in personnel selection and classification. By 1945, more than 13 million persons had been screened. Military aviation psychologists developed the Air-Crew Classification Test (ACCT) Battery, which was administered to more than 600,000 men during World War II. The ACCT Battery was found to be predictive of pilot and navigator training success (H. L. Taylor & Alluisi, 1994).

Until 1950, military psychologists developed service-specific selection and classification tests. In 1950, the Armed Forces Qualification Test was set up in recruiting offices to screen applicants for all services. It was succeeded by the Armed Services Vocational Aptitude Battery (ASVAB; H. L. Taylor & Alluisi, 1994; Wiskoff & Rampton, 1989). Computerized adaptive testing of the ASVAB is routinely assessed for cost-effectiveness, item pool development, and validity of performance prediction.

Computers have been used extensively in military settings to maximize the efficiency of person–job matching, analyze predictive models, handle large volumes of data, assess performance, and run simulations. Pilot selection is a significant task, given the high costs of training military pilots, the complexity of the technology and tasks, and the costs of the airplanes and training equipment. The trend in pilot selection has been toward use of apparatus-based performance tests (H. L. Taylor & Alluisi, 1994).

Performance

The military is concerned with the human factors affecting the performance of service members in their duties, particularly in military tasks and combat operations. Human factors include adjustment, human–machine systems, training and simulation, motivation, and morale. The objective of the extensive preparation and training programs is to improve and sustain the performance of service personnel in a variety of environments (Krueger, 1991a, 1998a). In 1984, the Army Research Institute for the Behavioral and Social Sciences asked the National Academy of Sciences to study techniques to enhance human performance. Results at the individual, group, and organizational levels were extraordinary in increasing the understanding of behavioral and social phenomena (Druckman & Bjork, 1991, 1994; Druckman, Singer, & van Cott, 1997; Druckman & Swets, 1988).

Adjustment and Mental Health of Military Personnel

At the beginning of the 20th century, mental illness in troops (and in civilian populations) was narrowly defined mainly as severe psychotic abnormality. In 1912, mental disorders were broadened from "insanities" to "mental alienation." The incidence of mental illness in military personnel rose from 1 to 2 per 1,000 troops per year to 3 to 4 per 100 troops per year (Glass & Bernucci, 1966). Because of the low rates, the military authorities found no need to expand military medical services. However, Civil War reports demonstrated situational causation of mental illnesses in military personnel, called the "nostalgia" syndrome (Barnes et al., 1870–1888; G. Rosen, 1975) and "soldier's heart" or "irritable heart syndrome" (DaCosta, 1871). The military attributed the nostalgia symptoms to the vulnerability of young troops. In the Russo–Japanese War (1904–1907), psychiatric specialists treated mental disorders and stress reactions of combat troops. Reports were published in military literature. These reports provided the first demonstration of rapid-intervention psychological support techniques under fire; the results were successful in returning troops to duty (for a bibliography, see U.S. Army Military History Institute, 2000; see also Richards, 1910).

In 1979, the health promotion publication entitled *Healthy People* was initiated to encourage health promotion and disease prevention policies. In 1981, the Department of Defense issued a directive on physical fitness and weight control for active duty personnel (Health Promotion Program, 1986). The Department of Defense population health improvement efforts paralleled the *Healthy People* goals (Bibb, 2002; T. W. Britt, Adler, & Bartone, 2001). The Fit to Win—Substance Abuse Prevention health prevention program was initiated in the Army in 1987, continuing the Army Alcohol and Drug Abuse Prevention and Control Program initiated in 1971. Service members were surveyed using the Health Risk Appraisal and then the Health Evaluation Assessment Review to determine risky behaviors. The behavior of service members has been monitored from 1980 to the present (R. M. Bray & Marsden, 2000) for trends in substance use (smoking, drug use, alcohol consumption) and stress. The results show a decline in substance use. Whether the changes resulted from the Department of Defense health prevention programs or from demographic changes cannot be determined.

The Department of Defense Force Health Protection program, initiated in 2003, maintains health and provides protection through (a) fitness and health promotion, (b) protection and prevention, and (c) medical and rehabilitative care (Winkenwerder, 2003). The armed services encourage healthy lifestyles; service members routinely practice physical training and mental fitness. Lifetime exposure to one or more traumatic events for active duty military personnel was estimated at 65%; victims of any traumatic event were at twice the risk of having two or more physical and mental health problems than nonexposed controls (Hourani, Yuan, & Bray, 2003). Baseline measures were collected before overseas deployments, during service abroad, and on return to allow assessment of changes in service members' health. Hourani et al. (2003) reinforces the notion that military experience may change individual health. Monitoring and assessing changes in health have been important in peacekeeping and humanitarian assistance operations since Operation Desert Storm in 1991 and in deployments in former Yugoslavia (1992), Afghanistan (2001), and Iraq (2003; Wright, Huffman, Adler, & Castro, 2002).

Nature of Warfare, Weapons, and Tactics

The nature of warfare changed dramatically in the 20th century; extended combat operations have involved thousands of troops using heavy artillery fire and resulting in large numbers of casualties. In World War I, machine guns, lethal gas, airplanes, and high-explosive munitions combined with armored tanks to change the lethality of land warfare (Brodie & Brodie, 1962). In World War II, science, technology, and improved tactics created even greater destructive capabilities; the intensity and devastation increased exponentially, and new lessons needed to be learned to meet the challenges. During both world wars, the resources of American and Allied scientists were pooled, permitting interchange of scientific information. The coordination among academic scientists, industry, and military research departments promoted significant achievements in military health and technology. Technological advances in World War II included airplanes, radar, submarines, rockets, and the atomic bomb. Unlike in World War I, combat operations during World War II were conducted in multiple theaters (Europe, North Africa, and the Far East), requiring different tactics, equipment, and training considerations to meet new threats. Service members and civilians faced more stressors.

Human Factors Engineering

The military is concerned with the human factors affecting the performance of troops in their tasks, particularly basic military tasks and combat operations. Human factors include human–machine systems, training and simulation, organizational factors, motivation, and morale. The diversity of military jobs is reflected in the number of different jobs involved and their unique training requirements; many military jobs have no civilian occupational equivalent. Military personnel must operate, repair, and maintain complex equipment and weapons platforms, often in hostile environments. Military personnel train to master basic skills, military-specific skills, and special job skills.

Military training during World War I was personalized and followed an apprenticeship model. In World War II, training became more group oriented and lecture based. During World War I and late into World War II, the Army Training Program relied on troops being trained by the units to which they were assigned. Training techniques varied considerably; there were no significant attempts to standardize occupational performance and outcomes. As the nature of the military tasks became more complicated with new weapons platforms (tanks, airplanes, ships), new environments (cockpits, high noise levels, work underwater, presence of hazards, extreme weather conditions), and more sophisticated equipment (radar, artillery, communications, lethal gas), training to specific standards became more important. During all of the wars of the 20th century, psychologists were tasked to apply scientific procedures to discover the effectiveness of the training programs and to recommend approaches to improve and sustain performance.

Military personnel train as they expect to operate; realistic training builds appropriate expectations, muscle memory, confidence in completing tasks under stress, trust in leaders, unit cohesion, and rehearsal of options. Training to standards assists overall performance. The National Training Centers, developed in the 1970s, allow units to rehearse against well-trained opposition forces to sharpen skills and build confidence. Practice using realistic training scenarios is standard for emergency, first responders, and security personnel preparing for crisis interventions. Under stress, performance decreases; rigorous preparatory training helps to diminish the decrement. Studies of sustained and continuous operations have defined the issues and examined the limits of human performance (Krueger, 1989, 1991a, 1998a).

Military psychological research has been concerned with improving military performance and operations. The American Soldier series on social psychology in World War II chronicled some of the military research efforts during World War II (Stouffer et al., 1950; Stouffer, Lumsdaine, et al., 1949; Stouffer, Suchman, DeVinney, Star, & Williams, 1949). The Army Research Institute for the Behavioral and Social Sciences is among the largest government employers of research psychologists (Uhlaner, 1977; Zeidner, 1986, 1987). The Navy's Office of Naval Research funds extramural research (Darley, 1957; Macmillan, 1951; Office of Naval Research, 2004). The Air Force Human Research Laboratories support both external and in-house research. The Rand Corporation (2004) employs a significant number of behavioral science researchers supporting military needs. The coordination between academic scientists, industry, and military research departments has led to significant achievements in health and technology with applications to first responders. Task analysis and part-task training principles are routinely applied to military settings (Gagné, 1962; Melton, 1947). The military has made many contributions to management sciences, including the use of computers; the development and systematization of the planning, programming, and budgeting system; simulations; project management; operations research; program evaluation and review techniques; systems management; line (combat) and staff (support and service) officers; executive development; Delphi forecast; and management of research and development (D. S. Brown, 1989; O'Connor & Brown, 1980; Smalter & Ruggles, 1966).

PSYCHOLOGY'S CONTRIBUTION TO PRESERVING THE FORCE

Preserving the fighting forces is a key consideration in military operational planning and is the foundation of the Department of Defense Force Health Protection program. The evolution of preventive medicine and public health in military operations has a long history of lessons learned and forgotten (Mangelsdorff, 1985). The availability of trained medical personnel and logistics support reassures soldiers going into combat although historically, during military operations, more losses occur from disease and nonbattle injuries than from fighting. Control of disease and nonbattle injuries is a command responsibility because commanders are responsible for the health and well-being of their personnel. Preventable casualties can be reduced through inoculations, high standards of field sanitation, preventive medicine, vector surveillance, stress management, and leadership (Bellamy & Llewellyn, 1990; Grau & Jorgensen, 1997; James, Frelin, & Jeffrey, 1982; Winkenwerder, 2003).

Occupational Health

Service in the armed forces is demanding and stressful. Military occupational health programs were started during World War I to protect service members and workers against the effects of chemicals (Deeter & Gaydos, 1993). During the early 1940s, the Army gradually expanded its industrial medical programs to protect all War Department employees. Infectious diseases were monitored in both service members and the civilian work forces regularly. Any delays in providing adequate coverage were generally due to lack of funds and trained health personnel.

Occupational health psychology (Adkins, 1999; Quick, 1999a, 1999b; Quick, Barab, Fielding, et al., 1992; Quick et al., 1997; Quick & Tetrick, 2003) became important with the passage of the Occupational Safety and Health Act (P.L. 91-596; 1970). Sauter, Murphy, and Hurrell (1990) framed a strategy to address occupational health risks and job stress. Collaboration between the American Psychological Association and the National Institute for Occupational Safety and Health (NIOSH) in the 1990s helped establish national health objectives regarding job stress (NIOSH, 1996). The focus of occupational health psychology proposed by NIOSH concerned the application of psychology to improving the quality of work life and promoting the safety, health, and well-being of workers (Sauter et al., 1990). The Department of Defense Force Health Protection plan targets improving the quality of work life for service members and promoting safety, fitness, health, and well-being. An occupational health psychology model is appropriate to examine work life, national security challenges, and attempts to improve the well-being of the armed forces.

Training Stress, Accidents, and Injuries

Military training is designed to be realistic and stressful to achieve a high degree of operational readiness; it is sometimes hazardous and even lethal. Adjusting to military life and training requires learning new coping skills. Psychological screening programs are used to predict which individuals will show difficulty adjusting to the basic training programs (Carbone, Cigrang, Todd, & Fiedler, 1999; Cigrang, Carbone, Todd, & Fiedler, 1998; Cigrang, Todd, & Carbone, 2000).

Helmkamp and Kennedy (1996) used the Department of Defense's Worldwide Casualty System (which has reported the cause of death of active duty service personnel since 1979) to classify the 27,070 deaths among active duty personnel from 1980 to 1993. Ninety-five percent of all military deaths occurred among men and most (84%) among enlisted personnel. Unintentional injuries were the leading cause of death (they are the fifth largest cause of death [62.3/100,000] in the U.S. population as a whole). Diseases, in second place, accounted for about 20% of deaths. Suicide was the third most common cause of death among men, followed by homicide; among women, this order was reversed, with homicide third and suicide fourth. About 2% of deaths resulted from combat operations. Among recruits, only 28% of deaths were classified as traumatic (unintentional injury, suicide, and homicide), in comparison to 75% in both the overall active duty military population and the U.S. civilian population ages 15 to 34 years (Scoville, Gardner, & Potter, 2004). During military recruit training, close supervision, emphasis on safety, and lack of access to alcohol and motor vehicles reduce deaths and injuries among trainees. Figure 1.1 presents active duty military deaths from 1980 through 2004.

Operational Stress Reactions

Because the lethality of combat has increased with improvements in tactics and weaponry, soldiers become stress casualties more quickly (see, e.g., Glass & Bernucci, 1966; Grinker & Spiegel, 1944; Mangelsdorff, 1985). Swank (1949) noted that the incidence of combat exhaustion was related to the casualty rate during World War II. Appel and Beebe (1946) recognized that psychological stresses increased with time in combat operations and that every soldier has a breaking point (Grinker & Spiegel, 1944). With the increase in operational tempo, service members are being deployed more frequently and for longer operations, and stress levels have increased correspondingly, particularly with the peacekeeping operations and the combat in Afghanistan and Iraq (Hoge et al., 2004; Stuart & Halverson, 1997).

Psychiatric casualties among military personnel during the Russo–Japanese War in 1907 (Richards, 1910) and later in World War I appeared dazed and unsteady; the condition was labeled *shell shock*, in reference to the

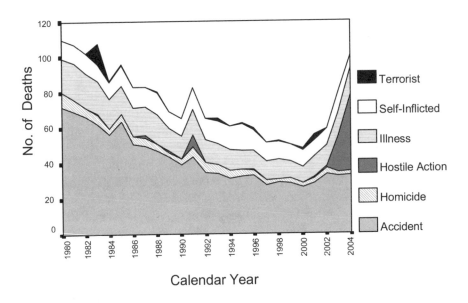

Figure 1.1. Active duty military deaths per 100,000 service members, 1980 through 2004. From Directorate for Information Operations and Reports (2005).

intense artillery shelling. Charles Myers, a psychologist in Britain's Royal Army Medical Corps, first used the term *shell shock* in 1915 to describe an inability to function in battle. The term *war neuroses* came into use later to explain the symptoms. The lack of a standardized psychiatric nomenclature to classify behaviors and symptoms was problematic. A taxonomy listing specific behavioral signs and symptoms permits observation and classification by many practitioners to compare different situations (against recognized criteria).

In 1914, British combat personnel with shell shock were evacuated back to England; unfortunately, their symptoms became fixed, and their conditions did not improve. The French evacuated shell shock casualties to rear facilities and likewise found unsatisfactory results for these soldiers. Later, advanced *neuropsychiatric* (i.e., neurological and psychiatric; permitted the physical basis of symptoms) hospitals were established near the front where the fighting occurred; rapid intervention with the expectation to return soldiers to their units proved more successful. In 1916, 66% of British soldiers and 91% of French soldiers with war neuroses were returned to combat duty following appropriate forward treatment of stress casualties (Glass, 1971). This natural experiment confirmed what had been learned in the Russo–Japanese War: Treating stress casualties near the fighting improved outcomes.

Before the United States entered World War I, Thomas Salmon (a psychiatrist) went to France; his mission was to discover how the combat stress casualties were recognized, treated, and rehabilitated in order to prepare the Army Medical Department for setting up U.S. operations in France. An integrated network of psychiatric services was developed by the Army Medical Department to establish treatment and rehabilitation centers for treating war neuroses. The principles of proximity (close to the fighting front), immediacy, and expectancy (rest, encouragement, then rapid return to duty) were developed and validated to restore troops to their units (see Artiss, 1963). These treatment principles effectively preserved scarce military manpower. The lessons learned from the military operations and interventions of World War I were chronicled and have shown continued applicability in wartime as well as in crisis and disaster situations (Ashburn, 1929; Bailey, 1918; Bailey et al., 1929; C. S. Myers, 1915, 1940). Unfortunately, the lessons learned about organization and treatment were forgotten and had to be relearned in World War II (Mangelsdorff, 1985).

Following World War I, psychiatric treatment facilities became a regular component of peacetime military medical care (Glass & Bernucci, 1966). It was believed that screening at induction or recruitment would eliminate psychiatric disorders and that psychiatric screening could be accomplished by any medical examiner. There was little planning or organization for prevention programs. It would take until the end of 1943 for an effective reporting system to be established to document the extent of mental disorders among admissions (Appel, 1966).

During World War II, psychiatric disorders were again the most important cause of personnel losses (Appel & Beebe, 1946; Lewis & Engle, 1954). In 1941, as combat divisions were reorganized to economize on medical personnel, psychiatrists were dropped from the divisional medical staff. Great numbers of psychiatric casualties were sustained in combat in North Africa in 1942, with few casualties returned to duty. In 1943, when a few psychiatrists were assigned to forward combat areas, the number of troops returned to duty increased to 50%, and in late 1943 Army psychiatrists were returned to staff forward combat units. In the North African theater in 1943, almost all soldiers who were not disabled or wounded eventually became neuropsychiatric casualties because they were not appropriately diagnosed, treated, and returned to duty (Appel, 1966; Appel & Beebe, 1946).

Evaluation of the effectiveness of treating neuropsychiatric casualties showed the importance of early intervention (Bailey et al., 1929; Glass, 1953) in improving cognitive functioning, overcoming stressful experiences, lowering the intensity of operational and environmental stimulation, and promoting social interaction (Mangelsdorff, 1985; McGrath, 1970). These findings had to be relearned during each of the world wars, in Korea (Glass, 1953), and in Vietnam (F. D. Jones et al., 1994, 1995). Treating casualties in theater using the principles of proximity, immediacy, and expectancy of returning, rather than evacuating, is the preferred intervention for returning troops to duty.

Debriefing

During World War II and in Korea, Marshall developed a method of historical group debriefing following combat (Marshall, 1944, 1947; Shalev, Peri, Rogel-Fuchs, Ursano, & Marlowe, 1998; F. D. G. Williams, 1990). The reconstruction group process gathered historical data on the events and reactions of those involved; this process helped group members learn from each other's experiences by validating individual experience and supportive acceptance. The historical debriefing technique and its variations, such as after-action reviews and critical incident stress debriefing (J. T. Mitchell, 1983), are still used by the military, police, emergency services, and other first responders after group critical incidents (Dyregrov, 1997, 2002; Gal, 1986; Mangelsdorff, 1985; Raphael, 1986; Shalev et al., 1998; Tedeschi & Kilmer, 2005). After-action reviews are still used as part of the training process following an accident to discover what happened, why, and how performance could be improved. There is some controversy over the effectiveness of debriefing techniques after extreme stress (Foa, Keane, & Friedman, 2000; Z. Kaplan, Iancu, & Bodner, 2001; Raphael & Wilson, 2000). From a teaching perspective, reviewing the events and analyzing what happened (preventive maintenance) can create a learning process that helps people understand what to do next time. Group dynamics, trust building, and the development of cohesive units contribute to protecting service members (Mangelsdorff, 1999a).

The military has assembled and trained operational unit and rapid reaction mental health teams that have deployed worldwide to provide assistance. The Army developed combat stress control teams of mental health professionals to monitor and intervene after critical events; the teams were deployed during Operations Desert Shield and Desert Storm (Bacon & Staudenmeier, 2003; F. D. Jones et al., 1994, 1995; U.S. Army, 1994a, 1994b). Navy Special Psychiatric Rapid Intervention Teams (SPRINT) were first deployed in 1978 to assist survivors of ship collisions (McCaughey, 1986, 1987). The Army Europe Stress Management Team has assisted survivors of terrorist attacks, such as the attack on the USS *Stark* in 1987, and hostage situations (Kootte, 2002; Sokol, 1989). Tyler and Gifford (1990) studied grief after fatal training accidents; timely interventions, on site, are the preferred mode after critical incidents involving injury or death. The USNS *Comfort*, a Navy hospital ship, provided relief to rescue workers in New York following the September 11, 2001, terrorist attacks (Reeves, 2002). Mental health support for survivors, rescuers, and family members following the Pentagon attack the same day was well coordinated (Cozza, Huleatt, & James, 2002; Hoge et al., 2002; Milliken et al., 2002; Ritchie & Hoge, 2002; Wain, Grammer, Stasinos, & Miller, 2002; Waits & Waldrep, 2002; Waldrep & Waits, 2002). At least two best practice guidelines have been developed for treating casualties of mass violence (American Red Cross, 1998); Ritchie and Hoge (2002) reviewed the lessons learned from the Pentagon attack, and the National Institute of Mental Health (2002) developed broad guidelines for brief interventions in mass emergencies but did not address chemical, biological, or radiological events (Ritchie et al., 2004). Coordinating and practicing intervention plans will be critical in preparation for future disasters.

CONCLUSION

This volume examines the contributions of psychologists and other social and behavioral scientists to creating healthy workplaces in the armed services to better serve the nation's security needs. The chapters offer perspectives on individual differences, personnel management, clinical and counseling psychology, and applied social psychology. There is some overlap in themes and presentations, but this overlap reinforces the maturation of the profession of psychology.

2

FOUNDATIONS OF NATIONAL SECURITY PSYCHOLOGY

WARREN R. STREET

In January 2002, 4 months after the World Trade Center was destroyed in New York City, U.S. Presidential Science Advisor John Marburger asked the American Psychological Association (APA) to suggest research studies for coping with the nation's most recent threat to its national security, terrorist attacks. Psychologists quickly provided over two dozen research questions focusing on the national emergency (APA Public Policy Office, 2002). Examples included studies of promoting vigilance in luggage screeners, negotiating with people from other cultures, training peacekeeping troops, designing communication systems to be effective in a crisis, and inventing intelligent systems for sifting through electronic documents. This rapid response was a recent instance in a long history of psychologists' adapting the skills of behavioral scientists and practitioners to national security needs. This chapter briefly reviews that history.

For the purposes of this chapter, national security psychology is divided into four themes: individual differences, learning and instruction, applied social psychology, and clinical and counseling psychology. In the 20th century, two world wars drew these thematic strands and their practitioners together to form the new field of military psychology. The final section of this chapter considers the history of professional organizations of military psychologists.

THE PSYCHOLOGY OF INDIVIDUAL DIFFERENCES

Explaining the nature and causes of differences between individuals is among the oldest puzzles of human inquiry. This review considers three lines of relatively recent U.S. national security applications: differences in performance abilities, in intelligence and aptitude, and in personality.

Performance Abilities

Studies of individual differences in reaction time were among the first studies in modern scientific psychology. Their origins can be traced to 19th-century attempts to eliminate errors in time estimates of celestial events for use in British naval navigation (Sobel, 1995). Astronomers discovered differences between themselves of more than a second in making these crucial celestial timings. The resulting techniques for studying differences in reaction time were among the earliest laboratory methods of *psychophysics*, the analysis of how humans experience physical events. Psychophysics was in the vanguard of the emerging science of psychology (Boring, 1950; Sanford, 1888). Reaction time tests and other measures of individual performance were included in military selection and placement batteries in World War I.

Some of the first military applications of psychophysical performance measures were to select candidates for airplane pilot training. The first U.S. military pilot selection tests were conducted by the Physical Examining Unit of the U.S. Army Signal Corps Aviation Section on May 15, 1917. By the fall of 1917, 67 units were conducting examinations. The 33-item battery included tests for stereoscopic vision, equilibrium, and color vision. One test for equilibrium measured a pilot candidate's ability to hold his arms horizontal after being spun several times in a swivel chair (U.S. War Department, 1919).

The testing program was later administered by the Air Service Medical Research Laboratory, established in October 1917 (U.S. War Department, 1919). This unit, now the U.S. Air Force School of Aerospace Medicine, was organized at Hazelhurst Field, New York, on January 19, 1918. John B. Watson was on the first medical research board, and Knight Dunlap headed the psychology department, which studied personnel selection, ability requirements, and the effect of mental state on pilot performance (Flanagan, 1948; Link & Coleman, 1955). Apparatus tests were expanded to include tests for reaction time, alertness, and eye–hand–foot coordination (Hilton & Dolgin, 1991).

Pilot selection methods became more sophisticated between the world wars. New reaction time tests and stick-and-rudder tests were created but did not accurately predict flight school performance. A complex coordination task developed by Melton (1947) was used throughout World War II. Measures of reaction time and steadiness were replaced by rotary pursuit measures, which require a person to maintain contact between a handheld stylus and a spot on a rotating disk. The testing program was administered by the Army Air Corps Aviation Psychology Program (Hilton & Dolgin, 1991). Modern pilot candidate screening and training methods make use of realistic computer simulations of the pilot's tasks. For a comprehensive history of techniques for military pilot selection, see Hilton and Dolgin (1991).

In World War II, interest in enhancing human performance expanded to include *human factors engineering* or *ergonomics*, the technology of designing machines for effective use by human operators. This field had its beginnings in the early 20th century with the work of industrial efficiency engineers (Hilgard, 1987). Military applications followed after World War I. For example, the first study of the effects of air flight conditions on human performance was of cold temperature on pilot efficiency, reported by Captain Harry G. Armstrong of the U.S. Army Air Corps on March 1, 1935 (Armstrong, 1936; Dempsey, 1985). Armstrong had established the Aero-Medical Laboratory early in 1935 and went on to command the Air Corps Physiological Research Laboratory, created later in 1935. In 1985, the laboratory was renamed the Harry G. Armstrong Aerospace Medical Research Laboratory (Alcott & Williford, 1986; Dempsey, 1985).

During World War II, behavioral performance studies at the Aero-Medical Laboratory became increasingly independent of physiological and medical studies, and the psychology branch of the laboratory was inaugurated on May 25, 1945, with Paul M. Fitts in charge. The first studies were of instrument legibility, movement of controls, instrument reading under acceleration, and shape coding of controls. The latter studies were undertaken by William Jenkins to prevent accidentally raising the landing gear while attempting to land the airplane (Benjamin & Baker, 2004; Dempsey, 1985).

Leonard Carmichael, Paul Fitts, Franklin Taylor, Leonard Mead, Alphonse Chapanis, and John Kennedy were among those who expanded the theory and research base of human factors engineering in World War II (Chapanis, Garner, & Morgan, 1949; Finch & Cameron, 1956; Hilgard, 1987; Stager, 1987). Finch and Cameron (1956), McCormick (1957), Link (1965), and Gal and Mangelsdorff (1991) have provided comprehensive summaries of research on human factors and environmental factors in aerospace and military performance.

Intelligence and Aptitude Differences

Tests of intellectual abilities appear alongside tests of physical performance in modern military selection and placement test batteries. The classic first example of intelligence testing applied to national security occurred during World War I. On January 23, 1917, APA president Robert M. Yerkes wrote to President Wilson to offer his services in the event America joined the war in Europe. On April 6, war was declared, and by April 22, the APA council had created 12 wartime committees, one of which dealt with the psychological examination of recruits (Evans, Sexton, & Cadwallader, 1992). On May 28, 1917, the eight-member committee first met at the Vineland Training School in Vineland, New Jersey, where H. H. Goddard was the director of research. Lewis Terman, whose Stanford–Binet intelligence test had been published in 1916 (Terman, 1916), was also a member of the committee. Testing all recruits dictated the use of group tests, and the committee supervised the construction of the Army Alpha test, for recruits who read and spoke English, and the Army Beta test, for those who did not. The first forms of the test were adopted by the War Department on December 24, 1917 (Routh, 1994). By January 31, 1919, more than 1.7 million recruits had been tested (Uhlaner, 1968). Comprehensive histories of the tests and the testing program have been provided by Yerkes (1921), Samelson (1977), Hilgard (1987), and Routh (1994).

After World War I, public reports of Army Alpha and Army Beta test data led to public debate about the low educational achievement level of American youths, the validity of intelligence tests, racial differences in intelligence, and intelligence differences in relation to the geographic origins of White soldiers' European ancestors (Harrell, 1992; Hilgard, 1987; Lippmann, 1922; Samelson, 1977). One by one, some alarming conclusions were rebutted by reasoned examination of the data.

Walter Van Dyke Bingham and Walter Dill Scott led another line of psychometric contributions to the war effort. With Bingham as head of the Carnegie Institute of Technology Division of Applied Psychology, Scott and his associates at the division's Bureau of Salesmanship adapted their Rating Scale for Selecting Salesmen to a military setting and produced the Rating Scale for Selecting Captains. It was approved for use by the Army on August 23, 1917, 18 days after U.S. Secretary of War Newton D. Baker created the Committee on Classification of Personnel in the Army at Scott's urging. For this work, Scott was awarded the Distinguished Service Medal in 1919 (Benjamin & Baker, 2004; Jacobson, 1951; U.S. Adjutant General's Office, 1919).

Yerkes and Bingham returned to national security work when World War II became increasingly likely. Bingham joined the Personnel Testing Section of the U.S. Army Adjutant General's office in April 1940 and headed the group that developed the U.S. Army General Classification Test (AGCT), eventually administered to 12 million service personnel (Harrell, 1992). Bingham's unit evolved into the Army Research Institute for the Behavioral and Social Sciences, created in 1972.

Some branches of the service used further specialized assessment. In July 1940, psychologist John C. Flanagan was commissioned in the U.S. Army Air Corps and became director of the Aviation Psychology Program (APP). Pilot selection examinations designed by the APP were first used at Maxwell Field, Alabama, on October 13, 1941. On September 15, 1942, the first APP training detachments to work at flying schools were ordered to air bases in Nevada, Texas, and Florida. The APP became the Air Force Human Resources Research Center in 1949 and the current Human Resources Directorate of the Air Force Research Laboratory (Hendrix, 2003).

Personality Differences

Apart from its application to psychotherapy, the study of personality differences has not been as prominent in the history of military psychology as that of intellectual and performance differences. Testing American military personnel for psychological fitness began in 1918 after General John J. Pershing sent a telegram to the Army Chief of Staff complaining that the "prevalence of mental disorders in replacement troops" called for an "intensive effort in eliminating [the] mentally unfit" (Ridenour, 1961, p. 51). In 1919, near the end of the war, Robert S. Woodworth developed the Personal Data Sheet in an attempt to predict which soldiers would be susceptible to "shell shock" (Benjamin & Baker, 2004).

The period between the wars appears to have generated only moderate progress in personality screening. In 1940, Medical Circular #1 of the Selective Service System provided a guide to minimum mental and personality inspection of draftees. Eight abnormal types were described: mental defect, psychopathic personality, mood disorder, syphilis of the central nervous system, psychoneurosis, grave mental and personality handicaps, chronic inebriety, and organic nervous system disease (American Psychiatric Association, 1944).

Beyond screening for personality disorders, the use of personality tests to predict military performance has borne few fruitful results. Hilton and Dolgin (1991), for example, summarized the history of finding the optimum airplane pilot personality, which began in World War I, when it was believed that coolness under pressure, ingenuity, and courage were qualities of the successful pilot. In World War II, American studies of personality predictors of pilot performance discouraged their use, but German military psychologists became devoted to defining the character type of the superior pilot. Fitts (1946) described German reliance on intuitive assessment, first impressions, and unreliable measures such as handwriting analysis. He observed that Nazi orthodoxy and maintaining upper class access to the officer corps may have been more important than considerations of reliability and validity in German military psychology. In the postwar era and up to the present, "despite decades of evidence to the contrary" (Hilton & Dolgin, 1991, p. 94), the belief persists that the most successful pilots have the "right stuff." Studies continue to be funded in the hopes that refinements will improve the predictive validity of personality measures.

LEARNING AND INSTRUCTION

The history of modern instructional technology and instructional design had its beginnings in World War II (Reiser, 2001a, 2001b). Contemporary instructional design emphasizes thorough analysis of knowledge and skill requirements and their hierarchical relationships in the task to be learned as well as holistic integration of learned skills through performance in a realistic context. Instructional technology assists training through automated presentation, practice, and assessment of knowledge and skills. The military environment has provided a demanding test bed for improvements at all levels of civilian education, because it requires training of large numbers of diverse people to high levels of error-free performance.

Modern instructional technology began with training films in World War II and has evolved into interactive and recursive computer presentations. During the war, the Army Air Force alone produced more than 400 training films and 600 filmstrips. Video media later replaced film, but training content continued to be presented in a linear fashion. Modern computer technology allows branching and repetition, fully implementing the principles of instructional design developed during World War II by Robert Gagné and John C. Flanagan (Reiser, 2001a, 2001b). Flanagan's work is described later in this chapter. Gagné developed tests for classifying air crew members in World War II and continued to study learning and instruction for the U.S. Air Force until 1991 (Gagné, 1962; Saba, 2002). In Gagné's (1962) APA Division 19 (Society for Military Psychology) presidential address, he warned against using learning theories to guide the design of military instructional programs. Learning theories deal primarily with the acquisition of new behavior, but the component behaviors of most military tasks are already in the soldier's repertoire. They need to be performed in the right order and with reliably high accuracy. Gagné urged military educators to rely on careful task analysis, ensure mastery of each component skill, and arrange the learning sequence to ensure optimal transfer from one component to another. Gagné's procedures readily generalized to public educational settings.

These procedures were also characteristics of programmed instructional methods refined by B. F. Skinner, Fred Keller, and other behavioral psychologists. Keller's first programs of instruction were adopted by the Army Signal Corps to teach Morse code to military personnel (Keller, Christo, & Schoenfeld, 1946; Root, 2002). Because of their well-defined performance outcomes and ability to track performance over time, military training procedures are among the most effective contemporary educational programs.

APPLIED SOCIAL PSYCHOLOGY

Although many lines of social psychological research existed before World War II, the war transformed this field into a mature scientific specialization. Research during and immediately after the war provided knowledge about persuasion and propaganda, group cohesiveness and morale, leadership, cooperation and conflict, bargaining and negotiation, conformity, compliance, and obedience. Communication studies by Carl Hovland and small group studies led by Kurt Lewin at the Research Center for Group Dynamics exemplified this period of growth in social psychology.

Hovland was at Yale University when the war began, and he joined a unit of the Research Branch of the Information and Education Division of the War Department headed by Samuel Stouffer. Hovland guided the work of 15 researchers studying the effectiveness of training programs and films. These studies were published in *Experiments in Mass Communication* (Hovland, Lumsdaine, & Sheffield, 1949), one of the four volumes of the influential series *Studies in Social Psychology in World War II*, under the editorship of Samuel Stouffer (Stouffer et al., 1950; Stouffer, Lumsdaine, et al., 1949; Stouffer, Suchman, DeVinney, Star, & Williams, 1949). After the war, Hovland continued his research at Yale and also played a crucial role in founding the Behavioral Research Center of the Bell Telephone Laboratories. This small division of the Bell Labs at one time employed five members of the National Academy of Sciences and made significant discoveries in bargaining and negotiation, multidimensional scaling, instructional technology, learning and memory, visual perception, and human information processing (Carroll et al., 1984; Shepard, 2000).

Kurt Lewin emigrated from Germany to the United States in 1933 during Hitler's rise to power. His studies of leadership, with Ronald Lippitt and Ralph White, seemed to U.S. readers to demonstrate the superiority of democratic over autocratic styles of leadership (K. Lewin, Lippitt, & White, 1939; M. A. Lewin, 1998). Lewin's concept of *action research*—problem-solving research in authentic situations—was put to use in stateside wartime campaigns to promote changes in eating patterns to reduce waste and food shortages (M. A. Lewin, 1998). Lewin's creation of the field of group dynamics opened up group communication, leadership, decision processes, cohesiveness, and cooperation and competition to experimental examination (Hilgard, 1987).

The most comprehensive product of applied social psychological research in World War II was the aforementioned *Studies in Social Psychology in World War II* series. The first data for these studies were gathered on December 8, 1941, the day after the attack on Pearl Harbor, and data collection continued throughout the war. The series is usually referred to as the American Soldier series from the titles of the first two volumes, published in April 1949. Many of the research themes of postwar experimental social psychology can be traced to these volumes.

CLINICAL AND COUNSELING PSYCHOLOGY

The historical trajectory of the treatment of mental illness is one of increasingly objective diagnosis, humane treatment, and public stewardship.

When World War I began, mental illness was treated by psychiatrists, and clinical psychologists worked primarily with schoolchildren. In World War I, psychologists in hospital detachments typically administered tests to soldiers admitted for medical treatment, but they did not provide direct therapeutic services. Major Bird T. Baldwin was the first psychologist assigned to a military hospital, Walter Reed General Hospital in Washington, DC, on April 17, 1918 (McGuire, 1990). By the end of the war in 1919, psychologists were stationed at each of the Army's 40 hospitals.

During World War II, psychologists were gradually added to treatment teams at military hospitals. The first clinical psychologists assigned to hospital duty appear to have been six first lieutenants in the Sanitary Corps of the Army Medical Department assigned to military hospitals in spring 1942. A 22-day course for clinical psychological training was begun at Fort Sam Houston, Texas, on September 9, 1944, and by July 1945, 250 commissioned psychologists were working in military hospitals (Uhlaner, 1968).

The end of the war presented American hospitals with an unprecedented number of veterans with mental illnesses as patients and a government that stood ready to pay for their treatment. J. G. Miller (1946) reported that patients with neuropsychiatric disorders made up 44,000 of the 74,000 patients in all Veterans Administration (VA) hospitals. In short order, the VA and the U.S. Public Health Service (USPHS) developed treatment team models that included clinical psychologists, the federal government created the National Institute of Mental Health (NIMH) from the USPHS Division of Mental Hygiene, and the APA produced a list of reliably high-quality graduate training programs in clinical psychology.

David Shakow headed the committee that published the first APA accreditation standards (Committee on Training in Clinical Psychology, 1947) and a list of graduate programs that conformed to them (Committee on Training in Clinical Psychology, 1948; also see Shakow, 1978). The APA, VA, and NIMH sponsored the Boulder Conference in 1949 (Raimy, 1950), resulting in the construction of the scientist–practitioner model of graduate training for clinical psychologists. The net effect of these events spread far beyond providing psychological services for veterans. Responsibility for psychotherapy, previously the domain of psychiatrists, was extended to clinical psychologists, and an independent profession and body of scholarship were created. In a 1999 APA survey, about 70% of doctoral psychologists said they were employed in health services professions, and about 50% were clinical psychologists (Kohout & Wicherski, 2003).

The VA has promoted research, innovative therapies, and graduate internships in clinical psychology for several decades. A recent training program in psychopharmacology has extended prescribing privileges to armed services clinical psychologists. The first prescription written by a legally trained psychologist in the United States was written by U.S. Navy Commander John L. Sexton on February 10, 1995 (W. R. Street, 2001). Some years would pass before prescribing privileges were extended to the first civilian psychologists, in Guam (1998), New Mexico (2002), and Louisiana (2004).

The current discipline of military psychology has been substantially influenced by issues of personal adjustment and mental health. Of the 21 symposia and paper sessions sponsored by APA Division 19 (Society for Military Psychology) at the 2003 and 2004 annual meetings of the APA, 11 focused on themes of personal adjustment and quality of life in the military. Topics included psychological interventions for families with a deployed parent, substance use and mental health, sexual harassment research, sexual orientation and military service, and dual relationships in the military.

PROFESSIONAL ORGANIZATIONS

Professional organizations for military psychologists have paralleled the development of the field. The National Research Council created its Committee on Psychology for the purpose of war preparedness on November 15, 1916, and the creation of the APA's war service committees in 1917 was described earlier in this chapter. The committees specialized in such matters as examination of recruits, aptitude testing, morale, training, motivation, emotional disorders, acoustics, vision, and aviation (Camfield, 1992). These committees and military units were disbanded at the end of the war, but research interests and professional alliances continued. Many military psychologists applied their skills to the problems of industry in peacetime America and joined the American Association of Applied Psychology (AAAP) when it was founded in 1937. The AAAP provided a forum for applied psychologists apart from the APA, which was dominated by academic psychologists.

World War II created a similar muster of psychologists to military service, but unlike World War I, mobilization preceded America's entry into the war. The National Research Council appointed a Committee on Public Service in April 1939. This group was soon renamed the Committee on the Selection and Training of Military Personnel, assuming that psychologists would resume the duties they performed in World War I (Capshew & Hilgard, 1992).

In 1940, the National Research Council appointed the Emergency Committee in Psychology to mobilize psychological skills for war service (Dallenbach, 1946). The Emergency Committee excluded women from its plans, so the National Council of Women Psychologists, later renamed the International Council of Psychologists, was formed to promote the employment of women in wartime service. Thirteen female psychologists met at the Manhattan apartment of Alice Bryan on November 11, 1941, to inaugurate this organization (Capshew & Laszlo, 1986; see also Scarborough, 1992). Hoiberg (1991) provided a review of women's roles in the military.

As more applied psychologists became involved in World War II, an AAAP section on military psychology, headed by Chauncey M. Louttit, was added in 1943 to earlier sections on clinical, consulting, industrial and business, and educational psychology. With the modern reorganization of the APA, these sections of the AAAP became APA Divisions 19, 12, 13, 14, and 15, respectively (Capshew & Hilgard, 1992; Gade & Drucker, 2000).

The modern APA is itself a product of World War II. In late 1940, the National Security Council convened representatives of psychological organizations to coordinate a response to the approaching war (Hilgard, 1987). The group that emerged, eventually named the Emergency Committee on Psychology, included representatives of the AAAP, APA, Psychometric Society, American Association for the Advancement of Science, Society of Experimental Psychologists, and Society for the Psychological Study of Social Issues. The Emergency Committee ultimately created work for 12 standing committees, 26 special subcommittees, and 18 other projects, including programs in such disparate areas as training in military psychology, lie detection, psychology in the child welfare program, high school psychology classes, marksmanship, length of the workweek, leadership, and interpersonal hostility (Dallenbach, 1946).

Harmonious relations among member groups led to an Intersociety Constitutional Convention of Psychologists in May 1943, where it was decided to unite under the umbrella of the APA. The bylaws of the APA were radically redrawn to create the modern division structure of the organization, and member societies were accorded division status in the larger organization. Nineteen APA divisions were thus created, and the AAAP division on military psychology became Division 19 (Society for Military Psychology). Hilgard (1987) and Capshew and Hilgard (1992) reported these events in detail.

John G. Jenkins was the first president of Division 19. Jenkins was chair of the National Research Council Committee on the Selection and Training of Military Personnel in 1939 and director of the U.S. Navy Bureau of Aeronautics, Medical Section, during World War II. The division began with about 100 members. The APA membership directory listed 150 fellows, members, and associates in 1948; 276 in 1960; 621 in 1985; and 453 in 1995. The count as of spring 2004 was 397. In this same time span, the membership of the APA had grown from 5,700 to over 80,000.

The stable and relatively small size of Division 19 has been a matter of concern to its members. Gade and Drucker (2000) speculated that a few issues, such as involvement in the war in Vietnam and policies toward gay and lesbian personnel, have separated the mainstream of the APA from the military and, by association, from Division 19. To preserve some organizational influence and vitality, members submit occasional proposals to merge Division 19 with Division 21 (Applied Experimental and Engineering Psychology).

Many American military units related to psychology were created in World War II and survive, in some form, to the present day. The Office of Naval Research, the Army Research Institute for the Behavioral and Social Sciences, and the Air Force Office of Scientific Research are examples of these units. Psychological applications to military life are also represented in the curricula of the service academies.

Civilian research organizations were spawned by talented veterans of World War II and continue to contribute to national security psychology through grant- and contract-supported research. The American Institutes for Research (AIR) and the Human Resources Research Organization (HumRRO) are two examples of this line of development. John C. Flanagan, whose role with the Aviation Psychology Program was described earlier in this chapter, founded the American Institute of Research, later renamed the American Institutes for Research, in December 1946. With colleagues at AIR, Flanagan developed the critical incident technique for personnel selection, inaugurated programs of test development to identify promising students, and produced human engineering test methods that permitted equipment evaluation in the design stage. AIR later turned its attention toward improving systems of education (AIR, 1996, 2003).

From 1950 to early 1952, Harry Harlow served as chief of the Human Resources Section of Research and Development, U.S. Army. His last staff study recommended the formation of a Human Resources Research Office. HumRRO was founded in August 1951 by an agreement between the U.S. Army and George Washington University, with Meredith Crawford as its first director. HumRRO became a civilian counterpart of the Army's Human Research Units (Crawford, 1984). The first HumRRO divisions were activated in Washington, DC (Training Methods), and Fort Knox, Kentucky (Armor Human Research; Uhlaner, 1968). HumRRO became an independent corporation in 1969 and continues to investigate personnel testing and training, leadership, and morale; evaluate programs; and implement development programs for military and civilian clients.

CONCLUSION

The role of psychologists in American national security continues to be defined by the threats and opportunities of the day. The historical development of systems for testing, assigning, and training recruits informs modern practices; clinical and counseling psychologists are being deployed to combat settings as well as serving in clinics and hospitals, and research on human behavior in military circumstances continues to be undertaken by the armed services, universities, and private consulting organizations.

Psychologists are active advocates for congressional support of their research and practice presence in the military. In 2003, APA's annual 3-day

Science Advocacy Training Workshop was dedicated to military psychology. APA also cosponsored a congressional briefing with the office of Senator John McCain on September 29, 2003, titled "Psychological Science in Support of the Soldier" (Kelly & Robinson, 2003).

The Coalition for National Security Research (CNSR), a consortium of more than 40 research universities and professional societies, was founded in 2001 to promote a national research effort equal to the demands of the modern world. The APA is a member organization of the CNSR. In June 2002, psychologist Stephen Zaccaro represented the APA before the U.S. Senate Committee on Appropriations Subcommittee on Defense on the subject of psychology's contribution to national security, adding the APA's weight to CNSR support for military science budget enhancements (Zaccaro, 2002). History emphasizes the importance of maintaining military readiness in all domains. Psychological science and practice grew through the 20th century to take their place in today's comprehensive array of tools that ensure U.S. national security.

II

INDIVIDUAL DIFFERENCES

Explaining the nature and causes of differences between individuals—in performance ability, in intelligence, and in personality—has been a continual challenge for military psychologists across the branches of the services. Researchers have examined the assessment, selection, classification, and adjustment of military personnel as well as human and systems performance and human factors. This applied research has helped shape the services' missions, policies, operational requirements, organizational structures, systems, and environments.

During World War I and World War II, most human factors and psychological research addressing national security challenges was conducted and managed by psychologists who served as military officers or civilians. Following World War II, this capability was augmented by contract research with universities, federal research laboratories, and other organizations. The common objectives were to enhance human performance, promote resilience, and optimize service members' adjustment under arduous demands and extreme environments.

The chapters in Part II provide snapshots of the organizations, people, and missions involved in research on individual differences. There is some overlap among these chapters (and others) because of the similarity in operational requirements across service branches. This research has evolved with technology development, increasing interdisciplinary emphasis, and changes in organizational structures. Lessons learned have applicability in homeland defense and fighting the global war on terrorism.

3

HUMAN FACTORS RESEARCH IN THE NAVAL SERVICE

PAUL D. NELSON

In his invited address at the First International Symposium on Military Psychology in 1956, Melton (1957) suggested that the roles of the military psychologist and the requirements for research in military psychology are essentially the same for the different branches of the armed forces. When one examines personnel and human factors research themes over nearly a century of psychology's applications to the military (Driskell & Olmstead, 1989; Nelson, 1971; Wiskoff, 1997), one finds ample evidence in support of Melton's contention. Distinctions one might observe between the ways psychologists in each service function or in the types of problems on which they work, Melton further suggested, are due mainly to different military service missions, policies, operations, and systems. Consequently, although the psychological research highlighted in this chapter may not be altogether different from that conducted in the other military services over comparable peri-

The author is currently Deputy Executive Director, Education Directorate, American Psychological Association (APA). Prior to joining the APA staff, he served 26 years on active duty as a commissioned officer of the Medical Service Corps, U.S. Navy, retiring in 1982 with the rank of Captain. Through most of his naval career he was engaged in the conduct and management of human performance research applicable to Navy and Marine Corps training and operational environments. In his final 4 years of active duty, he served as Chief of the Medical Service Corps, U.S. Navy.

ods of time, it was considered distinctly relevant to missions, policies, operations, and systems of the naval service.

Although some research was conducted in the Navy before the 1940s, it was not typically systematic or otherwise of a programmatic nature, as it was to become during and subsequent to World War II. This chapter therefore places primary emphasis on the nearly 50-year period of the Cold War following World War II. Typical military defense scenarios of this period involved preventing nuclear attack and deterring large-scale combat engagement of ground forces. For the naval service, control of the seas (including the airspace above them) was a strategic requirement. A corollary of this requirement was the ability to project power ashore, through naval aviation, surface and submarine weapon systems, and amphibious operations carried out by the Navy and Marine Corps, for the removal of threats to U.S. defense and for protection of allied nations. Within this envelope of operational requirements were strategic and tactical mission profiles for various operational units of the Navy and Marine Corps, together referred to as *the naval service*. This chapter provides historical highlights of human factors research in the naval service—that is, research applicable to human performance in Navy and Marine Corps operational systems environments.

OFFICE OF NAVAL RESEARCH, OTHER NAVAL LABORATORIES, AND CONTRACT RESEARCH

Although most psychological research during World War II was conducted and managed by psychologists who served as military officers or civilians in Navy Department in-house organizations dedicated to various mission requirements of the naval service (i.e., bureaus, laboratories, and training commands), during the decades that followed that war this capability was augmented significantly by contract research, most of which was with university faculty. This increase in contract research was funded largely by the Office of Naval Research (ONR), established by act of Congress in 1947, 3 years before the National Science Foundation was founded. Although the latter agency, together with the National Institutes of Health, is commonly credited for advancing science through university research grants during the past half century, one could easily argue that ONR contributed at least as much, if not more, to the advancement of psychological science during that period, especially during its first 25 years.

From its inception, however, ONR was subject to the skepticism of those who questioned the integrity of research sponsored by a defense department agency of the federal government, at least from the perspective of whether the mission of such an agency and the purpose of basic research might be in conflict. Rear Admiral T. Solberg, as chief of naval research, first spoke to this issue publicly in reply to an article in the November 12, 1950, issue of the *Washington Post*, reprinted in the *American Psychologist* on request of the editor by John Macmillan (1951), then director of the Human Resources Division at ONR. A variant of this issue had to do with the integrity or veracity of scientific reports that emanate from military psychological research, whether conducted by in-house laboratory or by contract through a defense department agency such as ONR (Older, 1947).

Darley (1957) described the origins of ONR, its mission, and its success in advancing psychological science during its first decade. In a gesture denoting the integrity and value of ONR-supported research for psychology over this time period, the American Psychological Association (APA) president Leonard Carmichael presented on behalf of the APA a certificate to Rear Admiral Rawson Bennett II, Chief of Naval Research, with the following citation: "Presented in recognition of the exceptional contributions of the Office of Naval Research to the development of American psychology and other sciences basic to the national welfare" (Darley, 1957, p. 305). Nonetheless, even 15 years later there remained controversies about mission-oriented basic research, addressed at that time by ONR Psychological Sciences Division director Glen Bryan (1972).

Despite the skepticism, the leaders in those years of psychological research at ONR demonstrated unequivocally that science of high quality could be and was advanced in a mission-oriented government agency, particularly one related to the military. ONR contractors produced scientific publications in refereed journals and scholarly books in virtually all areas of scientific psychology to advance the theory, method, and applications of the discipline as a relatively young science. If one were to list the psychologists who served as ONR contractors during the decades following World War II, it would read as a who's who in psychological science for that period. ONR contractors' research had an impact on the development of an entire generation of new psychologists who were privileged to work as graduate research assistants on the grants and contracts. From the practical side of meeting needs of the naval service, moreover, one also could trace to early ONR research in the psychological sciences such developments as job and task analysis methodologies, attitude measurement methods, leadership and small group theory, instructional technology, computer-based testing, and biomechanical and other human factors support technologies for the workplace. ONR was among the major facilitators in the development of such methods and technologies, from theory to practice, in psychological research.

Also clear from the history of psychological research sponsored by ONR is that most of the psychologists involved in its leadership and research program management were capable of bridging the worlds of science and of military operations, one of the critical roles Melton (1957) cited in his comments about military psychologists. This bridging was essential for at least two reasons. First, this ability enabled Navy program managers to argue the potential costs and benefits of their research to those who controlled the

Navy's research and development budget, and second, it helped leaders and program managers guide contractors to the needs of potential naval service users of their research. Most of the directors of the ONR Psychological Sciences Division and the managers of its research program areas had prior experience working with the Navy, either as civilians or as commissioned officers, and were thus acquainted with the mission and structure, culture, and operating platforms of the naval service. They were also familiar with the Navy's in-house laboratory structure and its research programs and worked collaboratively with colleagues in Navy bureaus or systems commands who were research and development program managers for those laboratories. On many occasions, ONR contract research was coordinated with Navy in-house laboratory research on the same or related problems.

By the end of the 20th century, the organization of ONR's research divisions was quite different from that at its inception; it had been tailored to mission modifications that evolved over 50 years and to a vision of the needs of the 21st-century naval service. There is no longer a Psychological Sciences Division; it has been renamed the Human Systems Department and encompasses divisions for medical and biological sciences and for cognitive, neural, and social sciences. There is much greater integration of research with technology development, more emphasis on interdisciplinary research, and an organizational structure appropriate to these efforts. Nonetheless, psychologists continue to be among the program management staff, as they have been since the commissioning of ONR, and psychological science continues to benefit and contribute to the mission of this organization.

Contract research in the Navy has been useful in advancing theory and methodological developments and in supplementing in-house laboratory efforts with a breadth of scientific capability that may not be available other than through the university or specialized contractor community. Research in Navy Department laboratories, however, offers the advantages of a distinctive mission focus related to naval service populations, systems, and operations; continuity of research and application relevant to those parameters over extended periods of time; and the capacity to respond to specific service needs on short notice, providing a scientific resource immediately at hand for training and operational commanders.

Navy Department laboratories are most commonly categorized as a function of the headquarters command to which they report (e.g., Naval Air Systems Command, Naval Sea Systems Command, Bureau of Naval Personnel, Bureau of Medicine and Surgery). Functionally, however, the psychological research conducted through major Navy Department in-house laboratories can be categorized as follows:

- fitness for naval service (behavior and health),
- manpower and personnel (recruitment, classification, training, and retention), and

- system environments (design, operational conditions, and human performance).

The first two of these topics are addressed in other chapters of this volume. Fitness for service research is summarized in chapter 4, which is focused on the Naval Health Research Center (and its predecessor laboratory). Manpower and personnel research is summarized in chapter 9, which is focused on the Navy Personnel Research and Development Center (and its predecessor and successor laboratories). The third category of psychological research conducted in Navy Department laboratories—system environments—is the principal focus of the present chapter.

SYSTEM ENVIRONMENTS RESEARCH

Most Navy Department laboratories have a mission focus on naval systems environments and operational platforms (e.g., shipboard, aviation, submarine, and diving). Psychological research in such laboratories has been focused accordingly on human performance under various conditions of system design and operation. Among the Navy laboratories in which research on such human performance issues has been common, the oldest is the Naval Research Laboratory (NRL), the Navy's first in-house research facility, established by Congress for the Department of the Navy in 1923 on the advice of scientist and inventor Thomas Edison. It likewise became the Navy's first laboratory for human factors engineering research. During World War II, NRL, in alliance with the National Defense Research Council, was a major research and development entity for the advancement of technology used in submarines and aboard ship.

Engineering Design

Following World War II and the establishment of ONR, NRL became the only in-house laboratory that reported directly to the chief of naval research. F. V. Taylor (1947), among the founders of the APA's Division 21 (originally the Society of Engineering Psychologists), described the history of engineering applications contributed by psychologists during the war, which served as a basis for the inclusion of psychology on a permanent basis among the sciences that have become integrated into the NRL mission to this day. Attesting to the continued contributions of significance that NRL psychologists have made to the scientific discipline of psychology is the fact that 50 years following F. V. Taylor's publication, Astrid Schmidt-Nielsen, then an NRL psychologist, was elected in 1996 to serve as president of Division 21.

Following World War II, applied experimental and engineering psychologists were similarly engaged at such laboratories as the Navy Electronics Laboratory Command, the Naval Air Development Center, the Naval Pacific Missile Center, and the Naval Training Devices Center (NTDC). Although early research on the evaluation of training devices had been done at NRL (F. V. Taylor, 1947), much more was to be done in the years following World War II with the advent and increasingly prevalent use of computer-driven simulation of work tasks or, more broadly, naval system work environments. It was largely to NTDC that this mission was assigned. Psychologists hired by NTDC in its early years identified key issues in training device research and development (Blaiwes & Regan, 1970), among which were transfer of training and the measurement of training effectiveness (Blaiwes, Puig, & Regan, 1973).

Among the higher cost training environments for which the use of simulators became cost-effective (not only in the military but in civilian life as well) is that of aviation. One application for naval aviation is the simulation of aircraft carrier landings, always a demanding task, especially at night (Ehrhardt, Cavallero, & Kennedy, 1975). Developed to train personnel in virtual environments, simulators also became useful as test beds for analysis of human factors engineering design (e.g., judgments of relative motion in tactical displays; Laxar, Beare, Lindner, & Moeller, 1983) and information processing in navigational displays in submarines (Laxar & Olson, 1978).

Operational Conditions

In addition to the effects of system design on human performance, the environmental conditions and operational schedules under which naval systems are deployed in the fleet can impose additional challenges to the sustained reliable performance of human operators. As technology advanced in sophistication following World War II, personnel who operated naval tactical systems were faced with increasingly complex sets of perceptual, cognitive, and psychomotor tasks. In these systems, or simulations of the same, psychologists were challenged to measure the effect on human performance of interactions between task complexity, system environments, and other operational conditions (e.g., time on task, work schedules, deployment periods). A few examples of such research conducted in Navy laboratories are

- the assessment of effects of physical environment stressors on performance (Bittner, Carter, Kennedy, & Harbeson, 1986),
- the development of critical tests for operational performance measurement (Kennedy, Turnage, & Lane, 1997),
- the development and use of a computerized human performance test battery for repeated measures application under different environmental conditions (Turnage & Kennedy, 1992),
- comparison of performance on visual and auditory monitoring tasks (Kennedy, 1971) and of tactical work stations in air and sea environments (Shannon & Carter, 1981), and

- effects of fatigue and sleep loss on performance (Elsmore, Hegge, Naitoh, & Kelly, 1995; L. C. Johnson, 1967, 1982; L. C. Johnson & Naitoh, 1974; Naitoh, 1989; Naitoh & Townsend, 1970).

The importance of work environment diversity in the naval service, with its undersea, surface shipboard, and aviation systems environments, was recognized by the Navy Medical Department before World War II when it established special training programs for Navy medical officers assigned to submarine and aviation duty. The establishment of these programs served as a precedent for the early roots of medical and psychological research related to these operational environments. It was within this context that what is now the Naval Aerospace Medical Research Laboratory was established in close proximity to flight training at naval air stations in and near Pensacola, Florida, and that what is currently the Naval Submarine Medical Research Laboratory was established in close proximity to the submarine base at Groton, Connecticut.

Human Performance in Naval Aviation and Flight Crews

Before World War II, there had been no formal psychological research program associated with naval aviation. Most of the screening of naval aviators was done by the aviators themselves, with the assistance of Navy medical officers qualified as flight surgeons. The cost in human life and equipment from military aviation accidents and the relatively high rate of failure in flight training prompted the Civil Aeronautics Administration, the Navy Bureau of Aeronautics, and the National Research Council to collaborate in establishing a program of research on naval aviation selection and training by the late 1930s in anticipation of a national defense buildup. Known as the Pensacola Project and staffed by psychologists who were commissioned through the Navy Medical Department, the research was based at the Naval Air Station, Pensacola, Florida. By the war's end, Fiske (1946) observed the following:

> On their own initiative as well as on request, aviation psychologists examined existing conditions and attacked those problems where they could help out, not caring whether the subject matter would ever be found in *Psychological Abstracts*. They asked questions and sought convincing answers, either by searching the official records or by counting the occurrences of simple events. (p. 548)

During the war, psychologists were also called on as consultants to examining boards of aviators (E. Weitzman & Bedell, 1944). The organization and nature of research conducted in aviation psychology during the war were very well chronicled by Fiske (1946, 1947) and Jenkins (1942, 1945, 1946).

What was often lacking was an ongoing and more systematic program of research through which reliable and meaningful performance criteria could be developed for validating aviation selection test batteries, decisions related to retention of trainees and their assignment to different types of operational aircraft, and training procedures in the different aviation pipelines. This charge became a major priority of psychologists, both commissioned officer and civilian, who conducted research at the Aviation Psychology Laboratory at the Naval School of Aviation Medicine in the decades immediately following the war. First of all, a factorial understanding of primary and basic training performance was achieved for aviation trainees that could then be used to predict subsequent training performance in what were essentially multiengine (propeller) and single-engine (jet) aircraft training pipelines at the time (Shoenberger, Wherry, & Berkshire, 1963; Wherry & Waters, 1960a, 1960b). Contributing to this more effective prediction of aviator performance was the conceptualization of secondary selection (Berkshire, Wherry, & Shoenberger, 1965), in which measures of student performance in the various stages of naval air training were combined progressively to augment the predictive validity of the initial naval aviator screening test battery and early training performance measures. This model proved cost-effective in the extensive and expensive training involved in military aviation. At any given stage of training, an actuarial forecast could be made for a student having difficulties on the basis of both normative data and on that student's performance to date, thus affording Navy line staff and medical review boards of student aviator progress an evidence-based estimate of outcomes if a particular student were to continue in training. Although the types of aircraft in which students train and naval aviators fly in the fleet are markedly different today from those on which this system of performance prediction was initiated, the principle of the data system remains valid and its implementation in effect.

Successful completion of flight training is necessary for naval aviators to be ready for operational assignment, but it is not sufficient as a criterion of performance in the fleet. Increasingly, with advanced aircraft systems, further training is required of naval aviators in preparation for fleet operations. Thus, performance measures are needed at the fleet level for use as meaningful criteria, just as they are in the various stages of basic and advanced training. In addition to critical incidents and squadron leader ratings, peer ratings were explored as performance measures (Bair, 1952; Jenkins, 1948). Peer ratings obtained during early stages of officer and flight training also were explored as predictors of subsequent performance in training and the fleet (Bair & Hollander, 1953; Berkshire & Nelson, 1958; Hollander, 1954). Positive outcomes led to the implementation of peer ratings during preflight training and, as a consequence, to their availability for use with initial selection test battery scores and secondary selection measures of training performance in the prediction of training and fleet performance effectiveness among naval aviators. In recent years, psychologists at the Naval Aerospace Medical Research Laboratory conducted large-scale fleet surveys on aviator performance so that the performance database could be updated and revalidated on the basis of more current operations.

One of the special challenges posed to naval aviators in high-performance jet aircraft is that of performing under extreme conditions of acceleration and high-speed maneuverability. As early as World War II, psychologists engaged in research on naval aviation worked with medical officers and biomedical scientists on aviator disorientation problems (Graybiel, Clark, MacCorquodale, & Hupp, 1946). Later, even more dynamic flight environments of naval aviation led psychologists to assess individual differences in motion sickness susceptibility (Hutchins & Kennedy, 1965; Lentz & Guedry, 1978), the relationship of perceptual style and reactivity to motion (Long, Ambler, & Guedry, 1975), the influence of visual display and frequency of whole-body angular oscillation on motion sickness (Guedry, Benson, & Moore, 1982), and factors related to disorientation among naval helicopter pilots (Tormes & Guedry, 1975) as well as in high-performance combat aircraft immediately following rolling maneuvers (Lentz & Guedry, 1982). Among the technology developments resulting from this type of research was the Brief Vestibular Disorientation Test for use in screening naval aviators (Ambler & Guedry, 1966) and other naval aircrew personnel (Ambler & Guedry, 1974). Fred Guedry, a leader of this research at the Naval Aerospace Medical Research Laboratory for most of the past 50 years, was also a scientific consultant to the National Aeronautics and Space Administration.

Human Performance in Submariners and Navy Divers

Although psychologists at the Submarine Medical Research Laboratory (SMRL) gave some attention to issues of selection for submarine duty during World War II, they gave far more attention to visual and auditory performance in submarine environments (Luria, 1990; Weybrew, 1979). Resulting from that early research was the development of what became the armed forces color vision test (Farnsworth, 1951) and the identification of optimal lighting patterns for performance aboard submarines (Farnsworth, 1952). In later years, in addition to continuing research related to visual performance of submarine crews on extended deployments (Kinney, 1963; D. O. Weitzman, Kinney, & Ryan, 1966), new requirements for Navy divers to perform underwater tasks in small submersibles and open seas led to additional research on visual performance under such conditions (Carter, 1978; Kinney, Luria, & Weitzman, 1969; Kinney & Miller, 1974; Luria & Kinney, 1970; Luria, Kinney, & Weissman, 1967; Moeller, Chattin, Rogers, Laxar, & Ryack, 1981). Jo Ann S. Kinney, who provided leadership for this research over many years at SMRL, also served on a National Research Council scientific advisory panel for visual research.

Auditory signal detection aboard submarines, especially for sonar operators, was likewise a major focus of SMRL research, as was underwater communications for Navy divers. Pitch discrimination under varying background noise conditions (Morse, Libby, & Harris, 1973), the interaction of pitch and loudness discriminations (J. D. Harris, Pickler, Hoffman, & Ehmer, 1958), monaural and binaural audible angles for detecting a moving sound source (J. D. Harris & Sergeant, 1971), and a factorial study of signal detection abilities (J. D. Harris, 1964) are but a few examples of auditory problems addressed at the laboratory for some 3 decades under the leadership of J. Donald Harris who, on two occasions, was invited to author chapters in the *Annual Review of Psychology* on hearing (Harris, 1958) and audition (Harris, 1972). In more recent years, extensions of that research emphasis have continued (Doll & Hanna, 1989; Doll, Hanna, & Russotti, 1992). In Navy diving environments, auditory problems to which researchers gave primary attention were speech intelligibility under conditions of open sea or saturation diving environments and the effectiveness of communications equipment; this research paralleled that on similar problems in manned space flight (Sergeant, 1966) and in earlier years of naval aviation environments (Tolhurst, 1957, 1959).

With the advent of nuclear technology, the capacity of submarines to remain submerged and at sea in general for longer periods increased. Correspondingly, the demands on submarine crew performance were increased through advances in electronics technology and engineering design, rendering human operator tasks much more complex and the risks for failures in human performance greatly enhanced. Coupled with longer periods of submersion at sea on patrols that precluded normal interaction by most of the crew with the world external to the submarine, the resulting stresses of isolation and confinement necessitated psychological research on human adjustment and effective performance in such an environment. In the face of these challenges, among the research themes of psychologists at the Naval Submarine Medical Research Laboratory was the adjustment of submarine crews to the confining and extended deployments aboard submarines (Beare, Bondi, Biersner, & Naitoh, 1981; Moes, Lall, & Johnson, 1996; Weybrew, 1959; Weybrew & Molish, 1979; Weybrew & Noddin, 1978).

During this same period psychologists studied the effects of sensory deprivation on human performance, primarily through contract research (e.g., Solomon et al., 1961; Suedfeld, 1968; Zubek, 1969). Stimulated in part by experiences reported following the Korean conflict by military personnel who had been captured and subjected to confined and sensory-deprived conditions (Schein, 1957), but also by anticipation of future naval systems that would require crews to function effectively under extended periods of relatively isolated and impoverished environmental conditions, this research became a focus of Navy psychologists as well. In addition to research conducted on the adjustment of Navy and civilian personnel in the relatively isolated and confined Antarctic stations, the Navy implemented a program of laboratory research on isolation, confinement, and sensory deprivation at the Naval Medical Research Institute (NMRI), in Bethesda, Maryland, from the late 1950s through the 1960s (Altman & Haythorn, 1967). In 1969, under the Navy's leadership, a major conference was sponsored by the North Atlantic Treaty Organization to summarize the results of a decade of such research, including related contributions of other laboratories and nations (Rasmussen, 1973).

In terms of more immediate military application, however, the NMRI program was deemed by then to be of limited benefit in relation to the cost of carrying it out, and thus the laboratory program did not survive beyond several field studies of the adjustment and performance of Navy aquanauts who were confined for extended periods in undersea habitats in Projects SEALAB and TEKTITE (Radloff & Helmreich, 1968); the latter was a joint venture by NMRI and ONR. By this time, the Navy was well into research on saturation diving, much of the biomedical research for which was conducted at NMRI. Thus, under the leadership of Arthur Bachrach, the focus of NMRI psychological research shifted from one on human performance under conditions of isolation and confinement to one on human performance in hyperbaric diving environments. Consistent with the research previously noted on performance effects of system environmental conditions, Bachrach's team of psychologists demonstrated that human performance on tasks for which reliable individual baselines are established can serve as an early alerting signal to human neurophysiological incapacitation in need of medical attention. Bachrach and Egstrom (1987) provided an overview of this research.

CONCLUSION

The research cited in this chapter illustrates the fact that what some refer to as "basic" and "applied" research can coexist and that the demarcations between them reflect more the intent of the investigator and the context in which the research is undertaken than the substance of the research itself (Bryan, 1972; Geldard, 1953). Basic research serves to advance theory, and applied research serves as a platform for the test of theory; one is reminded of scientific philosopher Kurt Lewin's (1951) dictum that "there is nothing so practical as a good theory" (p. 169). It is perhaps worthy as well to contemplate in the context of this chapter the antithesis of Lewin's notion, suggested some years later by Levy-Leboyer (1988), that "there is nothing so theoretical as a good application" (p. 785).

In applying an established theory to the solution of a problem in the field, and in cases in which there is no theory directly applicable to a problem, it is not uncommon for further discovery to occur among those well grounded in science and engineering. Such discovery, when replicated, can lead to modifications of existing theory or to the initial formulation of a theory when one does not clearly exist. It is through such iterative processes of conceptual formulation and experimental procedure or other disciplined forms of scientific inquiry that science and an understanding of its applications is advanced. In addition to providing consultation and technical solutions to problems of human performance in naval systems environments, much of the research highlighted in this chapter resulted in peer-reviewed scientific publications, the beneficiaries of which are others in the scientific and engineering communities as well as the general public. It is notable also that some of the human factors scientists who served in the Navy's research agencies also served on the basis of their research as consultants for organizations external to the naval service that likewise serve the public welfare through the advancement of science and technology essential to the security of this nation (e.g., National Research Council and National Aeronautical and Space Administration) and its international allies (e.g., North Atlantic Treaty Organization). Others have served as national leaders in the advancement of psychology as a science through their leadership in professional scientific societies and, as in the case of ONR, in the development of a generation of new psychological scientists through their association with Navy research grants.

Thus, by several measures, the examples of human factors research conducted through ONR and the Navy Department laboratories cited in this chapter confirm the thesis that science and related technology can be advanced in the context of an applied mission. The benefits of such research, moreover, extend beyond the context of the applied missions and environments in which it is conducted to benefit human society in general.

4

NATIONAL SECURITY INTERESTS AT THE NAVAL HEALTH RESEARCH CENTER

KARL F. VAN ORDEN AND D. STEPHEN NICE

Each of the armed services invests in and maintains a capacity to execute research focused on the health and well-being of its service members. The rationale for the investment is simple and straightforward: Military life is stressful. The stressors are many: frequent deployments, arduous training, separation from family, uncertainty about future assignments, and, of course, combat. Service members experience many of these stressors over the course of their careers, or even over the course of a 3-year enlistment, and they often occur simultaneously. They also occur within the context of a unique military culture where dedication to the mission is highly valued and physical and behavioral issues are often viewed as weaknesses and therefore suppressed.

Psychologists at the Naval Health Research Center (NHRC) and its associated laboratories have a distinguished history of examining, characterizing, and developing strategies to understand and ameliorate the stresses

This chapter was authored or coauthored by an employee of the United States government as part of official duty and is considered to be in the public domain. Any views expressed herein do not necessarily represent the views of the United States government, and the author's participation in the work is not meant to serve as an official endorsement.

that military life imposes on sailors and marines. The NHRC was established in 1959 as the U.S. Navy Medical Neuropsychiatric Research Unit with a mission to conduct research in the area of neuropsychiatry as it applies to the naval service. In 1974, the unit was redesignated the Naval Health Research Center, and its mission was expanded to study medical and psychological aspects of health and performance among naval service personnel. In 1999, the Naval Health Research Center Laboratory in San Diego assumed command and control responsibility for a number of subordinate commands and detachments, including the Naval Submarine Medical Research Laboratory in Groton, Connecticut; the Naval Aerospace Medical Research Laboratory in Pensacola, Florida; the Environmental Health Effects Laboratory at Wright–Patterson Air Force Base in Dayton, Ohio ; and the Directed Energy Bioeffects Laboratory at Brooks Air Force Base in San Antonio, Texas. The NHRC currently leads an organization that employs over 350 civilian, military, and contract personnel and has an annual budget of over $82 million.

The Navy's psychological and biomedical research attempts to improve human performance, protect service members from psychological and physical harm, and better integrate human capabilities with the systems personnel must operate. Although the contributions of military and civilian psychologists at NHRC are varied and have a distinguished history, this chapter focuses attention on several specific areas of current interest and importance: (a) psychological adjustment to military life, (b) understanding and enabling healthy behavior, and (c) understanding and improving cognitive performance. Military psychologists enable the Navy to enter the marketplace in the acquisition of new materials, devices, and systems as an operationally oriented buyer, with technical expertise and experience in the disciplines relevant to the development of such systems. The presence of Navy civilian psychologists ensures a corporate memory of past technical problems and their solutions. Crawford (1970) provided numerous examples of the benefits of organizational continuity derived from military research organizations.

PSYCHOLOGICAL ADJUSTMENT TO MILITARY LIFE

Because of its unique customs and traditions, transition to the military culture can be challenging for some individuals. Subcultures exist between, and even within, the armed service branches. Physical and mental requirements vary among specific military occupations. Appropriately selecting and monitoring the adjustment of service members to military culture is highly necessary because of both operational readiness and financial considerations.

Selection

Tests designed to screen candidates for particular military occupations have existed for decades. Sailors and marines are categorized for technical abilities by the Armed Services Vocational Assessment Battery (ASVAB).

Candidates for specialized training in the aviation and submarine communities are required to complete additional testing. In the case of individuals volunteering for the submarine service, these tests focus on personality variables that correlate with a sailor's ability to adjust and adapt to the unique stresses of living and working aboard a submarine for months at a time.

For submarine force duty, the Manual of the Medical Department of the U.S. Navy states that "the psychological fitness of applicants for submarine duty must be carefully evaluated, because of the unique nature of the submarine environment and the responsibilities placed upon each person in a submarine" (U.S. Navy, 2005, pp. 15–92). This requirement is currently being met by researchers at the Naval Submarine Medical Research Laboratory in Groton, Connecticut, who have developed a 240-item self-report questionnaire called SUBSCREEN that identifies prospective submariners with psychological traits that may hinder successful adaptation to the submarine environment (Bing & Eisenberg, 2003). SUBSCREEN is administered to each new officer and enlisted submarine school class. Recent modifications have enabled test administrators to calculate the probability of disqualification and separation. Enlisted students identified by the questionnaire as having an 80% or greater probability of negative fleet attrition are now referred to the mental health clinic for a mental health status interview. The overarching goal is to reduce psychological disqualifications and psychologically based medical evacuations during operational deployments.

There is a 40-year history of psychological testing for selection and classification focused on psychosocial and mental achievement factors (Plag & Goffman, 1968), attitudes and motivation (Hoiberg & Pugh, 1978), and mental health (Gunderson & Arthur, 1969; Gunderson, Looney, & Goffman, 1975). Attrition from military service was most often the result of a combination of several factors, including preservice demography and social background and in-service experiences such as service history, satisfaction, and job and training performance (LaRocco, Pugh, Jones, & Gunderson, 1977). More recently, NHRC researchers have completed one of the most ambitious assessments of medical and psychosocial factors associated with recruit retention (Booth-Kewley, Larson, & Ryan, 2002; Larson, Booth-Kewley, & Ryan, 2002). These studies of more than 66,000 recruits found that the Sailors Health Inventory Program questionnaire, a 40-item medical and psychosocial history questionnaire, was a considerably more powerful attrition predictor than either educational credentials or mental ability scores. These findings may have significant value as the Navy enters an era of profound changes in the force structure—namely, the initiatives to reduce personnel aboard ships and shore installations and to increase the frequency of operational deployments across the globe.

An area of increasing emphasis is the role that physical symptoms can play in psychological assessment. For example, Larson, Booth-Kewley, Merrill, and Stander (2001) reported strong associations between anxiety, depression, and total number of physical symptoms (e.g., headaches, back pain), and factor analyses indicate that emotional distress combined with certain physical complaints form a common factor that predicts basic training attrition (Larson et al., 2002). Physical symptom reports may constitute a potentially valuable but underutilized role in military selection screening, because acknowledgement of physical discomfort carries less stigma than does acknowledgement of emotional disturbance. Thus, respondents may be more honest on items measuring physical discomfort. The utility of physical symptom reports in psychological screening for military service is an area currently under active investigation.

Navy researchers have also increasingly sought to understand the role that positive psychological traits can play in lowering attrition risk. Historically, attrition studies have focused on negative traits or events (e.g., anxiety, depression, trauma history) and have ignored the potentially beneficial role of positive constructs such as optimism, hope, self-esteem, and related concepts. Balanced assessments of attrition risk that take into account both positive and negative characteristics have been rare in military selection research, despite the obvious need to assess individuals holistically. Recent NHRC exploratory studies have determined that various measures of positive-focused psychological traits reflect a common broad factor, *positivity* (Larson et al., 2004); validity analyses have indicated that positivity may have incremental validity over personality scores for predicting a positive outcome (adaptive coping) but not a negative outcome (physical symptoms). Thus, although military screening research has traditionally attempted to determine how negative traits increase risk for undesirable outcomes such as attrition, there may be equal value in further studies of how positive traits increase the likelihood of desirable outcomes.

Mental Health

Mental and emotional disorders have been a major, longstanding public health problem in the military, as they have been in the general population (Arthur, Gunderson, & Richardson, 1966; Gunderson & Hourani, 2003). Gunderson and Hourani (2003) studied all first hospitalizations for psychosis of active-duty enlisted persons in the U.S. Navy during the period 1980 to 1988. The largest number of psychotic cases were diagnosed as schizophrenia (57 per 100,000), followed by affective psychosis (38 per 100,000), other nonorganic psychoses (31 per 100,000), alcohol psychoses (16 per 100,000), drug psychoses (11 per 100,000), and paranoid states (4 per 100,000). Consistent with a previous NHRC finding (Kilbourne, Goodman, & Hilton, 1988), Gunderson and Hourani found that schizophrenics were most likely to have an indicated pre-Navy psychotic condition. Gunderson and Hourani also found that overall incidence ranged from a low of 3 per 100,000 for obsessive–compulsive disorders to a high of 58 per 100,000 for undifferentiated neuroses and miscellaneous minor subtypes. Unlike psychotic disorders, individuals with anxiety disorders in this study were disproportionately female, and this difference was pronounced for all subgroups except phobic and obsessive–compulsive disorder.

Although studies of active duty military personnel have usually shown that age-specific suicide rates in the military are lower than those of the general U.S. population (Chaffee, 1983), suicide remains the third leading cause of death, after accidents and illness (Hourani, Warrack, & Coben, 1997). In addition to identifying incidence rates and correlates of completed suicides, NHRC researchers are improving the Navy's suicide assessment instruments and methodologies (Hourani, Jones, Kennedy, & Hirsch, 1999) and contributing to the literature on suicide attempters. With the assistance of NHRC psychologists, the Navy and Marine Corps now use a standardized Department of the Navy Suicide Incident Report (DONSIR) to standardize the review and reporting process on suicides among active duty personnel (Hourani & Hilton, 1999). The comprehensiveness of the DONSIR database allows a thorough investigation and analysis of potential risk factors and intervention strategies. Suicide attempts represent a risk factor that NHRC is currently exploring. Approximately 30% to 40% of persons who complete suicide have made a previous attempt, and the risk of a person who attempts suicide becoming a suicide death statistic is 100 times greater than average in the first year following the attempt. Trent (1999) conducted a comprehensive study of more than 4,500 individuals in the Navy and Marine Corps who were hospitalized for a nonfatal suicide attempt (called *parasuicide*). She found that between 1989 and 1995, the incidence rate ranged from a low of 79 per 100,000 to a high of 107 per 100,000. Parasuicide rates for women were 2 to 3 times higher than those for men, in contrast to the suicide completion data. Although the parasuicide rate in the Marine Corps is generally lower than the rate in the Navy, the rate among Marine Corps women is dramatically higher than the rate among Navy women. Poisoning (63.7%) and cutting or piercing (30.6%) were the leading methods of attempt. Trent speculated that a focus on mental health issues, particularly among recruits and young enlistees, might enhance suicide prevention efforts.

HEALTH PROMOTION

Unlike many occupations within the general population, military personnel are required to maintain a high degree of physical readiness for the duration of their military careers. As is certainly the case with the civilian population, military health and readiness is dependent on many factors, the most important being the behaviors and attitudes of service members.

Behavior and Health

The current "epidemic" of obesity sweeping the United States has elevated interest in studies and strategies the armed services have used to improve and maintain healthy lifestyles among service members. The armed services are uniquely prepared to study health promotion for three reasons: (a) Service members are routinely measured for weight and body fat; (b) their physical performance is regularly tested; and (c) there is an essential requirement that members of this population be physically capable of performing their duties, and for many occupations, performance is positively correlated with physical fitness levels.

In an effort to better understand the behavioral, psychological, and demographic factors associated with various aspects of physical readiness, Conway (1989) conducted a study of 1,357 Navy men assigned to nine ships in the San Diego area. Controlling for exercise activities, she found that physical readiness was positively associated with wellness behaviors, believing in the importance of physical fitness, expecting to reach or maintain ideal weight, having been athletic as a youth, and education level. Fitness was negatively associated with tobacco use, "preventive/avoidance" behaviors, age, and ever being overweight. Similarly, Trent and Conway (1988) found that dietary factors such as between-meal snacking, overeating, and caffeine intake were negatively associated with physical fitness. However, other research has demonstrated only a modest negative association between body composition and fitness among Navy personnel (Conway, Cronan, & Peterson, 1989).

Health behaviors are actions undertaken to maintain or improve health (Kasl & Cobb, 1966). At the most general level, NHRC researchers have found that health behaviors form two broad categories or dimensions: preventive behavior and risk-taking behavior (Vickers, Conway, & Hervig, 1990). As described later in this chapter, much of the NHRC research in health promotion has addressed the encouragement of wellness behavior, a component of preventive behavior, and the reduction of substance use, a component of risk-taking behavior.

Nutrition and Weight Control

Nutrition and weight control are important preventive behaviors for obesity, which has high rates in the United States. Hoiberg and McNally (1991) found that between 1974 and 1984, obesity as a primary diagnosis accounted for 10,000 days of hospitalization in Navy facilities at an estimated cost of $2,115,000. This estimate did not include the contribution of obesity to other disorders such as diabetes mellitus, digestive diseases, coronary heart disease, and cerebrovascular disease. Although recruits must meet Navy body composition standards on entry into the service, NHRC studies have estimated that 10% (Conway, Trent, & Conway, 1989) of Navy personnel failed to meet body fat standards in the late 1980s and that this number had grown to 19% (Graham, Hourani, Sorenson, & Yuan, 2000) by the end of the 1990s. However, it is difficult to estimate how much of this increase was due to similar patterns in the nation as a whole or to changes in body fat standards in the Navy. The equations used throughout the Department of Defense to estimate body fat from anthropometric measurements actually were developed at NHRC by Hodgdon and his colleagues (Beckett & Hodgdon, 1984; Hodgdon & Beckett, 1984).

As the Navy moved to incorporate nutrition education into the recruit training program, NHRC researchers assisted by assessing nutrition knowledge among Navy recruits. Using an abbreviated version of the National Dairy Council's Nutrition Achievement Test 4 that was developed for high school students, Conway, Hervig, and Vickers (1989) found that only 2% of recruits answered 90% or more of the questions correctly. Questions answered incorrectly by more than 50% of recruits involved assessing nutrient needs and whether those needs are being met, identifying the four major food groups and recommended servings, and describing the effects of alcohol and drugs on nutritional status. As the Navy expanded nutrition education Navy-wide, Trent (1992) assessed baseline nutrition knowledge among active duty personnel. Although this study found that Navy personnel appeared fairly strong relative to the criteria of national dietary guidelines and objectives, there was room for improvement. Trent recommended that the Navy's nutrition education campaign continue to reach beyond remedial weight control programs to engage entire commands. In addition, she recommended emphasizing the role of complex carbohydrates in a healthy diet, widely disseminating of guidelines for the use of nutritional labels on food products, and developing a Navy-wide campaign to label food items and menus in military dining facilities.

NHRC researchers have also contributed by evaluating Navy treatment programs. Hoiberg, Berard, Watten, and Caine (1984), for example, conducted an evaluation and a 1-year follow-up of 531 women and 155 men participating in a Navy weight-loss treatment program. Results of this study found an average weight loss of 22 pounds for women and 28 pounds for men. Correlates of weight-loss maintenance included a change in eating behavior, a self-reported improvement in health status and feelings toward dieting, adult onset of obesity, and physical exercise participation. More recently, Trent and Stevens (1995) evaluated the effectiveness of the Navy's three-tiered obesity treatment program, which includes Level I (command-directed remedial conditioning program), Level II (weight management counseling), and Level III (inpatient obesity treatment). Height, weight, and body circumference measurements were obtained from 624 program participants at the beginning of the program and at 6 weeks, 6 months, and 12 months after the start of the program. There was a significant reduction in percent of body fat after 1 year in all three program tiers. Results demonstrated a sustained downward trend through the 6-month data point, then a plateau between 6 and 12 months. The number of participants meeting the Navy's body fat criteria improved from 1% to 27%, and the number of participants classified as obese dropped from 63% to 43%. However, absolute losses (mean percent body fat) were small: –3.6% fat for men and –4.5% fat for women after 1 year. Of the sample, 4.6% were discharged from the Navy for obesity. Level III, which uses diverse treatment techniques, was the most effective program in helping participants to reduce body fat. Level I, which is primarily an exercise program, was the least effective. An aggressive and supportive aftercare program was recommended to enhance weight loss among program graduates.

Smoking

Cigarette smoking is of particular concern in military populations because in addition to the health hazards it poses, this behavior has a demonstrated effect on the physical readiness of U.S. Navy personnel. In a study of 241 men enrolled in Navy Underwater Demolition Team training, Biersner, Gunderson, and Rahe (1972) found that smoking was negatively correlated with both sports interests and physical fitness. In a study of 1,357 shipboard Navy men, Conway and Cronan (1988) found that smoking was clearly associated with poorer fitness. Men who had never smoked were leaner, could do more sit-ups, and scored higher on the overall physical fitness rating than current smokers and former smokers. Former smokers also performed better than current smokers on the 1.5-mile run and sit-up tests.

Similarly, Conway and Cronan (1992) studied the relationship between smoking, exercise, and fitness among 3,045 Navy personnel. Smoking was clearly associated with lower exercise levels and lower physical endurance (cardiorespiratory and muscular) even after controlling for exercise. The relatively high prevalence rates of smoking among Navy men and women compared with the general population and the health and physical readiness consequences of smoking prompted NHRC psychologists to further examine the correlates of smoking behavior and evaluate potential interventions.

Cronan, Conway, and Kaszas (1991) followed a group of recruits for 1 year after they entered the Navy to determine when, where, and why men started smoking during their 1st year in the service. Because many individuals began smoking soon after joining the Navy, the authors suggested that effective prevention programs needed to be implemented in recruit training and repeated in early training schools. These findings were supported by another study that found that the Navy was not attracting a higher than expected percentage of smokers from the U.S. population. According to self-reports, 28% of the recruits were current smokers, whereas 50% of a shipboard sample were current smokers. In 1989, NHRC researchers were asked to evaluate prototype smoking prevention and cessation programs in recruit training. In this study, four groups of incoming recruits were compared: an education group, a no-smoking group, a health risk appraisal feedback group, and a no-treatment control group (Cronan, Conway, & Hervig, 1989). Relative to the control group, recruits in the education and no-smoking groups perceived the Navy and company commanders as more discouraging of tobacco use and were less likely to start smoking. These NHRC studies contributed substantially to the Navy policy to ban cigarette smoking during recruit training. A subsequent NHRC evaluation of this policy demonstrated a meaningful impact of the Navy's no-smoking policy during recruit training in reducing smoking prevalence at a 1-year follow-up (Hurtado & Conway, 1996).

Despite significant improvements resulting from the tobacco ban during basic training (a policy that all services have now implemented), military smoking rates remain high, particularly among younger service members and particularly in the Marine Corps (R. M. Bray et al., 2003). In 2002, NHRC researchers developed a brief tobacco cessation intervention for Marine Corps recruits. Baseline data indicated that nearly 50% of incoming recruits were either smokers or smokeless tobacco users, and many used both (Trent, Hilton, & Melcer, 2004). A two-part video training program was designed to build on the existing tobacco ban by providing information, advice, graphic imagery, testimonials, and cognitive–behavioral skills to help educate, motivate, and support recruits in choosing to maintain their abstinence from tobacco after graduating from basic training. As of this writing, a randomized controlled design has been planned to evaluate the effectiveness of the intervention in terms of tobacco use and quit attempts in the 12 months following recruit training.

Alcohol Abuse

Alcohol consumption, particularly for recreation or celebration, has long been an accepted part of military life; however, following the dramatic rise in illicit drug use among armed forces personnel during the late 1960s, national and military leaders recognized that substance abuse, including alcohol abuse, was a serious public health problem (Kolb & Gunderson, 1985). In a 13-year study of problem drinkers in the U.S. Navy, almost two thirds were hospitalized for reasons other than alcoholism one or more times compared with less than half of the control participants. The higher rates of illness noted for the problem drinkers in nearly every period beginning with their 1st year in the Navy suggested that they were more subject to illness and accidents from the time they entered the service. A more recent study of drinking behavior among marines showed that they drank an average of 6 days per month, consuming an average of almost six drinks per drinking day, and they reported more than three times per month in which they consumed six or more drinks per occasion (Shuckit et al., 2001). The authors concluded that the prodigious level of alcohol intake and associated problems indicated that the Marine Corps personnel were at especially high risk for alcohol-related life problems.

During the 1970s, the Navy responded to the alcohol problem by developing and expanding treatment and rehabilitation programs for alcohol abusers (Kolb et al., 1983). Over the years, the Navy's inpatient alcohol treatment program has operated with an excellent cost–benefit ratio (Kolb, Baker, & Gunderson, 1983; Trent, 1998). In a comprehensive evaluation of 2,823 active duty inpatients for alcohol treatment, Trent (1998) assessed the efficacy of a 6-week program versus a 4-week program. For this comprehensive study, she used patient demographics, family background, clinical profile, and treatment characteristic data obtained by counselors. One-year follow-up data concerning alcohol use, behavior problems, retention on active duty, reason for discharge, career status, job performance, and quality of life were obtained from participants, their work supervisors, their drug and alcohol program advisors, and automated Navy personnel master files. In this study, Trent found that the single best predictor of success at 1 year was months of aftercare attendance. She concluded that a reduction in length of stay from 6 weeks to 4 weeks in the Navy's inpatient alcohol treatment program would not have an adverse effect on outcome.

In addition to treatment and rehabilitation, which can involve cost and maintenance issues, more recent interest has been given to the prevention of alcohol abuse among young military members. NHRC developed and evaluated a cognitive–behavioral alcohol abuse prevention training program in a pilot study with Marine Corps infantry battalions in Okinawa, Japan (Hurtado et al., 2003). This study found that implementation of the program was feasible and that there was a greater decrease among the intervention group in the percentage of some self-reported alcohol-related problems in comparison with the control group. Although the program did not have a significant overall effect on drinking behavior, the training prompted a short-term decrease in drinking, suggesting potential for the prevention program for marines. This study led to modifications of the training program and a similar evaluation among another military sample (Hurtado, Simon-Arndt, Patriarca-Troyk, & Highfill-McRoy, 2004). Although participants who received the full training program showed a tendency to drink less and had a higher readiness-to-change level and fewer alcohol-related productivity losses, there was an overall lack of program effects on drinking and alcohol-related problems, which suggests that additional efforts are still needed. These studies underscore that substance abuse behavior change is a complex and difficult challenge and suggest that multimethod approaches that focus on social and environmental factors, in addition to individual training and education, hold the most promise for reducing risky alcohol misuse behavior among military personnel.

HIV/AIDS

In 1986, the Navy and Marine Corps began routine testing of all personnel for the presence of antibodies to HIV (Herbold, 1986). At the beginning of this program, NHRC was tasked to develop and maintain a registry of all tests and test results (Garland et al., 1989). Within the first 3 years of the program, the Navy had administered 1,795,578 ELISAs (the HIV antibody test) to 848,632 active duty enlisted personnel. Because there is no vaccine or cure for HIV at this time, behavioral change is currently the primary means of preventing HIV/AIDS. In 1997, a National Institutes of Health consensus panel concluded that behavioral interventions to reduce HIV/AIDS were effective and should be widely disseminated (National Institutes of Health, 1997). In collaboration with colleagues at the University of California, San Francisco, NHRC researchers developed a prevention intervention for sexually transmitted diseases (STDs) and HIV in Navy personnel based on cognitive–behavioral principles (Boyer et al., 2001). In this controlled study of 619 marines aboard four ships deployed to the western Pacific, the authors found that the intervention significantly decreased the risk of alcohol abuse, a contributing factor to STDs and HIV.

A second controlled study using this cognitive–behavioral intervention was conducted among marine security guards (Booth-Kewley et al., 2001), and similar positive results were obtained. Based largely on the success of this work, NHRC was selected by the Department of Defense as the executive agent for HIV/AIDS prevention among foreign militaries. Since its inception in 2000, the Department of Defense HIV/AIDS Prevention Program has expanded globally and is currently working with the militaries of more than 50 nations in an effort to reduce the incidence of HIV/AIDS through primary prevention activities.

COGNITIVE PERFORMANCE

The process of appropriately integrating human physical and cognitive abilities with the machines military personnel use, called *human factors engineering* or *human systems integration*, has evolved over time from a focus primarily on safety toward a greater emphasis on improving overall system performance. The evolution of command and control systems has led to a requirement for systems to provide decision support (or at least some protection from decision bias) to enable effective human decision making in dynamic and information-intensive settings. The areas described in this section represent two extremes in the human factors engineering continuum: (a) situational awareness within a complex system to make more effective and timely decisions and (b) performance enhancement in the presence of fatigue. Strategies to improve physical performance in extreme environments (heat, cold, altitude, depth) have always been of major concern to the military and are discussed in this section because of the strong effects on cognition of both extreme conditions and of amelioration strategies.

Situational Awareness

In 2002, the fast-attack nuclear submarine USS *Greeneville* surfaced beneath a Japanese vessel (*Ehime Maru*) carrying a group of tourists off the coast of Hawaii. The collision sank the Japanese vessel, damaged the submarine, and left nine civilians dead. Investigations indicated that loss of situational awareness (SA) was a significant contributing factor. SA was first used to describe the ability of fighter pilots to maintain an understanding of their relationship to other aircraft in aerial battles. More formally, SA refers to an understanding of a complex, dynamic system. SA is multifaceted, relying on the ability of the person to perceive the relevant elements in the environment, to integrate and comprehend the meaning of these elements, and to predict future system states based on this understanding (Endsley, 1995). In addition to the USS *Greeneville* incident, several other incidents involving submarine collisions and groundings have been attributed to lack of SA.

As a result of these incidents, design engineers are beginning to appreciate the concept of building information systems that will support SA. The Operational Requirements Document of the USS *Virginia* class of submarines has specifically stated that the control room will increase the SA of the commanding officer. Researchers have begun developing and testing metrics of SA for submarine operational testing and evaluation. Currently, work is ongoing on the underlying cognitive abilities that support SA for submariners, with an emphasis on working memory and long-term working memory. It is thought that long-term working memory ability is an important component of expertise (Ericsson & Kintsch, 1995). Although previous accounts focus on the SA of the individual, research is also focused on team SA. Team SA is particularly important for surface navigation, which requires the commanding officer to successfully coordinate four separate teams (Shobe & Severinghaus, 2004).

New research at NHRC has begun to examine SA among marines involved in close combat situations. Soldiers and marines are at significant risk when they lose SA in combat situations. Many current operations involve placing these individuals in urban patrol situations, which can become hostile fire events in a matter of seconds. Maintaining individual and team SA is recognized as critical to team effectiveness and survival. Current work is focused on understanding the factors that contribute to and degrade SA in marines during urban war-fighting training. The primary focus of SA research is to use the results to build better systems, better procedures, and better concepts of operations that optimize the operating environment, leading to increased mission effectiveness and safety.

Fatigue

Performance enhancement, whether it be physical endurance or cognitive sharpness, remains an area of keen interest to the armed forces. Fatigue represents a central area of concern for the Navy, as arguably the largest employer of shift workers in the world, and research has been underway for decades. Early research was focused on understanding the basic neurophysiology of sleep, including its distinct stages and electroencephalographic correlates. More recent fatigue and sleep deprivation research at NHRC includes a variety of efforts ranging from basic science laboratory investigations to operationally applied methods and technologies.

Research in the 1980s began to focus on lapses in attention and how brain activity (as measured by electroencephalography) covaries and can be used to model moment-to-moment task performance. The goal was both to understand complex brain dynamics and to develop an approach that could eventually be used in a real-world system to monitor a human operator. This work showed that the continuum between wakefulness and sleep is much more dynamic—occurring on a minute scale—than had been previously thought (Makeig & Inlow, 1993). It also demonstrated that the changes occurred at specific EEG frequencies and that individual differences in spectral changes were evident and needed to be considered if models sensitive to fatigue-related performance changes were to be developed (Jung, Makeig, Stensmo, & Sejnowski, 1997). Complex modeling of eye activity measures (blink rate and duration, saccade frequency and duration, pupil diameter) was also found to be sensitive to dynamic changes in performance produced by drowsiness (Van Orden, Jung, & Makeig, 2000). These studies were instrumental in demonstrating that EEG and eye data could be used in real time to model task performance levels on a moment-to-moment basis with a high degree of sensitivity. The development of dry electrode technology and remote eye activity recording devices coupled with inexpensive processing units will make feasible alertness monitoring systems for use in vehicles or workstation consoles within the next decade.

NHRC's more basic science investigations of fatigue and sleep deprivation have focused on elucidating mechanisms behind individual differences in resilience to the effects of sleep deprivation. Functional magnetic resonance imaging data collected on the Psychomotor Vigilance Test, the first such data ever recorded, show a difference in neural activation as a function of fast and slow responses (Drummond, Salamut, Brown, Dinges, & Gillin, 2003). This study, conducted in collaboration with the University of California, San Diego, and with the Veterans Administration Hospital in La Jolla, California, demonstrated that fast responses are associated with greater activation, as expected, except in the medial prefrontal cortex, where slow responses are associated with greater activation. The mechanistic explanation of this apparent pattern is still under consideration. Another notable finding is that naturally occurring habitual sleep length influences performance during sleep deprivation. Specifically, "short sleeper" subjects (i.e., persons who have a habitual pattern of 6 hours of sleep per night) show relatively faster degradation of ability during sleep deprivation and faster recovery of ability after sleep obtained, suggesting that resilience to sleep loss needs to be regarded in the context of differing effects on sleep-deprived ability and sleep deprivation recovery ability.

Operational work has focused on the nature of sleep and rest aboard ships. NHRC researchers have used actigraphy to measure sleep and activity over the course of many days. An *actigraph* is a wristwatch-like motion detector that records individuals' sleep–wake activity in their everyday environments over extended periods (e.g., 1 month). This approach has been validated by the American Academy of Sleep Medicine (Ancoli-Israel et al., 2003). In one study, NHRC was asked to investigate the potential for fatigue and safety problems among aircraft carrier flight deck personnel (Carr, Phillips, & Drummond, 2003, 2004). Data collected in response to this request revealed some fatigue and sleep restriction, as expected, but not to an extent that would warrant critical safety concerns or follow-up investigation. In the future, actigraphy may be a key component of a system that continuously monitors sailors to ensure that they receive adequate rest and maintain a high state of readiness.

Cold and Altitude

Changes in cognitive and physical performance in extreme environments have been well researched. Hypo- and hyperthermia affect muscular endurance and coordination and result in significant loss of cognitive capability, producing a state of dementia. Research at the Naval Medical Research Institute demonstrated that deficits in short-term memory could be observed even under nonhypothermic conditions. J. R. Thomas et al. (1990) found significant deficits in a delayed matching-to-sample task in individuals exposed to 4 °C air for 70 minutes. Although these individuals experienced no changes in core temperature, circulating catecholamines were significantly elevated. Furthermore, brain evoked responses had shorter latencies, indicating hyperactivity, compared with ambient temperature conditions (Van Orden, Ahlers, House, Thomas, & Schrot, 1990). More recently, Schrot and colleagues showed that ingesting tyrosine mitigates the cold-induced memory deficit (Schrot, Thomas, & Shurtleff, 1996; Shurtleff, Thomas, Schrot, Kowalski, & Harford, 1994). Cold is believed to disrupt catecholamine processing in the brain. Administering tyrosine (a catecholamine precursor) appears to alleviate the cold-induced memory impairments caused by cold exposure. Current research is examining the levels of tyrosine necessary to protect against cold-related cognitive deficits.

Rapid ascents resulting in altitude changes of 8,000 or 9,000 feet place individuals at risk for acute mountain sickness (AMS), which is associated with headache, nausea, lethargy, and sleepiness. AMS has occasionally become an issue of operational concern, most recently among Special Forces personnel deployed to the mountains of Afghanistan. Confusion, dementia, and loss of psychomotor ability can occur in more severe AMS states. Military personnel with AMS can quickly become casualties and place their units' mission effectiveness at risk. More troubling is that AMS is often a precursor to the life-threatening conditions of high-altitude cerebral edema and high-altitude pulmonary edema. Avid hikers and mountaineers have long dealt with AMS by carefully controlling exposure and ascent rates. The expedition community has also relied on drugs like acetylzolamide (Diamox) to reduce the likelihood of AMS onset.

Recently, NHRC began investigating the effectiveness of drugs to prevent AMS at the Marine Corps Mountain Warfare Training Center in the Sierra Nevada mountains in Bridgeport, California. Aside from the prevention of AMS, Parker and colleagues were interested in whether there were any cognitive effects of altitude and acetylzolamide or ginkgo biloba (Parker, Griswold, & Roberts, 2003; Parker, Walsh, & Roberts, 2003). They used a weapon assembly task and computer-based cognitive tasks. Although study participants found some side effects of acetylzolamide unsettling (e.g., tingling of fingers and lips), there was no effect on any performance measure.

CONCLUSION

Military service will continue to place people in situations that are extremely demanding, both physically and psychologically. Even as weapon systems become more technically advanced, researchers are continually reminded that individual sailors and marines are the most important "systems" of all. Personnel continue to be of greater value to the military as selection requirements become more stringent, training becomes more extensive, and deployment schedules continue to strain personal lives. There is no doubt that sailors and marines will continue to work in far from optimal environmental conditions. Fatigue will continue to be a rate-limiting factor in individual and unit performance and to affect both physical and cognitive performance. Applied psychologists will be directly involved in ensuring that individuals are appropriately selected and trained for specific military occupations and for the unique stresses of military life in general. There will continue to be problems associated with adjustment to military life, particularly the abuse of both legal and illegal substances, and researchers will undoubtedly be tasked with understanding root causes of adjustment and substance abuse problems and devising amelioration strategies.

Undoubtedly, the psychologists who work to improve the lives and performance of military personnel will continue to advance the general field of psychology and the interests of national security. It is noteworthy, however, that many of the psychologists working within the NHRC laboratories contribute to multidisciplinary teams or are conducting research in areas unrelated to psychology. These areas include epidemiology, medical modeling and simulation, operations research, force protection, deployment health, electromagnetic radiation, laser bioeffects, underwater sound, submarine escape and rescue, undersea medicine, hearing conservation, toxicology, human performance physiology, aerospace medicine, and research administration. Although this chapter did not cover these research areas, the unique ability of psychologists to work outside their primary discipline to contribute to a broad range of national security issues is a powerful testimony to the rigor of their training and their commitment to the organization.

Although psychologists have made many significant contributions in the interest of national security and to science, new challenges will emerge as the nature of security, warfare, and society itself evolve over time. The insights and innovations born from research and development will inevitably benefit the military establishment and its people at the individual level. Adjustment and coping with the stresses of deployment and military life, as well as cognitive performance under dynamic and information-intensive conditions, remain at the forefront of issues facing Department of Defense psychologists today. Research in these and related areas remains the surest means to protecting military personnel, improving their performance, and improving the quality of their lives.

5

U.S. ARMY RESEARCH
IN HUMAN PERFORMANCE

GERALD P. KRUEGER

After World War II, hundreds of research psychologists working for the national defense returned to academia for less militaristic civilian endeavors. But the Korean War (1950–1953) rekindled the military services' appetite for behavioral sciences research. During the 1950s and early 1960s, the U.S. Department of Defense established a sizeable number of government-owned and -operated military research laboratories on military bases. Some of them had a mission to do behavioral sciences research, often in conjunction with other scientists and engineers in multidisciplinary laboratories. For the 60 years after World War II, the U.S. military also continued to sponsor and financially support behavioral sciences research and applications work via extramurally funded research grants to university-based research programs as well as at various other individual contract research centers that focused their work on military requirements. U.S. Army–sponsored programs fostered numerous behavioral science research projects at prestigious universities by issuing support grants to prominent professors to do research, including funding the doctoral dissertations of hundreds of graduate psychology students and thereby producing some of the personnel "seed stock" who matured into the leaders of today's applied, experimental, and engineering psychology programs. Many military laboratory scientists began in graduate student roles sponsored by the Department of Defense.

Through decades of excellent work in military labs and military-sponsored academic programs, these scientists have produced behavioral research findings that have contributed significantly to the understanding of psychological science and to psychological databases. Much research instrumentally helped to resolve unique problems of soldiers, sailors, airmen, and marines of the time, in training and in combat, by helping establish advanced military doctrine, deployment and operational procedures, and new personnel policies to ensure force readiness to protect the nation's security interests. The work of Army behavioral scientists, often participants in multidisciplinary programs, has helped preserve and protect the physical and psychological health and sustain the performance of countless combatants who have participated in U.S. military overseas deployments since World War II (Gal & Mangelsdorff, 1991; Krueger, 1998a, 1998b).

Much of the military psychological research was pacesetting, and it was sufficiently generalizable (often going well beyond the unique needs of the military) that it offered significant applications to the civilian populace as a whole. New psychological knowledge from military research programs has been adapted to several national psychological and social programs and movements. In particular, Army lab–based psychological and human engineering research has influenced national programs in government and industrial workplaces by setting preferred operational procedures for jobs; placing safety restrictions on environmental and occupational exposures in workplaces; integrating women into jobs traditionally filled by men; improving the design and use of public transportation systems; improving the design of hundreds of user-friendly consumer products; contributing new adventure and leisure sporting activities and related equipment, clothing, and even field food products; and improving civilian lives in countless other applications. Now, after the turn of the millennium, and especially after the terrorist attacks of September 11, 2001, the contributions of these military psychologists to national homeland security and the war on terrorism within U.S. boundaries has been brought into focus.

HUMAN FACTORS ENGINEERING AND PSYCHOLOGY

Like its sister services, in the 1950s the U.S. Army embraced the emerging new discipline of *engineering psychology* or *human factors research* (Chapanis, Garner, & Morgan, 1949; Parsons, 1972) to assist in the design of user-friendly, safe-to-operate man–machine or human–materiel systems. The U.S. Army's Human Engineering Laboratory (HEL), founded at Aberdeen Proving Ground, Maryland, in 1951, paved the way for good human engineering and ergonomic design of complex Army systems such as tanks, armored personnel carriers, artillery pieces, helicopters, missiles, weapons of various sorts, communication systems, and so forth. HEL's contributions to broadening engineering psychological research methodologies are still significant today.

As just one among many examples, HEL scientists developed classic full systems approach methods for evaluating human performance in complex scenarios involving artillery weapons by identifying and independently studying each of the numerous sources of human error variance on the battlefield, leading to systematic improvements in each area of concern and dramatically improving the first-round hit probabilities of artillery teams. Previous studies demonstrated that more than 50 years of improving artillery weapon tubes, shells, and explosive charges collectively had barely improved first-round hit probability in Army artillery missions. HEL human factors researchers, in a series of battalion artillery field tests (Horley, Dousa, Phillabaum, Lince, & Brainerd, 1978), enacted a paradigm shift by pointing out that examining each of the contributions of human error variance in large team efforts in the field could assist the artillery community significantly by implementing man–machine interface fixes necessary to improve artillery accuracy. HEL's field artillery studies set the methodological stage for other human factors and ergonomics studies of complex man–machine interface problems in other military and civilian applications as well (van Cott & Kinkade, 1972).

HEL's basic science studies elucidated many psychological principles for problems as diverse as determining how equipment operators process sensory inputs and handle large amounts of information presented rapidly, how computer display symbologies and icons can be made easy to use in command and control centers such as those in missile and communication center hubs, and how people communicate in the face of significant ambient acoustical noise. HEL's engineering psychology and human factors research methodologies established from the 1950s through the 1980s later enabled researchers to understand the best human use and proper design of complex information, intelligence, and communication systems, many of which are being used extensively in tracking terrorist activities and engaging in counterterrorism security measures.[1]

PSYCHOLOGY OF OCCUPATIONAL, ENVIRONMENTAL, AND PREVENTIVE MEDICINE

The Army formed two occupational and preventive medicine research labs employing psychologists. The U.S. Army Research Institute of Environ-

---

[1]HEL is now called the Human Research and Engineering Directorate, a part of the corporate Army Research Laboratory. For an extensive list of published reports, see U.S. Army Human Engineering Laboratory (1990).

mental Medicine (USARIEM; see http://www.usariem.army.mil) was formed at Natick, Massachusetts, in 1961 to provide environmental medicine research support for soldiers deploying to harsh environments. USARIEM later added the missions of nutritional research to ensure the safety of field feeding systems and to explore ways to enhance soldier performance through nutritional supplements and of research to measure the efficacy and safety of soldier clothing and equipment to improve soldier performance. The U.S. Army Aeromedical Research Laboratory (USAARL; see http://www.usaarl.army.mil) was established at Fort Rucker, Alabama, in 1962 to provide direct aviation medical research support to all Army aviation and airborne activities; later research on vision and acoustics and on health hazard assessments and countermeasures for air and tactical ground vehicles and weapon systems were added to their mission.

Over the decades, the missions of both USARIEM and USAARL expanded to include a wide swath of occupational medicine research in support of the four U.S. military services as personnel at these two labs worked to understand human behavioral and physiological responses to significant environmental stressors in military occupational settings (e.g., extreme heat and cold, high terrestrial altitude, sustained performance requirements, intense noise and blast, vehicle vibration and acceleration, low-altitude flight, excessive pilot workloads, flight with sophisticated night vision systems; see Gal & Mangelsdorff, 1991). Psychologists, along with scientists and researchers in a variety of other disciplines (e.g., biophysicists, optometrists, audiologists, biomedical engineers, nutritionists, biochemists, sports medicine specialists, flight surgeons, aviation safety officers, occupational and preventive medicine specialists) at these two laboratories established some of the nation's best occupational, preventive, and environmental medicine guidance for dealing with the harsh conditions of military training, deployment overseas, and combat.

Research psychologists, both uniformed service members and Army civil servants, have led multidisciplinary teams of laboratory scientists who conducted numerous experiments and extensive field studies to examine human stress physiology and psychological response to unique military environments and how to predict behavior and operational performance at work, both on an individual basis and for small groups, crews, and teams (for lists of USARIEM publications, see Krueger, Cardinal, & Stephens, 1992; USAARL, 1991, 1996). Pioneering work at both labs on the psychology, health implications, and performance of women in tough military jobs pointed the way to integrating women into career fields traditionally filled by men, and this work carried over into many civilian work settings (K. E. Friedl, 2005; Vogel & Gauger, 1993).

Aviation psychologists at the USAARL pioneered work with heavily instrumented test helicopters and flight simulators to examine the following helicopter pilot workload issues, among others:

- pilot performance at low-level, nap-of-the-earth flight;
- helicopter flight with night vision goggles (Crowley, 1991; Rash, Verona, & Crowley, 1990);
- use of helmet systems coupled with night vision sensors on the fronts of helicopters (LeDuc, Greig, & Dumond, 2005; Rash, 1999);
- navigation at nap-of-the-earth levels using hand-held maps versus automated tracking systems (Cote, Krueger, & Simmons, 1985);
- extensively long flying hours with and without night vision goggles;
- search and rescue and aeromedical evacuation missions;
- performance under the influence of various chemical substances, such as atropine sulfate (Simmons et al., 1989), sedating and nonsedating antihistamines, stimulant drugs (J. A. Caldwell & Caldwell, 2005), and sleeping compounds (J. A. Caldwell, Caldwell, Smith, & Brown, 2004); and
- use of cumbersome chemical protective suits (J. L. Caldwell, Caldwell, & Salter, 1997).

The pilot workload studies of USAARL aviation psychologists made significant contributions to staffing decisions among Army aviation unit personnel. They helped to determine how many aviators should be trained to staff combat aviation battalions. They also contributed to decisions about numerous other personnel, aviation medicine, and operational procedural matters affecting military and civilian helicopter operations.

For decades, research psychologists at USARIEM conducted psychological and performance assessments as part of multidisciplinary teams examining many physiological and psychological parameters of soldiers' work in harsh natural environments of high heat (both dry and humid air), severe cold (Kobrick & Johnson, 1991), and high terrestrial altitudes (Banderet & Burse, 1991). Their work ensured that the military forces understood the advantages of realistic training and preacclimitization to climatic extremes before deployment to combat zones (Krueger, 1993). Thirty years of pioneering physiological and psychological research at USARIEM paid big dividends in a matter of days when U.S. military forces were rapidly deployed in 1990 to the harsh desert environments of Southwest Asia to face Iraqi forces in a large operation that became known as Operation Desert Shield in the Persian Gulf War of 1991. Research psychologists and preventive medicine specialists at USARIEM teamed overnight to prepare preventive medicine pamphlets for distribution to hundreds of thousands of U.S. troops deploying to the deserts of Southwest Asia from August to December 1990 (see Glenn et al., 1991). The U.S. Army Surgeon General acknowledged the value of disseminating such preventive medicine guidance for preserving the health and performance of combatants in the desert, particularly in terms of prevention of anticipated heat stress casualties. This successful USARIEM theme of incorporating years of research findings into such pamphlets, delivered on time to troops deploying to harsh environments, was repeated for soldiers and marines sent to Somalia in 1993, to Rwanda in 1994, to Bosnia in 1996, and to other harsh locations as well. Subsequently, the Army Medical Department institutionalized the USARIEM process and transferred the responsibility for producing preventive medicine guidance pamphlets for the troops to the Army Center for Health Promotion and Preventive Medicine (see http://chppm-www.apgea.army.mil).

Other psychological studies of note at USARIEM have included assessments of soldier performance while wearing chemical protective clothing, which demonstrated the importance of such variables as bulkiness and fit, visual distortions, and heat stress imposed on wearers of the uniforms (Bensel, 1997; R. F. Johnson & Kobrick, 1997; Krueger & Banderet, 1997); studies of various uniform camouflage designs and target detection, anticipating advanced digital and chameleon patterns; and examination of the effects of high stress levels on Army Rangers and Special Forces soldiers in training and acquisition assessment testing (Marriott, 1993). Behavioral scientists who teamed with exercise physiologists doing load carriage, forced march, and parachutist studies served the infantry, but they also contributed to modern sports psychology, particularly for high-adventure, extreme performance individual sports such as mountaineering and ultramarathoning.

Foundation behavioral and medical science work at Army research laboratories identified and made recommendations to prevent or control health hazards attributable to high levels of acoustical noise, blast overpressure, toxic fumes, electromagnetic radiation, whole body vibration, and heat stress that often accompany operation of advanced development materiel systems. These health hazard assessment programs, spearheaded by research psychologists in several Army labs, were the forerunners of many civilian workplace safety programs watched over by the Occupational Safety and Health Administration (OSHA). Frequently, technological advancements cause human operators to make adjustments and enact compromises in how they must train or operate. For example, the advent of sophisticated night vision systems enables people to work through darkness in forests and deserts; in another example, exposures to toxic fumes, excessive noise, and vibration of heavy vehicles become the norm. Army occupational behavioral scientists set the pace for this burgeoning field of workplace safety by assisting in formalizing the health hazard assessment program described in Army Regulation 40-10 (written in parts and driven by Army psychologists; U.S. Army, 1983), which requires assessments of the developmental weapons and materiel systems the Army plans to procure. The health hazard assessment program was joined with six other human assessment domains (including manpower, personnel, training, human factors engineering, system safety, and soldier vulnerability) in the Army Manpower and Personnel Integration (MANPRINT) program for ensuring that human variables are all properly accounted for in materiel system acquisition programs (see http://www.manprint.army.mil/). This Army program was expanded to become the Department of Defense's Human Systems Integration program and the Navy's Systems Engineering Personnel Integration (SEAPRINT) program. The principles espoused in these human factors–related programs are now are of considerable interest to the OSHA and the Department of Labor (for health hazards, see Krueger, 1983).

SOCIAL AND PSYCHOLOGICAL COPING WITH EXTREME STRESS

The Walter Reed Army Institute of Research (WRAIR) in Washington, DC, launched its Division of Neuropsychiatry in 1951 to do medical psychology research. At its height, the division had a staff of more than 100 research psychologists and other behavioral scientists, including military psychiatrists and social anthropologists. Basic neuroscience research using both animal and human studies elucidated mechanisms of human stress reaction and responses to a wide variety of stresses encountered by military personnel. In several high-technology innovations, WRAIR's research programs were the neuropsychological pacesetters for decades. For example, one WRAIR department first used positron emission tomography scans or magnetic resonating imaging to elucidate mechanisms of human reaction and responses to a wide array of stresses, especially sleep deprivation, encountered by military personnel (Balkin et al., 2000, 2002; M. Thomas, Balkin, Sing, Wesensten, & Belenky, 1998; M. Thomas et al., 2000).

In WRAIR's Department of Military Psychiatry, groups of social experimental psychologists and social anthropologists examined somewhat more holistic stresses facing soldiers deployed to various regions of the world, and they pioneered research on stress syndromes associated with the harshness of military occupations and lifestyle. The popular book *Boys in the Barracks* by two prominent uniformed research psychologists (Ingraham & Manning, 1984) helped Congress and military decision makers in Washington, DC, understand the personnel turmoil in the U.S. Army after the Vietnam War. Prominent WRAIR research psychologists and anthropologists studying team and unit cohesion positively influenced Army unit staffing and personnel policies that affected all soldiers in the Army.

For decades, WRAIR's military psychophysiologists pioneered research on assessments of soldier performance in sustained-duration missions that entail significant loss of rest and sleep. As early as the 1950s and 1960s, seminal work on this and related research topics established the WRAIR as a center of expertise on sleep loss and performance; military psychologists of those days examined hundreds of volunteer participants confined to the laboratory for long, sustained cognitive work bouts to clarify expectations during continuous around-the-clock operations (Krueger, 1986, 1989, 1991b; H. L. Williams, Lubin, & Goodnow, 1959). In the 1980s and 1990s, WRAIR's staff began directing significant research to explore a variety of stimulant drugs and hypnotic sleep aids for military use in the field (Newhouse et al., 1992; O'Donnell et al., 1988; Wesensten et al., 2002, 2004); their research on the effects of caffeine (Committee on Military Nutrition Research, 2001; McLellan et al., 2005) contributed significantly to an understanding of the nation's most ubiquitous stimulant, found in coffee, tea, soft drinks, and chocolate and, more recently with military impetus, inserted into chewing gum (Kamimori et al., 2002; McLellan, Bell, Lieberman, & Kamimori, 2003–2004).

The research programs at WRAIR formed perspectives on anticipated social–psychological human responses to the extreme stresses of combat and to conditions of anxiety and panic, on the diagnosis and treatment of choice for those suffering from posttraumatic stress syndrome; and on handling the grief reactions of first responders in mass casualty situations (Bartone, 1998). Military psychologists at WRAIR's Department of Military Psychophysiology pioneered the use of cognitive and neuroscience computerized test batteries, beginning with the WRAIR Performance Assessment Battery (Thorne, Genser, Sing, & Hegge, 1985); progressed toward standardization with colleagues in other U.S. military laboratories using the Uniformed Tri-Services Performance Assessment Battery (Englund et al., 1987); and then developed the Automated Neuropsychological Assessment Metrics, an automated cognitive test battery available on CD-ROM (O'Donnell, Moise, & Schmidt, 2005; U.S. Army Medical Research and Materiel Command, 2002). Widespread use of these research tools in dozens of military and university behavioral research laboratories around the world has helped standardize the methodologies and data collection procedures used in psychological research programs. In turn, this standardization has fostered easier comparison of data findings, replication of results, and collaboration and cooperation among hundreds of behavioral scientists worldwide.

Renamed the Department of Human Behavioral Biology in 1984, this same department also pioneered the development and use of other sophisticated high-technology research tools and techniques. Among these are wear-and-forget wrist-worn activity monitors, based on piezoelectric chip technology, to permit assessments of how much sleep an individual obtains in military field exercises or deployments (Redmond & Hegge, 1985; Russo et al., 2005). Department researchers configured prototype versions of a wrist-worn activity monitor into a soldier's "sleep watch," which contains a personal sleep management algorithm that permits the soldier to monitor the "reservoir of cognitive alertness" remaining before he or she must stop working and obtain needed sleep. This sleep watch has been deployed on an experimental basis with soldiers and with commercial long-haul truck drivers (Dinges, Maislin, Krueger, Brewster, & Carroll, 2005; for publications of WRAIR's Department of Human Behavioral Biology, see Krueger, 1986).

Army psychological research on sustained soldier performance has had impacts on the regulation of public transportation operators' hours-of-service rules (for commercial aviation, rail, shipping, trucking, and municipal transit systems), operator fatigue countermeasure programs (e.g., McCallum, Sandquist, Mitler, & Krueger, 2003), and fitness-for-duty rules at nuclear power plants. All of these are based in part on seminal behavioral science work performed over the decades at the Army's WRAIR (Balkin et al., 2000), USAARL (J. A. Caldwell & Caldwell, 2003, 2005), USARIEM, and sister service labs in the Navy and Air Force. The civilian community has not yet determined how to apply the significant behavioral research on stimulant (J. A. Caldwell & Caldwell, 2005; Newhouse et al., 1992; Wesensten et al., 2004) and hypnotic (Balkin et al., 2002; O'Donnell et al., 1988) drug compounds and other nutritional and chemical substances (Lieberman, 1990) explored by Army and other military psychologists in search of ways to sustain military performance.

PSYCHOLOGICAL PREPARATION FOR WEAPONS OF MASS DISRUPTION

The U.S. Army Medical Research Institute of Chemical Defense (USAMRICD) was established in 1979 as the Army transitioned from developing offensive chemical weapons to devoting significant amounts of research to medical defense against them. Research psychologists have conducted numerous medical and behavioral experiments using animal or human models to develop pretreatment, prophylaxis, and treatment drugs as well as medical antidotes for those who might be exposed to battlefield chemical threats. The research psychologists at USAMRICD are physiological or pharmacological psychologists, neuroscientists, and comparative behavioral biologists (Romano & King, 2002). This work has provided the military with predictions of the performance and health effects of vaccines and pretreatment prophylactics before exposure to weaponized chemical threats and also determines the effects on performance and biology of medical treatment antidotes after exposure of combatants to threat agents (e.g., research on chemical compounds like pyridostigmine as a pretreatment and on atropine sulfate as an antidote given after nerve agent exposure; see Headley, 1982; McDonough, 2002; Romano & King, 2002).

This work provides equally significant payoffs in preparing the civilian populace and emergency response teams to cope with mass exposure to terrorist attacks with chemical and biological weapons. Most behavioral science work on chemical threats has been ongoing at the USAMRICD at Edgewood, Maryland, and research on protection against biological threats, at the U.S. Army Medical Research Institute of Infectious Diseases (USAMRIID), was reformulated in 1973 from offensive biological weapons research at Fort Detrick, Maryland. Significant medical research on both these topics is also being accomplished at the WRAIR.

Behavioral scientists with extensive practical experience in military medical laboratories have much to contribute in readying the civilian populace to inoculate themselves with immunizations such as those developed for smallpox and anthrax and to prepare first responders to analyze terrorist activities and select the proper response in dealing with a panic-stricken civilian populace (see Pastel, 2001). Many occupational and preventive medicine findings now are being used to train civilian disaster, first responder, and rescue teams around the country to deal with harsh circumstances of possible terrorist attacks, especially those that might involve weapons of mass destruction or, as military psychologists refer to them, weapons of mass disruption; the term *disruption* encompasses the wider impact terrorist attacks have on people who just want to be assured they have not been contaminated by a chemical or biological agent (Pastel, 2001). Research by military psychologists on prediction of human performance while wearing chemical–biological protective clothing and using decontamination equipment are helping the nation prepare for weapons of mass disruption and determine the terrorist attack response readiness posture (Krueger & Banderet, 1997).

CONCLUSION

Over the past 60 years, the U.S. Army has been fortunate to have teams of dedicated uniformed and civil servant psychological and behavioral science researchers. They have frequently collaborated with colleagues in Air Force, Navy, and allied foreign laboratories and universities. Together, their research helps enable U.S. soldiers, sailors, airmen, and marines and soldiers of U.S. allies to fight better, longer, stronger, smarter, and safer. These scientists, working as part of multidisciplinary teams, adhere to laboratory mission mottos such as "Conducting Research for the Soldier," "Conserve the Fighting Strength," and "Protect and Sustain a Healthy and Medically Protected Force." Their stated goals include ensuring that military forces are deployed in a state of superb health and are equipped to protect themselves from disease, injury, and the many health hazards that accompany operation of complex military equipment in high-tempo deployments in harsh environments. But most of all, military psychologists work to permit combatants and support personnel to engage in optimal performance in the face of arduous battlefield operations.

This chapter has described only snippets of citations from the broad research of the military research laboratories and numerous sponsored extramural civilian research programs. This chapter, indeed this book, briefly describes only a sampling of their work. Significant findings of such dedicated military psychological research studies have been published in peer-reviewed scientific journals and periodicals. Many other findings are documented in government laboratory technical reports; most of those reports are archived and available from the Defense Technical Information Center at Fort Belvoir, Virginia, or from the laboratories themselves. The millions of soldiers, sailors, airmen, and marines who have worn the uniforms of the U.S. armed forces since World War II have benefited from the work, dedication, and significant accomplishments of psychological science patriots and their service to this nation.

6

THE HISTORY OF SPECIAL OPERATIONS PSYCHOLOGICAL SELECTION

L. MORGAN BANKS

The history of special operations psychology is relatively recent. Although what is now called *special operations* can be traced to the very beginnings of the United States, it had its modern start in the 1940s, during World War II. During that war, such units as the U.S. Army Rangers, Merrill's Marauders (the 5307th Composite Unit [Provisional]), and the 1st Special Service Force began the long road to institutionalizing the concept of special operations in the U.S. military. One of the better known U.S. special operations organizations of the war was the Office of Strategic Services (OSS). In late 1943, tapping into the patriotism and dedication of some of the best psychologists in the United States, the OSS established the first psychological assessment center in the United States (MacKinnon, 1980; Morgan, 1957; OSS Assessment Staff, 1948). Ten years later, when U.S. Army Special Forces was created, halting steps were taken to use psychological assessment as part of the selection process. Over the years, this program waxed and waned, and it was eventually eliminated during the Vietnam War.[1]

---

[1]Although I have been unable to document the reason for this elimination, the success of Special Forces operations in Vietnam and the need for greater numbers of Special Forces soldiers than a strict selection program could supply likely were the major causes.

Following the Iran hostage crisis (November 1979–January 1981), an Army aviation unit was organized in 1981 to provide aviation support to U.S. special operations ground forces. These aviators flew difficult missions (e.g., pioneering flight under night vision goggles), and they consequently had a high accident and fatality rate. Partly because of this, the unit, which is now referred to as the 160th Special Operations Aviation Regiment (Airborne), began a formal psychological assessment for all of its pilots. This assessment program continues to this day.

In 1988, the use of psychological assessment in Special Forces was reborn. That year, a formalized assessment program was created that had screened, by late 2005, over 40,000 soldiers for assignment. In 1994, the 75th Ranger Regiment instituted a formal selection program involving psychological assessment for all officers and NCOs being considered for assignment to the regiment.

The use of psychologists in U.S. Special Operations Forces (SOF) has become well entrenched, and their numbers continue to grow. The origin of all Army psychology is deeply rooted in assessment and selection (Banks, 1995), and it has historically had great (arguably, its greatest) success in this area. Army SOF psychology has expanded to include a multitude of services within SOF (e.g., training, organizational consultation, direct support to combat operations, research), but all of the current positions have as their basis the assessment and selection of soldiers for critical tasks.

OFFICE OF STRATEGIC SERVICES

A discussion of the development, utilization, and success of the OSS is beyond the scope of this chapter. However, because of its seminal role in psychological assessment, a brief description is included. The OSS was in many ways a unique organization for the United States. Although intelligence sections of the various military services had existed since the founding of the country, and although the Federal Bureau of Investigation conducted counterintelligence activities in the continental United States, at the beginning of World War II there was no unified agency responsible for organizing all foreign intelligence. Not too surprisingly, Army intelligence did not communicate well with Navy intelligence, and both tended to have a distinct tactical military focus. In 1941, President Roosevelt created the OSS and placed William Donovan, a well-known (and well-connected) attorney and World War I hero, at its head. Roosevelt gave it the mission of being the major national-level agency responsible for intelligence collection, espionage, subversion, and psychological warfare. As such, it was one of the forerunners of the modern SOF. The tremendous contribution of the OSS has been documented elsewhere (see Bank, 1986; Smith, 1972); as one example, General Eisenhower stated that in Europe alone, the resistance movement they had fostered was equivalent to 15 infantry divisions (Bank, 1986). Unfortunately, when the OSS was created, there was little screening of candidates before acceptance into the organization. One common nickname for the organization was "Oh-So-Social," because many of the members were well-to-do friends (or friends of friends) of Bill Donovan.

By mid-1943, having undergone tremendous wartime expansion, the OSS had developed a significant personnel problem. Because of the high-threat environments of some of the missions, many who were deployed overseas had difficulty adjusting to the danger and stressors encountered in OSS operations (OSS Assessment Staff, 1948). The United States considered using a psychological–psychiatric assessment unit for selecting officer candidates similar to that of the British (OSS Assessment Staff, 1948, p. 4; see also Ahrenfeldt, 1958). This idea led to the development of the first psychological assessment center in the United States (MacKinnon, 1980). Over the next year and a half, more than 5,000 prospective candidates were evaluated before acceptance into the OSS. This assessment was performed at no small cost and was the precursor to both the civilian personnel assessment center movement and to the current special operations selection programs.

A large number of prominent psychologists were responsible for the creation and operation of the OSS selection program. Donald K. Adams, Staff Sergeant Urie Bronfenbrenner, First Lieutenant John W. Gardener, Joseph Gengerelli, O. H. Mowrer, Edward C. and Ruth S. Tolman, and Robert Tryon were among the long list of distinguished participants. Five individuals, Lieutenant Donald W. Fiske, Eugenia Hanfmann, Donald W. MacKinnon, Captain James G. Miller, and Lieutenant Colonel Henry Murray, assumed overall responsibility for completing a remarkable book that discussed their selection program, *The Assessment of Men* (OSS Assessment Staff, 1948). Although many were deserving of recognition, Murray was probably the individual most closely associated with the development and philosophy of the program. He and his colleagues developed a 3-day assessment program that attempted to measure a candidate's suitability for assignment to the OSS. Although a solid evaluation of the success of the program is difficult, the rate of stress-related problems reported from the field dropped dramatically following the initiation of the selection program (Banks, 1995; OSS Assessment Staff, 1948).

The psychologists working in the OSS were used primarily for assessment and selection, although one psychologist did manage to deploy operationally (Morgan, 1957). At the end of the war, the OSS was disbanded. In 1947, its intelligence collection functions were given to the newly created Central Intelligence Agency, whose first chief of psychology was a former OSS psychologist (W. J. Morgan, 1994, personal communication).

SPECIAL FORCES ASSESSMENT AND SELECTION

U.S. Army Special Forces was created in 1952, initially as part of the Psychological Warfare Center at Fort Bragg, North Carolina.[2] On the basis of the experiences of the OSS, the 1st Special Service Force, and other similar units in World War II, and in anticipation of the Cold War, this new organization had a variety of missions that revolved around guerilla warfare (now also called *unconventional warfare*) and sabotage. In fact, the current missions of Special Forces include unconventional warfare, *foreign internal defense* (training countries how to fight insurgencies), *direct action* (specific, aggressive, tactical operations, usually with strategic implications), *special reconnaissance* (the collection of tactical military information, often deep within enemy territory), and counterterrorism. Initially, many of the Special Forces soldiers were prior OSS members or foreign nationals gaining their U.S. citizenship by serving on active duty under a program referred to as the Lodge Act (C. M. Simpson, 1983, p. 26).

In 1952, Colonel Aaron Bank, a former member of the OSS who had parachuted into occupied France and Indochina, was assigned to Fort Bragg, North Carolina (C. M. Simpson, 1983). On June 20, 1952, the 10th Special Forces Group (Airborne) was created under his command. Initially, Colonel Bank designed and ran the Special Forces training program without including a formal psychological assessment program. Instead, he relied on selecting only very fit and motivated soldiers (Bank, 1986). He did, however, rearrange each 12-man team for the best compatibility. Later, a psychiatrist was used briefly to screen candidates, and a psychological test battery, the Special Forces Selection Battery, was developed and used (Department of the Army, 1961). This battery measured a number of areas, including visual–spatial ability. This battery was dropped sometime during the 1960s, probably for the same reasons as discussed below.

Special Forces, which was born with the idea of helping to defend western Europe, instead got its biggest boost from the Vietnam War. It expanded greatly during the 1960s, first with President Kennedy's personal support and then because of a greatly increasing number of missions around the world, mostly involving counterinsurgency (C. M. Simpson, 1983). This expansion, however, generated some selection and assessment problems. Initially, the training for Special Forces lasted over a year and had a high attrition

---

[2]*Special Operations Forces* is a term that includes a number of distinct types of military units, and includes Special Forces units, Civil Affairs units, Special Operations Aviation Units, and Psychological Operations units. An interesting side note is that historically there has been limited involvement of military psychologists in army psychological warfare, or *psychological operations* as it is currently called. Although positions have existed at various times for psychologists within active duty psychological operations units, they have been difficult to fill. Perhaps this is because there have been limited numbers of psychologists who have been interested, but I suspect that it is a combination of that and limited knowledge about the existence of the positions. The reserve component psychology positions within psychological operations units have, conversely, usually been filled.

rate. The requirement for increased numbers of Special Forces soldiers implied that more candidates would have to be brought into, and turned out of, the training program. In turn, this meant that fewer candidates could be screened out if the increased requirement was to be met. Eventually, the need for psychological screening became moot and screening was discontinued. As the quote below describes, the emphasis was on getting Special Forces soldiers into the field, which was not without some controversy:

> At Fort Bragg, the Special Warfare School . . . dutifully shifted into high gear in an effort to meet the new manpower goals. . . . [the] Special Warfare School increased its output of graduates from something under 400 a year to almost eight times that. Attrition through training standards had always been a valuable tool when the Forces were small; it served to sort out those who belonged from those who did not. By 1962, attrition had fallen to about 70 percent from its earlier rate of near to 90 percent. By 1964, it was down to 30 percent, and more ominously, the "numbers merchants" at the Special Warfare Center and in the Pentagon were applauding the improvement. (C. M. Simpson, 1983, p. 66)

It was in Vietnam that Special Forces was to prove itself on the battlefield. Although the U.S. strategic policy used in Vietnam may have been flawed (see Summers, 1982), the tactics and techniques used by Special Forces units during Vietnam were extremely successful (Schemmer, 1976; C. M. Simpson, 1983; Stanton, 1985). According to one author, Special Forces had

> trained or retrained large portions of three standing armies for frontline combat . . . trained most of the indigenous special warfare contingents in South Korea, Taiwan, Thailand, the Philippines, and South Vietnam . . . [and] created, trained, and fielded the Civilian Irregular Defense Group program troops that fought a large share of the war throughout the most threatened regions of Vietnam. (Stanton, 1985, pp. 291–292)

However, after the post-Vietnam drawdown in the mid- to late 1970s, Special Forces began, like the rest of the Army, a significant decline in both number and quality of personnel. Additionally, selecting and training Special Forces soldiers required great expense in both manpower and money. (All Special Forces soldiers are men, because Special Forces is a combat arms specialty, and Congress has required that all combat arms specialties be composed of men, excluding women.) All assessment and training of Special Forces soldiers were, and continue to be, conducted by the John F. Kennedy Special Warfare Center and School (SWCS), the descendent of the old Psychological Warfare Center, at Fort Bragg, North Carolina.

By 1988, Special Forces training, technically called the Special Forces Qualification Course (SFQC), consisted of three phases. (For an excellent description of the final phase of the SFQC, see Waller, 1994.) The first phase included training soldiers in basic infantry skills, including patrolling, raids, and ambushes, but also advanced land navigation and some beginning Special Forces skills. There was limited prescreening of soldiers before their assignment to the SFQC. For this reason, the first phase was also used to screen out soldiers who were not considered appropriate candidates for eventual assignment to an operational Special Forces unit. Conceptually, the purpose of this first phase was ambiguous. The purpose of training soldiers up to a particular standard was confused with the task of selecting out soldiers who would probably not be successful in Special Forces. As might be expected, this phase had a high attrition rate, often up to 60%. Most soldiers were permanently moved to Fort Bragg, North Carolina, from other locations around the world before beginning this training. Therefore, when they were dropped from the course, they either had to be assigned somewhere else at Fort Bragg or moved, at significant expense, to another Army base. This caused morale problems and was not very cost-effective for the Army.

To improve the quality of Special Forces soldiers and reduce the overall expense of their selection and training, then-Colonel (later Brigadier General) Richard Potter proposed and successfully defended the concept of a 3-week selection course. Now referred to as the Special Forces Assessment and Selection (SFAS), the first iteration was conducted in 1988. Candidates would be brought on temporary duty to Fort Bragg, go through SFAS, and then be returned to their parent units. If they were successful in the selection, they would be given orders to report for training and reassignment into Special Forces. If they were not successful, they would have no derogatory information placed in their files and would, in fact, be thanked for attending. This philosophy of not denigrating nonsuccessful candidates had, and continues to have, several positive effects. Good soldiers who are not successful do not leave with a bad impression of Special Forces, and it encourages candidates who might not otherwise apply to do so. This philosophy also reinforces the standard that Special Forces soldiers should always treat people with respect, because most of the jobs in Special Forces involve working with people of other cultures, often in politically sensitive situations. Their treatment in the selection begins their own acculturation process.

To assist in the assessment, a psychologist position was created in the SWCS, and a uniformed psychologist was assigned in 1989. He conducted a screening of each individual and viewed the candidate's performance during the entire course. This type of psychological assessment, which is still used today, begins with the administration of a number of routine tests, including the Minnesota Multiphasic Personality Inventory (MMPI; a measure of psychopathology, now the MMPI–2; Greene, 1991) and the Wonderlic Personnel Test (a measure of intelligence; Wonderlic, 1992), soon after the candidates arrive. The candidates are then put through a grueling series of tasks, all of which are designed to measure their motivation, fitness, practical intelligence, and ability to work with others under stress. They are tested both individually and in groups, and carefully designed behavioral observations are taken during each task.

At the end of approximately 3 weeks, the psychologist reviews the psychological profiles of the remaining candidates. Soldiers with unusual profiles or with profiles that have historically been associated with poor performance are then individually interviewed by the psychologist, who attempts to assess each soldier's suitability for success in training and ultimately for assignment in Special Forces. Of necessity, this process focuses on selecting out unsuitable candidates rather than selecting in the most suitable ones. In particular, the psychologist attempts to identify potential problems that are likely to interfere with the soldier's success, either academically during the training course or behaviorally at any time. For example, a history of poor academic performance coupled with poor test scores on the Wonderlic Personnel Test indicates a poor likelihood of success in the more academically challenging courses. Similarly, a history of arrests or of nonjudicial punishment while on active duty is a good indicator of future problems with authority. The psychologist attempts to evaluate the "whole man," incorporating not only the test scores, but also the complete background and history of each individual, and then prepares recommendations for each questionable candidate.

Finally, a board of experienced Special Forces officers and sergeants major meets to review each candidate's performance. The senior officer on the board, usually the Training Group Commander, a Special Forces colonel, is the president of the board. The psychologist presents relevant information on questionable candidates to this board. The board president is the decision maker on selection, and the psychologist's role is that of an advisor. This is a major distinction between the role of the physician medical advisor, usually referred to as a "surgeon" in the Army, and the psychologist. Ordinarily, the surgeon finds an individual either qualified or not qualified, medically, for a particular job in accordance with specific regulatory guidelines. In contrast, the psychologist gives a descriptive analysis of a particular individual's strengths and vulnerabilities, and the president of the board makes the selection decision. As of summer 2005, about 40,000 soldiers had been screened and assessed.

From the inception of SFAS, Army Research Institute (ARI) psychologists were intimately involved in its design. ARI assisted by tasking research and personnel psychologists to assist in the development of this program. They provided insightful and carefully documented recommendations, always helping to ensure that the program was on solid empirical and legal ground. Many of these reports focused on larger issues, such as a major needs analysis of Special Forces (Brooks, 1992) and a detailed job analysis of Special Forces positions (Russell, 1994). The integration of clinical psychologists performing the assessments with personnel psychologists continually reviewing the entire selection program led to a close working relationship that continues to this day. This symbiotic relationship was so successful that in 1994, ARI created a field office at Fort Bragg. An industrial–organizational psychologist with extensive experience in personnel selection was permanently assigned to work with the Army special operations community. To the author's knowledge, not since the OSS program has such an integration been so successful. By virtue of the tremendous number of assessments conducted, the clinical psychologist assigned to the SWCS, along with the ARI psychologist, have been able to conduct extensive normative and predictive research on profiles associated with success and nonsuccess. They have developed new screening instruments and validated currently used tests. This emphasis on validating, through research, the effectiveness of the screening process began with the initiation of the program and continues to this day.

The SWCS is also responsible for training soldiers in a variety of advanced skills required of Special Operations Forces soldiers. These areas include psychological operations, civil affairs, foreign language training, and various high-risk training courses such as military free fall (parachute operations used to infiltrate enemy areas under the cover of darkness) and underwater (dive) operations. The SWCS psychologist provides instruction in many of these courses, teaching areas such as cross-cultural communications, target audience analysis, and even stress management. He has also been involved in preparing and debriefing personnel involved in many Special Forces deployments and is on call to provide support to the SOF community.

Since September 11, 2001, Special Forces soldiers selected using this process have been involved in combat operations around the world in support of the global war on terrorism. Many of these operations, such as the warfare conducted in Afghanistan, have required extremely high levels of stress tolerance and independence of action and are a testament to the success of the overall Special Forces selection process.

160TH SPECIAL OPERATIONS
AVIATION REGIMENT (AIRBORNE)

The unit currently referred to as the 160th Special Operations Aviation Regiment (Airborne; 160th SOAR) was created after the failed attempt to rescue the U.S. hostages in Iran in 1980. One of the identified problems in that attempt was that the United States had no dedicated, properly trained, special operations rotary wing aviators (Ryan, 1985). Consequently, a task force was initially formed, and later, in October 1981, the unit was formally designated as Task Force 160. One of the critical tasks that these aviators had to master, and that had not been well developed until that time, was the use of night vision goggles to fly helicopters at night. This was extremely dangerous work, and the unit experienced a number of fatalities while developing the techniques required for use of the goggles. Partly because of this, and to assist in selecting these men, a psychologist was assigned in 1984.

Selection into the 160th SOAR requires that each candidate aviator meet both rigorous physical standards and even more rigorous flying standards. The current selection program includes a complete psychological evaluation by the unit psychologist. Each candidate is fully tested and then interviewed for suitability. The results of that evaluation are presented to a board of senior leaders, and this board makes the decision on selection.

As with the other psychologists in SOF, teaching and training are an important role for this psychologist. As might be expected, these aviators undergo extensive and intensive training before they are considered operational members of this unit. The psychologist provides subject matter expertise to assist in aviator training. The psychologist also provides consultation to the training personnel and students on a case-by-case basis as required.

Because of the high operations tempo of the 160th SOAR, the psychologist also plays a crucial operational role. Not only is he able to provide typical combat stress or battle fatigue treatment and command consultation, but he is also trained in survival, evasion, resistance, and escape (SERE) debriefing (described in the next section). Following the 1993 capture and subsequent release of Chief Warrant Officer Michael Durant, the assigned psychologist provided critical support during Durant's transition back to the United States. Since the September 11, 2001, terrorist attacks, this psychologist has also deployed extensively in support of the global war on terrorism.

SURVIVAL, EVASION, RESISTANCE, AND ESCAPE

In mid 1980s, the Army created a program designed to train soldiers how to successfully survive captivity by training them in a simulated prisoner of war environment. At that time, the U.S. Air Force and Navy had permanent programs but the Army did not. Consequently, the Army SERE Department was created in the SWCS at Fort Bragg to train soldiers Armywide in a safe but intense environment. Lieutenant Colonel (later Colonel) Nick Rowe, who had been a prisoner of war for 5 years (Rowe, 1971), was brought in to create this training course. A requirement learned from past experiences with such programs was that a psychologist must be assigned to any SERE program. As described by Zimbardo, Maslach, and Haney (2000), programs that simulate captivity can become dangerous and can cause both physical and psychological damage to students. I was assigned in 1985 to provide psychological screening of all instructors providing such training and to ensure that such training was conducted in a psychologically safe manner. Since then, a SERE psychologist has been assigned and present at each training event.

To be effective, the training must simulate captivity in a realistic manner. As Zimbardo et al. (2000) pointed out, it is extremely easy for soldiers who are role-playing enemy guards and interrogators to act inappropriately, possibly causing harm to students. The original reason for assigning a psychologist was to screen out instructors who would be likely to endanger students during the training; by performing a detailed evaluation of each potential instructor, the SERE psychologist would help prevent the assignment of personnel who might injure students in a simulated prisoner of war environment. Although screening has certainly played a role in ensuring quality instructors, perhaps the most important value of SERE psychologists has been in providing behavioral expertise to enhance the design of the program to ensure its safety. In other words, they have made sure that psychologically effective safeguards are built into the training, including constant monitoring of instructors; treatment of stress reactions among students; and, most important, constant monitoring of the training process and environment. SERE psychologists have also been heavily involved in research on the effects of high-stress training and on prevention of these effects.

Because of their expertise in the behavior of soldiers in captivity, SERE psychologists have been involved in debriefing and helping Americans who have been held captive by foreign powers. For example, Chief Warrant Officer Bobby Hall was shot down and detained by the North Koreans in 1994. The SERE psychologist was sent to Korea to assist in his debriefing and transition back from captivity. In another example, when Chief Warrant Officer Michael Durant was released following his captivity by a Somali warlord, he was first met in Mogadishu by a SERE-trained psychologist from SOF who was supporting operations there. This psychologist accompanied Durant to Landstuhl, Germany, where he was met by the psychologist assigned to Durant's unit, the 160th SOAR. The unit psychologist then accompanied Durant to the United States and assisted in his debriefing and homecoming. SERE-trained psychologists have assisted in all repatriations in the past 20 years. In the current series of conflicts, they are deployed to assist any U.S. prisoners of war on their repatriation.

As might easily be imagined, most mental health professionals do not receive training in this area. Consequently, it is often easy for persons with limited experience to do more harm than good. For this reason, the Joint Personnel Recovery Agency, the staff element of the Joint Forces Command that oversees such training and debriefing, has established standards for specialty training and expertise in this area. At present, the ability to support repatriation activities is a specially credentialed skill for Army psychologists. Recently, the Army Medical Command identified 38 psychology positions worldwide that require this specialty and created a skill identifier to track psychologists with this training.

SERE psychologists, like the other psychologists working in SOF, also provide classroom instruction in the psychology of hostage and prisoner of war survival, captor behavior, and stress reactions. Because of their combined expertise in human behavior and in the actual history of prisoner of war behavior, they are often involved in helping to develop Army, and joint, doctrine both on SERE training and on how to best manage the care of former prisoners of war. SERE psychologists have been instrumental in helping write the current Army doctrine on how to survive captivity, *Field Manual 21-78* (Department of the Army, 1989) and the current chapter in the Army's survival manual, *Field Manual 21-76* (Department of the Army, 1992) on the psychological aspects of survival.

RANGER ASSESSMENT AND SELECTION PROGRAM

The 75th Ranger Regiment, the country's premier light infantry organization, has one of the most distinguished lineages of any U.S. Army unit. This began at least as early as the French and Indian Wars and continued through the American Revolution, the Civil War, and World War II (Center of Military History, 1990; Hogan, 1992; Lock, 1998) and has continued during the vicious street fighting in Somalia in 1993 (Bowden, 1999) and in numerous actions in the global war on terrorism. Today, the 75th Ranger Regiment is a flexible, highly trained, and rapidly deployable light infantry force that can be used against a variety of targets. It can conduct offensive operations against targets of strategic importance, such as seizing airfields or other key facilities and performing raids, or other crucial missions of national importance, such as evacuating noncombatants from a hostile situation. Additionally, it may conduct other sensitive operations in support of national policy objectives. All of these missions require tremendous dedication from the Rangers, often because the training requirements for such missions are extremely rigorous and because these missions can be some of the most difficult and dangerous the nation must perform.

The regiment has always selected its soldiers carefully, but in 1994, based partly on the success of the psychological selection programs discussed in this chapter, it added a psychological component to its selection process for leaders. All soldiers with the rank of sergeant and above must complete a Ranger Orientation Program before taking a position in the regiment. This orientation includes not only a physical assessment of the Ranger leader but also classes on professional ethics, Ranger standards of behavior, and the law of land warfare. In 1994, the psychological component of the assessment was added, referred to as a Ranger Assessment and Selection Program. The technical aspects of the assessment are similar to those of Special Forces and include personality testing and measures of intellectual functioning. In addition, a leader's ability to handle various challenging situations is assessed.

A board then meets to consider each candidate. The Regimental Deputy Commander is ordinarily the board president, and the members consist of the Regimental Battalion Commanders, sergeants major, and other field-grade officers of the regiment. The board evaluates each individual as a whole man, not focusing on any one particular trait. Again, the psychologist provides input to the board in the form of strengths and weaknesses of each individual and functions only as an advisor. The board gives a recommendation regarding each candidate to the Regimental Commander, who makes the final determination on assignment.

Since the beginning of the global war on terrorism, the 75th Ranger Regiment has been extensively involved in combat operations. The assigned psychologist has accompanied these soldiers throughout the world, providing direct psychological support where it is needed most.

PSYCHOLOGICAL APPLICATIONS DIRECTORATE

The U.S. Army Special Operations Command (USASOC), a Major Command, was created in 1989 as the headquarters element for all Army special operations units. USASOC currently includes the units discussed above (i.e., Special Forces, the 160th SOAR, the 75th Ranger Regiment, and the John F. Kennedy SWCS) as well as the U.S. Army Civil Affairs and Psychological Operations Command and the Special Operations Support Command. There are approximately 25,000 soldiers assigned to USASOC. As the use of psychologists grew within USASOC, a need developed for a special staff office to provide technical support and supervision of the use of psychology within the command. In 1994, the Psychological Applications Directorate was established to meet this need, composed of one colonel as the director and a small staff. As the senior psychologist within special operations, the director has technical responsibility for all the active duty psychologists within USASOC. The director reports directly to the commanding general of USASOC through the USASOC chief of staff. Specifically, this position is responsible for ensuring the provision of all psychological support to USASOC, including the selection and assessment programs discussed in this chapter and a variety of leadership development programs, command consultation, and operational psychology support.

CONCLUSION

Throughout the initiation and development of special operations psychological programs, several consistent themes have emerged. Psychologists function in an advisory role. In selection programs, they provide the command with the best possible assessment but leave the decision in the hands of the commander, who is the individual responsible for living with the consequences of the selection decision.

These psychologists have been most successful when they have been assigned to the lowest possible level. Because of the sensitivity of the information psychologists provide, they must be trusted agents. To be successful, they must have the trust of both the unit commanders and the unit's soldiers. To gain this trust, the psychologist must walk a very careful line ethically, being constantly aware of the potential for and avoiding improper dual relationships. At the same time, the psychologist must be intimately familiar with the job requirements of the unit and must be seen as supportive of the organization's mission. Although not without its own set of problems, the role of an internal consultant, assigned to the supported unit, has been successful.

Work in this area requires specialized training and experience. Although not inordinately difficult to acquire, in most cases it is critical to success. In some areas, such as SERE support, specialized training is an ethical and regulatory requirement. In others, such as providing assessments of average- to high-functioning soldiers, this work requires a shift in the internal norms that psychologists use from assessing psychopathology to assessing each individual's strengths and vulnerabilities. Psychologists must provide feedback in a useful and understandable manner.

The use of psychologists in the selection and assessment of soldiers for Special Operations Forces has shown tremendous growth over the past 15 years. Screening drill sergeants, providing support to basic trainees, and assisting with the selection and training of recruiters are just some of the areas currently being developed. This model has a bright future.

III

PERSONNEL MANAGEMENT

Personnel management includes the assessment, selection, classification, assignment, and training and education of personnel as well as occupational analysis, human factors engineering, performance assessment, and career management. Military jobs are highly varied and involve unique training requirements; many military jobs have no civilian occupational equivalent. Military personnel train to master basic skills, military-specific skills, and special job skills. As military tasks have become more complicated with new and more sophisticated weapons platforms, environments, and equipment, training to specific standards has become increasingly important.

Military psychologists are tasked to use scientific procedures to evaluate the effectiveness of training programs and to recommend approaches to improve and sustain performance. The research programs they have established have often been at the cutting edge in applying technology to meet service needs. Cooperative endeavors are common among the services, professional organizations, universities, and security organizations. The military is a demanding environment for training large numbers of diverse people to high levels of performance, especially following the advent of the all-volunteer force in 1973. Thanks in part to the efforts of psychologists, military training procedures are among the most effective occupational education programs.

7

THE HUMAN RESOURCES RESEARCH ORGANIZATION: RESEARCH AND DEVELOPMENT RELATED TO NATIONAL SECURITY CONCERNS

PETER F. RAMSBERGER

In 1950, the U.S. Army formed a study group to determine how re-search in the behavioral and social sciences could be better integrated into personnel, training, and policy efforts. The Research and Development Board of the Office of the Secretary of Defense established a Committee on Human Resources. Their report, *An Integrated Program in Human Resources Research*, was approved by the undersecretary of the Army in June 1951. One of the recommendations this report made was that

> a major contract be awarded to a recognized educational institution to provide for the formation of a Human Resources Research Office, which would have primary responsibility for conducting research in the areas of training methods, motivation and morale, and psychological warfare tech-niques. (Crawford, 1984, p. 1267)

The Department of the Army awarded this contract to George Wash-ington University on July 27, 1951. Six days later, on August 2, 1951, the

Human Resources Research Office (HumRRO) was opened on the campus of the university, with Meredith Crawford as its first director and employee. Crawford was dean at Vanderbilt University at the time. He was recruited to the position of HumRRO's director by Harry Harlow, who, although best known for his work on learning and attachment, was consulting with the Army and was one of the leaders of the study group that led to the recommendation that HumRRO be created.

HumRRO quickly expanded and fulfilled part of its mission by establishing field units at major Army installations across the country, including the Armor Research unit at Fort Knox, Kentucky; the Infantry unit at Fort Benning, Georgia; and the Air Defense unit at Fort Bliss, Texas. The size of the staff also grew, reaching a high of around 300 in the early 1960s.

In 1967, HumRRO's contract with the Army was modified to allow it to take on other sponsors. Two years later, another change came; HumRRO separated from George Washington University and became an independent, nonprofit research and development center. Although the Army continued to be the organization's principal sponsor, the percentage of its revenues generated from this source declined steadily throughout the 1970s. HumRRO's final sole-source contract with the Army came in 1975. Since that time, the organization has maintained a close relationship with the Army but has branched out to do work for the other services, government agencies, and private sector clients.

THE HUMAN RESOURCES RESEARCH ORGANIZATION'S CONTRIBUTIONS RELATED TO NATIONAL SECURITY

Over the years, HumRRO researchers have conducted a wide variety of studies and analyses in such areas as organizational development and program and policy evaluation. However, their contributions to national security issues have been primarily in the areas of personnel selection and classification, training, assessment, and force utilization. The relative emphasis on these various topics has not been consistent, reflecting both the interests of HumRRO staff and the needs of the clients they serve. This chapter highlights some of the major contributions of HumRRO and its sponsors to each of these areas and addresses some relevant but lower profile endeavors.

Selection and Classification

Over time, technology has played an increasingly important role in the defense of the nation. This is true both from a military and civilian perspective, as new weapons and other tools are used to protect U.S. interests at home and abroad. In the final analysis, however, it is the men and women who use those tools who are ultimately responsible for the country's well-being. Therefore, selection of individuals who will be able to meet the demands of training (where applicable) and the job itself is critical to ensure that organizations tasked with national security functions are able to perform them adequately.

During HumRRO's earliest years, selection into military service was not as paramount an issue as it is today, given that there was a draft that allowed the Army to select the required number of individuals from a large pool of American youths. A variety of classification studies were carried out, however, that were far reaching in scope and methodology as the Army sought to assign thousands of new recruits to appropriate jobs. Classic examples of such research are the Fighter studies, carried out in the 1950s and early 1960s. (Specific references are presented with the detailed descriptions below; there are 42 Fighter citations in the HumRRO bibliography.) Following World War II and other conflicts, a much-discussed topic was the incidence of ineffectual performance on the battlefield. Some estimates suggested that only 15% of soldiers actually fired their weapons during firefights. Although these figures were disputed, the fact remained that the failure of a significant proportion of soldiers to contribute to the efforts of their comrades was cause for concern among Army leadership. As a result, the degradation of behavior in combat was one of the first areas of investigation HumRRO was tasked to examine. Project Fighter I was designed to isolate characteristics of individuals who perform well under combat conditions and those who do not in the hopes of improving selection, classification, and training practices to overcome the noted deficits (Egbert et al., 1958).

HumRRO researchers went to Korea shortly after the cease-fire agreement was reached in 1953. There they interviewed members of units that had recently engaged in combat. A total of 647 soldiers were asked to name two or three men they would most like to have beside them in action and two or three they would least like to be at their side. They were then asked to describe actual incidents from past close combat to support their choices. Soldiers were designated *Fighters* when two or more comrades had identified them positively or one comrade identified them and some commendation had resulted from the action described. *Non-Fighters* were similarly selected. From the 1,000 men mentioned, 345 were selected as Fighters or non-Fighters.

The incidents themselves were used to generate categories of behavior that described the two classes of soldiers. Generally, Fighters were found to provide leadership, take aggressive action, perform supporting tasks while under fire, or any combination of these. Non-Fighters withdrew from the scene (either physically or psychologically), malingered, overreacted, or became incapacitated.

The selected soldiers underwent a week of psychological tests, including personality and interest inventories, intelligence tests, and life history measures. In all, some 27 questionnaires and 60 objective tests were administered, yielding more than 400 separate scores. The data were then analyzed in an attempt to isolate differences between the two groups. HumRRO researchers found that Fighters and non-Fighters could be distinguished on several dimensions. Fighters were more intelligent, socially mature, and emotionally stable as well as healthier; exhibited greater leadership potential; and possessed more military knowledge. Egbert et al. (1958) suggested that the findings be used to develop classification and assignment procedures that would permit Combat Arms to identify and select potential Fighters with maximum effectiveness. In addition, the authors suggested that training be examined to isolate practices that might inculcate the traits and attitudes that characterize soldiers who perform better under combat conditions. In the end, these results provided the most scientifically based portrait available at the time of soldiers who rose to the challenge of effectively performing under combat conditions.

Another line of research that concerned performance in combat focused on stress as a major contributor to skills degradation. The goal of Project Fighter IV was to achieve an understanding of the relationship between stress and performance and perhaps identify ways to reduce its impact (Berkum, Bialek, Kern, & Yagi, 1962). Over several years, the numerous HumRRO researchers examined ways to induce stress under experimental conditions (a problem not yet solved at that time) while embedding realistic performance dimensions so that the relationship between the two could be examined. In addition to contributing to the methodological literature regarding the study of stress, these experiments yielded a model of the concept with implications for the way training is conducted. These implications included the need for stimulus fidelity in the cues critical to the desired job performance, a high correspondence between the actions needed in the real world and those in training, a high degree of fidelity in regard to the outcomes that result from performing the required actions, and repeated practice with such situations to minimize the impact of stress after transfer from a training environment. The results of these experiments were published in Psychological Monographs and other civilian publications and documented in presentations at various conferences.

With the advent of the all-volunteer force in 1973, the selection issue became much more paramount for military manpower experts as well as for Congress and other interested observers. Over time, the military services had used a variety of measures to determine who would enter and who would not. As early as World War I, the Army administered the Alpha and Beta tests to informally assess individual strengths and weaknesses for the jobs that needed to be filled (Zeidner & Drucker, 1987). These efforts were expanded and formalized in the ensuing years until, in 1976, the Department of Defense mandated that all services use the same test for screening and classification purposes—the Armed Services Vocational Aptitude Battery (ASVAB; Eitelberg, Laurence, & Waters, 1984). A good deal of evidence showed the

ASVAB to be a valid predictor of performance in training. What was lacking was a systematic assessment of the relationship between scores on the test and actual performance on the job. A congressional directive in 1981 mandated that the services fill this gap. With that, the Joint-Service Job Performance Measurement Project was born (Wigdor & Green, 1991). Project A, the Army's contribution to this groundbreaking research, became the broadest effort. HumRRO was the lead contractor for Project A, working with the American Institutes for Research and the Personnel Research Decisions Institute under the aegis of the Army Research Institute for the Behavioral and Social Sciences (J. P. Campbell & Knapp, 2001).

Project A was a large-scale research effort designed to improve the selection and classification of Army enlisted personnel. The basic goal was to provide the Army with the tools to ensure the best possible match between soldiers' abilities and the Army's needs. To do this, Project A was devoted to validation of the ASVAB against job performance and success in training. Moreover, other cognitive and noncognitive tests were developed to supplement and enhance the ASVAB's predictive utility. These tests were validated against a full array of existing and newly developed soldier performance measures. As a longitudinal study, researchers were able to examine the relationship between early soldier performance (e.g., in-training measures) and later performance (e.g., job performance ratings and hands-on tests of job performance) so that more informed reassignment, reenlistment, and promotion decisions could be made.

An important distinction between this project and all past efforts to establish the validity of the Army's selection and classification was that this research was based on extremely comprehensive measurement of performance. Because of this focus, the results of this research were sure to stand up under operational, as well as legal and scientific, scrutiny. The Army needed to be able to guarantee that credible, valid, reliable, and fair measures were developed.

In addition, the HumRRO team developed and tested new selection and performance measures. The set of measures, called the Trial Predictor Battery (Zook, 1996), assessed a number of psychomotor, perceptual, and cognitive abilities that the ASVAB did not. The scope of the project made it possible to examine virtually the entire domain of information that might prove valuable to enlist soldiers, sample from that domain, and investigate the incremental value of each major piece of information. The performance measures included hands-on tests, rating scales, paper-and-pencil job knowledge measures, and training knowledge measures. Over the course of the project, thousands of soldiers were tested around the world, many at multiple points in their careers.

Several journal articles and books resulted from Project A research. In one such volume, *Exploring the Limits in Personnel Selection and Classification* (J. P. Campbell & Knapp, 2001), John Campbell, the principal scientist for

Project A, summarized the contributions made by this work. Among the areas of impact identified, he included job and occupational analysis, performance measurement, the study of individual differences, the estimation of classification efficiency, and the prediction of future performance from past performance. Campbell noted,

> The Project A data base, the models of performance and performance determinants that were developed, and the measurement and data collection procedures used are all playing critical roles in virtually all current and proposed research efforts that are attempting to deal with this more dynamic [human resource management] environment. (Campbell & Knapp, 2001, p. 577)

Using Project A data, as well as that collected by the other services as part of the Joint-Service Job Performance Measurement Project, HumRRO researchers undertook a large-scale effort to develop a method for linking recruit quality requirements and job performance data and to develop a cost–performance trade-off model for setting enlistment standards. This provided an empirically grounded basis from which the services could report to Congress the distribution of ability required in a given year's recruits that would achieve satisfactory overall performance levels at a minimum cost. This distribution standard was then used to develop a trade-off model for evaluating the relative costs and effectiveness of various quality mixes for reaching performance goals. Work from this project was featured in a book, *Modeling Cost and Performance for Military Enlistment*, published by the National Academy Press (Green & Mavor, 1994).

Starting in the late 1970s, the military services began development of a computerized adaptive version of the ASVAB. Computerized adaptive testing (CAT) is a technique in which test items are presented to individuals depending on their answers to previous items. If a person demonstrates a low level of knowledge through his or her responses to the test, the computer subsequently chooses items from the range of questions that fit this level. This method of administering tests is much more efficient, because time is not wasted presenting easy items to those who would have no problem with them or difficult items to those with no clue how to respond.

Over a 5-year period, HumRRO personnel studied the impact of CAT–ASVAB on military screening. Alternatives for both computerized and paper-based testing were examined. On the basis of this research, the Department of Defense converted from paper-and-pencil to computer testing in its 65 Military Entrance Processing Stations across the country starting in 1996. HumRRO personnel were also involved in the production of a book that documented the research and development of CAT as a means of administering the ASVAB. The result, *Computerized Adaptive Testing: From Inquiry to Operation*, tells the history of the CAT–ASVAB, including the practical lessons learned along the way (Sands, Waters, & McBride, 1997). The authors also provided reference information for practitioners developing computerized testing systems.

In addition to these large-scale efforts for the U.S. military, HumRRO personnel have developed valid entry-level selection processes for other federal and state law enforcement agencies, including selection tests for entry-level Federal Bureau of Investigation special agents (Schmitt et al., 1994) and Virginia State Police troopers and commercial vehicle enforcement officers (Keenan, Felber, & Dugan, 1997). For the Drug Enforcement Agency, HumRRO personnel conducted a job analysis for all special agent positions and used these results to develop content-valid job simulation exercises to identify qualified candidates for supervisory and managerial positions (Pulakos, Tsacoumis, & Reynolds, 1992). In all of these efforts, and many others not described, the emphasis was on applying proven scientific methodologies to collect data in a systematic and valid manner and then use such data to ensure that personnel selected for entry-level positions or for promotion within the organization have the requisite knowledge, skills, abilities, and other characteristics to perform at the maximum level.

Training

Training is obviously an important function to all organizations, including those responsible for the safety of the U.S. public. The military has a unique training situation in that the vast majority of enlisted personnel enter service with little or no relevant background in the jobs to which they are assigned. As a result, the Department of Defense oversees the largest training apparatus in the world. Over time, HumRRO has contributed in significant ways in this realm in the development both of specific programs to bring service members up to speed in their military jobs and of more generic methods for analyzing training problems and creating the systems to address them. In fact, HumRRO's founder, Meredith Crawford, was responsible for one of the earliest explications of the systems approach to training (Crawford, 1962).

One of the first major undertakings by HumRRO in this arena may have been responsible for the survival of the organization in the long run (McFann, Hammes, & Taylor, 1955). On June 25, 1953, President Dwight Eisenhower received a letter from Howard C. Sarvis of New Meadows, Idaho. Sarvis had some new ideas about "rifle shooting" that he shared with the President. His letter was referred to the chief of the Human Relations and Research Branch of the Army who, in turn, sent it to HumRRO. On October 14 of that same year, Sarvis met with representatives of the Department of the Army and HumRRO to discuss his ideas. HumRRO, with Sarvis serving as a consultant, was directed to study the issue in order to devise training methods that would more realistically simulate combat conditions and more thoroughly integrate marksmanship training into the overall basic training experience. The approach used in this endeavor included studying the conditions under which soldiers used their weapons in combat, developing training methods that would more closely approximate such conditions, devising measures to assess marksmanship ability, and evaluating the revamped training program using newly derived proficiency tests.

The study of combat conditions led to several conclusions that were to have a strong impact on subsequent efforts. In combat, researchers found, enemy targets are rarely visible except in close assault, indications of targets are usually fleeting, the prone position is often precluded by conditions under which firing takes place, and elevation is a crucial element of aiming that is complicated by the conditions under which the skill must be used and the training practices in effect at the time.

In examining the extant training program, HumRRO researchers concluded that there were several problems. Chief among them was the lack of realism inherent in the training, often the result of a greater concern with safety than with the efficacy of the training itself. The major changes introduced included the introduction of moving targets that appeared for brief intervals and that were created to resemble opposing forces. Changes were also instituted in the position trainees assumed, with a greater emphasis placed on firing while kneeling or standing as opposed to being prone. New aiming strategies were also introduced to correct for common errors that resulted from the conventional methods.

These innovations, called TrainFire exercises, were found to significantly improve trainees' ability to detect and hit targets. As a result, marksmanship training was substantially revised throughout the Army, with the innovations copied by professional organizations in law enforcement agencies throughout the country. Every soldier who passed through the Army took part in TrainFire exercises. The success of this project was extremely important to the still nascent HumRRO. Senior military leaders often failed to see how a group of civilian psychologists could achieve results that had a significant positive impact on the Army. TrainFire gave them tangible proof that they could. Further, it was an example that HumRRO leadership could point to over time as other efforts matured and provided benefits of a similar magnitude.

Over the years, HumRRO researchers have tackled systemic problems by developing generic models for the development and implementation of training programs. For example, in the past the Army has faced challenges associated with training individuals of lower aptitude and with developing instruction that can suit the needs of the entire range of aptitude levels when people are trained in a group. Perhaps the most extensive research and development projects aimed at the training of lower aptitude individuals were associated with Project 100,000. The goal of this initiative, authorized in 1966 by Secretary of Defense Robert McNamara, was to bring into service 100,000 men a year who would have otherwise been turned away on the basis (for the most part) of their failure to meet the minimum aptitude requirements at the time (Laurence & Ramsberger, 1991; note that only men, not women, were accessed under Project 100,000). The rationale was that military training and discipline would provide a leg up for these largely disadvantaged individuals. HumRRO was tasked with a number of research efforts at this time, including examining the roles that lower-aptitude men could successfully perform in the military, investigating the importance of literacy skills in training and job performance, creating remedial reading and comprehension programs that could be implemented efficiently and effectively, and developing generic training methods that could be used by others to develop instruction suitable to this particular population (J. E. Taylor & Fox, 1967; Vineberg, Sticht, Taylor, & Caylor, 1971).

A variety of important conclusions emerged from this body of work that are relevant to issues faced by many employers across the country today. For instance, researchers found that depending on the particular job, performance differences between higher and lower aptitude personnel faded over time. Reading skills could be significantly improved relatively quickly when the materials used in the instruction were job related. Targeted training strategies for those of differing aptitude levels proved to be more cost-effective in the long run; lower aptitude soldiers benefited from increased levels of personal interaction during instruction, whereas autonomy and self-pacing were most effective for higher aptitude men.

The work done in conjunction with Project 100,000 was to benefit researchers down the road. One of the difficulties encountered soon after the introduction of the all-volunteer force was the emerging need to train soldiers of varying aptitude levels in a group, as described above. As a result, HumRRO researchers were tasked with identifying or devising a set of instructional methods that would meet the needs of trainees of various aptitudes and to develop and test a complete training system based on those methods (Weingarten, Hungerland, Brennan, & Alfred, 1971). To accomplish this, a sample occupational specialty was chosen that incorporated a wide spectrum of aptitudes, and a variety of alternative instructional methods were devised and tested. Researchers concluded that peer instruction brought the best results in terms of mastery of the material by persons of varying aptitudes. They also discovered that this technique significantly decreased training time. The final five-step method included having soldiers act as observers, instructors, and assistants in an iterative fashion until 100% proficiency was obtained. The Army found that trainees made significant gains in proficiency when the model—known as APSTRAT (derived from *aptitude* and *strategies*)—was applied. Further, academic attrition was reduced, leading to overall cost savings. This success led to the expansion of the model to other occupations, exploratory studies that applied APSTRAT procedures in two public school systems, and inquiries from a wide variety of institutions facing similar training problems.

HumRRO, along with the military services and other civilian and contract agencies, have conducted—and continue to conduct—research and development efforts that advance the state (and art) of training for the personnel who form the core of the U.S. national security system. Of particular note are applications using training simulators in a wide variety of tasks and occupations, in conjunction with the development of state-of-the-art technology. Use of simulators provides opportunities to train the entire range of skills needed with repetition of tasks and functions to enhance attainment, all accomplished in environments that are characterized by the highest levels of training-to-job fidelity.

Performance Assessment

After determining who is most suitable for the available jobs within an organization and training those individuals to perform those jobs, there remains the issue of assessing subsequent performance to ensure ongoing quality. This is essential to overall organizational performance, not just because of the potential for skills degradation but also because of the constantly shifting nature of work in a world of unending technical advances. Employees must not only master a skill set to do their jobs but also adapt those skills to maintain maximum performance. HumRRO has been involved in the development of performance assessments since its inception, but one such effort deserves particular mention because of its scope and impact.

Starting in 1974, HumRRO worked with the Army on the concept of Skill Qualification Tests (SQTs; R. C. Campbell, Ford, & Campbell, 1978; McCluskey, Trepagnier, Cleary, & Tripp, 1975; Osborn, Campbell, & Ford, 1977). The idea behind the SQT was simple: Each soldier's job proficiency would be evaluated annually using valid and reliable tests. For the first time, Army testing would be a training function rather than a personnel function. The execution of the idea, however, was not so simple. There were some 800,000 soldiers in 200 different jobs at five different skill levels located throughout the world. In addition, there was little or no infrastructure to support the test development and implementation process. HumRRO's involvement was primarily centered on the creation of procedures for everything from task selection and analysis to validation and administration. This activity was accomplished by a small group within HumRRO and the Army under stringent timelines.

HumRRO researchers first worked with Army subject matter experts (SMEs) in eight combat military occupational specialties to develop performance objectives for "critical" and "important" job tasks. Four service branches took part in this effort: Infantry, Armor, Field Artillery, and Air Defense. Overall, project staff and SMEs identified some 1,800 performance objectives. HumRRO then conducted a field investigation to determine the potential reliability, validity, and feasibility of prototype performance tests developed from these objectives. The results of the investigation included a series of recommendations for improving hands-on performance testing. This work was extended further to explore the feasibility of performance-based tests in high-complexity military occupational specialities.

The move to performance-based testing required an extensive rollout of a concept that was largely unfamiliar to training and testing personnel in the field. HumRRO was integrally involved both with the development of the concept in the Army and with the creation of materials to guide the construction of the new tests so that the results would be reliable and valid. This work continued as HumRRO staff examined the feasibility of extending the test development work they had done for infantry jobs to other occupational series, which required that researchers work with staff from a number of Army schools, including Armor, Air Defense, Signal, Military Police, Ordnance, Institute of Administration, Engineers, and Quartermaster. The goal was to examine the content of instructional materials at each school and determine the degree to which performance-oriented tests could be developed. The primary outcome of this work was the volume *Handbook for the Development of Skill Qualification Tests* (Osborn et al., 1977). This manual provided detailed instructions to test developers for the creation of the three types of assessments that grew out of earlier work: hands on, written, and performance certification. The authors set out a systems approach to SQT development, including determining performance measures, identifying test conditions, preparing instructions for examinees and scorers, and conducting field tests of the instruments and procedures developed.

HumRRO followed up this work by conducting workshops on constructing and validating SQTs. Training specialists successfully developed instruction that met the Army's primary criteria of being exportable, self-paced, and efficient. HumRRO presented the training to 20 representatives of 13 Army schools who, in turn, delivered it to 213 participants across the country. HumRRO personnel also took the workshops on the road, with three researchers presenting them at 13 different locations over a 6-month period.

In 1978, the Division of Military Psychology of the American Psychological Association presented its Military Psychology Award to HumRRO for the organization's combined work in the development and implementation of SQTs. Among the accomplishments cited were the cooperation among the various organizations involved, the generalizability and applicability of the methodology developed, and the level of resulting organizational impact and change. As it turned out, SQTs were abandoned in the 1980s because of cost considerations. However, once again, lessons learned will not be lost; HumRRO is currently working with the Army Research Institute for the Behavioral and Social Sciences to identify viable approaches for the development of a useful yet affordable operational performance assessment system (Knapp & Campbell, 2004).

CONCLUSION

This chapter can highlight only a few of the thousands of studies HumRRO has conducted over the past half century to address real-world problems that influence the performance of personnel and organizations. As is often the case with applied research, the eventual impact of the results is often subtle in that it contributes to a body of knowledge without resulting in major changes to policies or procedures. Time moves on. Leadership changes. The nature of the challenges evolves.

It is the continuing building of knowledge that may, in the long run, make the greatest contribution to an understanding of and ability to deal with issues such as national security. The work done by researchers during Project 100,000 to determine the best means of utilizing and training lower aptitude personnel may have lost relevance with the ending of McNamara's contribution to the war on poverty, but it was to regain that relevance some years later with the discovery that the norms for the ASVAB had been incorrectly calculated and many below-average individuals had been admitted into the Army. The lessons learned during Project A are being well applied as HumRRO researchers work with the Army Research Institute to identify the characteristics and abilities that will be important to soldiers and leaders into the 21st century (Ford, Campbell, Knapp, & Walker, 1999). In addition, the work done for one client benefits others, as lessons are learned, improved strategies discovered, and more efficient processes developed. Finally, the sharing of information across entities and organizations benefits all who are striving to meet the many human resource challenges will be faced in the years ahead.

8

PSYCHOLOGY IN AIR FORCE TRAINING AND EDUCATION

HENRY L. TAYLOR

Military psychology is the area of applied psychology that focuses on the determination and application of the behavioral principles relevant to military problems. Military psychology had its beginning on the day the United States entered World War I, April 6, 1917. After World War I, from 1921 to 1939, there was minimum activity in this area. During World War II, between 1939 and 1945, military aviation psychologists in the Army Air Force (AAF) used applied experimental psychology methodology to solve aviation problems. The largest single program of the AAF was under medical auspices and focused on personnel selection problems (Melton, 1957).

Research and development (R & D) on aircrew selection and classification began in 1942 under the responsibility of the AAF Flying Training Command. This work led to the development of two testing programs: the AAF Qualifying Examination and the Air-Crew Classification Test Battery, which consisted of 20 tests. These personnel research programs made a substantial contribution to the war effort by determining the best fit of person to job in the largest single mobilization effort in the U.S. history.

Human engineering also had its beginning as part of the AAF Aviation Psychology Program. Paul M. Fitts is regarded as the founder of this field. In

August 1945, the AAF transferred Fitts and a small group of military psychologists to the Aero Medical Laboratory at Wright Field, in Dayton, Ohio, to establish the Psychology Branch with Fitts as its director (Pew, 1994). The mission of the Psychology Branch was to conduct psychological research on equipment design problems of interest to AAF. The R & D of the branch was organized into three areas: (a) cockpit displays and controls; (b) psychological problems in the design of radar, navigation, communication, gunnery, banking, and pilotless equipment; and (c) flight testing (Fitts, 1947). These R & D efforts helped make operational systems more effective by focusing on the human working within these systems.

The U.S. Air Force became a separate service in 1947. In rapid succession, the Air Force established the Human Resources Research Center under the Air Training Command, the Human Resources Research Laboratories under Air Force Headquarters, and the Human Resources Research Institute under the Air University Command (Lavisky, 1976). These actions clearly demonstrated the Air Force belief in the value of military psychology R & D and its contribution to operational effectiveness and to national security. These organizations were combined to form the Air Force Personnel and Training Research Center (AFPTRC) at Lackland Air Force Base in San Antonio, Texas, under the direction of Arthur W. Melton. The personnel and training research within the Air Force was conducted by the AFPTRC, and engineering psychology R & D was conducted by several different agencies, including the Wright Air Development Center, Rome Air Development Center, Air Force Cambridge Research Center, Air Force Special Weapons Center, and Air Force Flight Test Center (Melton, 1957). The AFPTRC was dismantled in 1957, but the Personnel Research Laboratory remained at Lackland Air Force Base and continued to conduct R & D and analysis for the deputy chief of staff for personnel within Air Force Headquarters. Over the years, this work has provided significant support to the decision-making process for Air Force personnel managers.

On the basis of a recommendation by the Air Force Scientific Advisory Board, the Air Force Human Resources Laboratory (AFHRL) was established in 1968 at Brooks Air Force Base in San Antonio, Texas. AFHRL was located near operational personal and training organizations to help define requirements for the R & D and to facilitate the implementation of research products. The Armstrong Laboratory, headquartered at Brooks Air Force Base, was established in December 1990 when the Armstrong Aerospace Medicine Research Laboratory, the Human Resources Laboratory, the Drug Testing Laboratory, and the Occupational and Environmental Health Laboratory and research functions of the School of Aerospace Medicine were merged. In October 1997, the four existing Air Force laboratories—the Wright Laboratory, the Phillips Laboratory, the Rome Laboratory, and the Armstrong Laboratory—were combined into one laboratory: the Air Force Research Laboratory, headquartered at Wright–Patterson Air Force Base in Dayton, Ohio.

This chapter focuses on the experimental aspects of Air Force military psychology that are relevant to the operational Air Force. I address three major areas—personnel research, human engineering, and training and instruction—and discuss the R & D efforts and the implementation of research findings in each area.

PERSONNEL RESEARCH

This section discusses the development of selection and classification batteries and assignment algorithms by Air Force military psychologists. The first classification battery for Air Force enlistees, the Airman Classification Battery, was developed and used for assignment purposes in the late 1940s. The use of personality inventories was begun in the 1960s when the five-factor model was developed. Air Force military psychologists initiated the development of an ability test using cognitive theory, the Learning Abilities Measurement Program (LAMP), which began to be administered in 1982. This section also discusses the specialized area of pilot selection and the involvement of Air Force military psychologists in the assignment of military personnel.

Selection and Classification

Soon after the U.S. Air Force became a separate service, the Airman Classification Battery was developed and implemented. The Airman Classification Battery, an aptitude test battery, was administered at Lackland Air Force Base to all Air Force enlistees and was used as the basis for making assignments (Weeks, Mullins, & Vitola, 1975). In 1958, the Air Force replaced the 5.5-hour Airman Classification Battery with a 2-hour Airman Qualification Examination, which was used for both selection and classification. During the 1950s, ability research in the Air Force was conducted by the AFPTRC until it was disbanded in 1957. In the mid-1950s, selection and classification of Air Force officers were transferred to the Air Force Reserve Officer Training Corps. In 1968 the Air Force became executive agent for R & D on the ASVAB and maintained this role until the mid-1980s, when the responsibility was reassigned to the office of the Secretary of Defense. The ASVAB became the test battery used by services for all enlisted personnel selection and classification.

In addition to ability testing, the Air Force also used personality inventories in its testing program. In the 1960s, Tupes and Christal of the Air Force Personnel Laboratory conducted a series of studies that identified five recurring dimensions of personality (Tupes & Christal, 1961), given the names of *surgency*, *agreeableness*, *conscientiousness*, *emotional stability*, and *culture*. During the late 1980s and early 1990s, a consensus developed among personality theorists and applied practitioners that these five factors constituted a consensus model of personality called the Big Five personality theory. A special issue of the *Journal of Personality* edited by McCrae (1992) was devoted to the five-factor model. The 1961 article by Tupes and Christal was reprinted in the issue, along with other articles dealing with issues and applications of the model. McCrae and John (1992) described the five-factor model of personality as a hierarchical organization of personality. Today the five factors are usually termed Extraversion, Agreeableness, Conscientiousness, Neuroticism, and Openness to Experience. McCrae noted that the most impressive feature of Tupes and Christal's work was its conceptual sophistication. The five-factor model of personality played a secondary role in selection and classification of Air Force personnel.

Beginning in 1985, the Armstrong Laboratory began a research project with the objective of producing the next generation of ability tests using cognitive theory. The project combined cognitive science, psychometrics, and computer-based testing. The result was a concept of cognition based on an information processing model. Kyllonen (1995) discussed a new approach to cognitive abilities testing called the *dimensional analysis* of cognitive abilities that was used to define how individuals differ cognitively. This work developed from the LAMP, which was first funded in 1982. Initially, data were collected from subjects on 30 cognitive tasks. In 1988, the LAMP used 30 microcomputer-equipped testing stations to collect data from Air Force enlisted personnel (Christal, 1988). In 1996, the test center contained eight testing bays and 250 computer testing stations. Approximately 24,000 enlisted personnel were tested during 1996 and 1997 (Kyllonen, 1996).

During the 1990s, the LAMP research team initially developed Cognitive Abilities Measurement (CAM), which was a framework for measuring cognitive abilities. Subsequently, a computerized test battery, the CAM battery, was developed using that framework (Kyllonen, 1995). The CAM battery consists of tests of processing speed, working memory, declarative knowledge, and procedural knowledge and was assembled to measure cognitive abilities in the verbal, quantitative, and spatial domains. The LAMP team has used the CAM battery of 59 tests to conduct a number of studies concerned with the degree to which performance on the CAM battery predicts success on a series of learning tasks. The results of these studies indicate that the CAM battery predicts performance on the learning task accurately and learning success on the tasks more accurately than the ASVAB. A 16-test subset of the CAM battery, the Advanced Personnel Test, is being compared with the ASVAB in terms of validity and fairness as a predictor of technical school outcomes for enlisted personnel. More than 9,000 airmen have been tested, and performance data from three Air Force technical schools and 15 Air Force technical specialists have been examined (Kyllonen, 1996). In 1997, the Air Force Research Laboratory decided to substantially reduce its investment in this area, which resulted in substantial reduction in the ability to conduct R & D in selection and classification.

By the end of World War II, Air Force pilot selection methods were well established. The use of apparatus tests that measured coordination and decision speed as part of the Air Crew Classification Test Battery was discontinued after World War II because of problems in maintaining the test apparatus in good working condition (H. L. Taylor & Alluisi, 1994). Recently, as computers have become more powerful, faster, and more economical, research on pilot training has increasingly used computer-based performance tests.

The tasks that Air Force military pilots perform in combat have shifted from predominantly manual control tasks to a highly automated cockpit with more emphasis on resource management tasks. Indeed, in current high-technology Air Force cockpits, many of the flight control tasks are automated. There has also been an increased use of digital displays that provide information to the pilot and of fly-by-wire systems to control of the aircraft. As pilot tasks have changed, the use of computer-based performance tests has increased. In addition, most Air Force military psychologists agree that substantial increases in the predictive value of paper-and-pencil tests (i.e., of intelligence and aptitude) above that of the current tests are extremely unlikely. This influenced the decision to reduce investment in R & D of selection and classification.

Assignment

In 1957, Air Force Headquarters directed that an occupational research project be established. In 1967, after a decade of R & D, an operational occupational analysis program using the Comprehensive Occupational Data Analysis Program (CODAP) was implemented in the Air Force (J. L. Mitchell & Driskell, 1996). Current job incumbents complete job task analysis surveys, and the surveys are analyzed to determine task frequency, importance, criticality, and difficulty of performance. Demographic information is also collected to permit the analysis of factors of job experience, military rank, and organizational level of job incumbents. CODAP consists of a set of computer programs that process, summarize, and display substantial sets and subsets of complex occupational data obtained from the surveys. CODAP is able to obtain and analyze detailed job task information from extremely large numbers of job incumbents. CODAP has been adapted by the U.S. military services and a number of additional organizations, including Canadian forces, Australian forces, the U.S. Coast Guard, and the National Security Council; in addition, several U.S. universities and corporate entities, such as AT&T and McDonnell–Douglas, have adopted CODAP (Mitchell & Driskell, 1996). For more detailed coverage of military occupational analysis methodology and application, see Bennet, Ruck, and Page (1996).

HUMAN ENGINEERING

Air Force military psychologists have played a major role in the area of human engineering for over 50 years. Initially, from the mid-1950s, the R & D focus was on cockpit display and controls, psychological problems in design, and flight testing, but in the mid-1990s, the program encompassed information management and display, performance enhancement, and design integration (R. J. Green, Self, & Ellifritt, 1995). This section will focus on anthropometry and workplace accommodation.

Anthropometry

Air Force military psychologists have collected human body surface data since the 1970s. Since 1985, there has been major technical advancement in the field of anthropometry in the areas of database systems, statistical methodology and applications, and 3-D anthropometric data collection. High-resolution typography on the human head and face has been possible since 1987.

In the mid-1980s, an automated 3-D scanner was first used as an anthropometric scanning system in the Armstrong Aerospace Medicine Research Laboratory (a predecessor of the Armstrong Laboratory) to collect human body size and shape data at the millimeter level of accuracy (Ratnaparki, Ratnaparki, & Robinette, 1992). The 3-D scanner changed the field of anthropometry by providing a fast and reliable method to collect anthropometric data. The advantages of 3-D surface anthropometry include a decrease in ambiguity, less dependency on the axis system, the ability to merge images of the human head and equipment, and reduced observer error in measurement (Robinette & Whittestone, 1992). In 1995, a fully automated, whole-body scanner that captures the shape of the human body with a single scan to an accuracy of 3-millimeter resolution was installed by the Armstrong Laboratory at Wright–Patterson Air Force Base. The Air Force currently uses the whole-body scanner to collect data for female accommodation; to improve the fit of helmets, military clothing, and antigravity suits; and to redesign cockpit layouts and escape systems.

Workplace Accommodation

Air Force R & D on workplace accommodation spans more than 50 years, starting with measurement of the strength required for aircraft controls in 1946. Body size surveys were conducted in 1950 that measured 146 dimensions of 4,000 male pilots. This led to the first general sizing system for flight clothing in 1959. In 1972, STICKMAN, the first computerized human stick figure, was developed and applied to first-generation computer graphic displays. This beginning focus on anthropometric modeling led to the development in 1975 of COMBIMAN, an interactive 3-D ergonomic computer graphics model of a human at a workstation. This engineering tool is used to evaluate operator capabilities and spatial accommodation. The technology has been used by the aerospace industries since 1978 (McDaniel, 1990). The model simulates both men and women, six types of military clothing, protective equipment, and three types of seat belt and shoulder harness restraints. Design engineers are able to use the model to vary size and proportion parameters for male and female U.S. Air Force and U.S. Army pilots and for U.S. Air Force nonpilots. Each of these populations has different size, weight, and proportion parameters and thus requires different accommodations. Strength parameters include stick, wheel, lever, pedal, and ejection seat variables, and reach parameters for a specific control device or reach envelope can be determined by varying both clothing and harness factors. By 1994 a "virtual" COMBIMAN had been developed that placed the viewer inside a 3-D cockpit drawing during landings (R. J. Green et al., 1995; McDaniel, 1990, 1991).

The development of CREW CHIEF, a computer graphics model of an aircraft maintenance technician, began in 1984 with the purpose of providing an ergonomic interface to aerospace computer-aided design systems (McDaniel, 1988, 1990, 1991). CREW CHIEF is a computer simulation of a fully articulated aircraft repair technician and was derived from COMBIMAN. It automatically simulates aircraft maintenance technician activities such as using hand tools and handling materials (lifting, pushing, and carrying) to determine the feasibility of performing a maintenance activity (Annis, McDaniel, & Krauskopf, 1991). The CREW CHIEF software has an expert system to create a 3-D human model that contains a full range of both male and female body sizes. It was first transferred to the aerospace industry in 1988, and currently the model is widely used to evaluate maintainability. Design engineers use CREW CHIEF to provide an initial mock-up of the system, and then they refer to published specifications concerning human anthropometric dimensions, range of motion, and strength capabilities to determine the feasibility of a maintenance technician interacting with the system during troubleshooting and repair. Components that are inaccessible or exceed weight limitations are redesigned. The value of CREW CHIEF is that it permits a simulation of the technician to be fully integrated with graphic design software so that ergonomic considerations can be assessed early in the design phase of a weapon system.

When system designers consider the physical characteristics of personnel during weapon system design, the 5th and 95th percentiles are used as design constraints. It has been known since the 1960s that the use of these percentiles induces gross errors, but the practice has continued because of the complexity of other design constraints. Improved analytical methods for representing anthropometric multivariate data developed by Air Force military psychologists provide alternatives to the use of the percentiles as cockpit design constraints (Zehner, Meindel, & Hudson, 1992). The multivariate approach has been applied to a number of areas. For example, the mandate by Congress to accommodate women in combat resulted in the use of the approach to evaluate the standard for pilot selection height and sitting height. In addition, anthropometric accommodation was one of the two selection criteria for the Joint Primary Aircrew Training System program (Zehner, 1994; Zehner et al., 1992).

The development of visually coupled helmet systems for military use requires R & D efforts in optical systems and electronic circuit design, hardware, and associated software. Air Force military psychologists have played a key role in the development of helmet-mounted displays and helmet-mounted sights. Air Force R & D in visually coupled helmet-mounted technologies began in 1966, and the first visually coupled system was demonstrated in 1969 by coupling a helmet-mounted tracker with a helmet-mounted display. It is generally accepted that this demonstration was the precursor of subsequent virtual reality systems. The Visually Coupled Airborne System Simulator (VCASS) project, whose purpose was to investigate visually coupled technology, was initiated to develop a fixed-base virtual environment simulator (Kocian, 1976). The VCASS had a large field of view that contained a binocular helmet-mounted display that generated 3-D stereo images. A head tracker with six degrees of freedom permitted subject head movements to control the visual scene presentation. The result was a simulation that immerses the subject within a computer-generated visual world that can be displayed in relation to the subject's head movement.

Recent focus of visually coupled systems R & D has been on critical systems and component tests and studies designed to solve the problem of integrating visually coupled systems with the human visual systems on advanced weapon systems such as the F-22 (Kocian, 1996). Visually coupled systems have military applications in the areas of within visual range operations, off-bore sight for air intercept, and air-to-ground interdiction missions. In 1993, the helmet-mounted display and head-tracker systems were successfully integrated and flown in two F-15Cs. The results indicated significant reduction in the times to locate the target, secure missile lock-on, and launch the missile. Analysis of the data from the flight test predicted an improvement in aircraft kill ratios.

TRAINING AND INSTRUCTION

Air Force military psychologists' involvement with training and instruction has focused on the development and use of aircraft simulators and computer-based testing. These areas have received substantial funding to solve training problems and to enhance human performance in airplanes. The R & D on simulators dates from the 1940s, and R & D on computer-based testing dates from 1982.

Aircraft Simulators

The Link instrument flight trainer, named after Ed Link, was used during World War II by the Army Air Force to train more than 500,000 pilots (H. L. Taylor & Stokes, 1986). The military services currently are the largest user of aircraft simulators. Air Force simulators range from relatively simple mock-ups and part-test trainers to sophisticated multimillion-dollar, full-mission weapon system trainers. Since the late 1980s, design trends of aircraft simulators have changed from large, complex, and costly systems to more affordable and flexible designs. These changes are the result of the need for more simulation systems located at a squadron organizational level. This section discusses the development and use of an advanced simulator for pilot training, transfer of training, and distributed mission training.

Advanced Simulator for Undergraduate Pilot Training

In the 1970s, the AFHRL contracted with the Singer–Link company of Binghamton, New York, to build the Advanced Simulator for Undergraduate Pilot Training (ASUPT), a high-technology simulator, to be used for research. The ASUPT design consisted of two independent cockpits that simulated the T-37B, the Air Force's primary jet pilot training aircraft (F. E. Bell, 1974; Gum, Albery, & Basinger, 1975). The cockpits were mounted on six-degree-of-freedom motion platforms that simulated the onset of kinesthetic cues. The sustained cues were provided by a "g-seat" (a seat containing bladders that inflate to simulate sustained g-forces). Each simulator had a wraparound visual display system consisting of seven video channels, each of which had an in-line infinity image optical display. The visual image was generated by a digital computer. The ASUPT was installed for research use at Williams Air Force Base in Mesa, Arizona, in 1975 (H. L. Taylor & Stokes, 1986).

Subsequent modifications to the ASUPT provided for the simulation of two tactical fighter aircraft, the A-10 and the F-16, instead of the T-37B trainer. The simulator was renamed the Advanced Simulator for Pilot Training (ASPT; Alluisi, 1981). The cockpit configuration and the name change resulted from a change in the research emphasis of AFHRL from undergraduate pilot training to advanced training, which included transition, continuation, and combat mission training.

A number of studies conducted by AFHRL demonstrated that the ASPT provided a significant contribution to combat training. A-10 pilots flying the ASPT learned offensive and defensive combat maneuvers in a simulated hostile environment (Kellogg, Prather, & Castore, 1980). A later study demonstrated that aircrew members trained in the ASPT to deliver air–surface weapons improved their initial weapons delivery performance in the A-10 aircraft (T. H. Gray, Chun, Warner, & Edwards, 1981). Hughes, Brooks, Graham, Sheen, and Pickens (1982) found that prior training in the ASPT by experienced A-10 pilots significantly increased survivability in the A-10 aircraft in a Red Flag combat exercise, which involves a simulated hostile threat.

Transfer of Training

The benefit of using flight simulators was further quantified in studies conducted at the University of Illinois. A. C. Williams and Flexman (1949) computed the first transfer effectiveness ratio, which they termed the *efficiency ratio*. In the 1970s, the Air Force Office of Scientific Research supported additional research at the Aviation Research Laboratory at the University of Illinois at Urbana–Champaign on the effectiveness of aircraft simulators. Povenmire and Roscoe (1973) conducted a study to determine the relationship between the number of training hours in the Link GAT-1 and incremental savings of flight time in the Piper Cherokee aircraft. They found that the mean flight time saved increased with the number of training hours in the aircraft simulator and that the differences were statistically reliable.

Air Force military psychologists conducted a number of experiments in the ASPT during the late 1970s using a transfer of training design (Martin, 1981; H. L. Taylor, 1985). The maneuvers investigated included basic contact flight, aerobatics, all phases of flight in the T-37 undergraduate pilot training syllabus, and air-to-ground bomb deliveries. The students showed significant positive transfer of training from the simulator to the aircraft. Wagg (1981) reviewed studies concerned with the performance implications of visual simulation and reported that the "overwhelming finding is that visual tasks learned in the simulator show positive transfer to the aircraft" (p. 17).

The follow-on to the ASPT as an aircrew training research simulator was the Display for Advanced Research and Training (DART; M. Thomas, 1993). The DART was used during the late 1980s to explore the effects of using low-cost display devices with sufficient fidelity for aircrew training. The DART has nine channels of imagery and uses a rear-screen projection technology; the result is wraparound display of real imagery. Research conducted during the early 1990s using the DART indicated that tactical pilots can perform their combat mission using low-cost, real imagery devices (M. Thomas, 1993). Recently, the DART has been used as an F-16 multitask trainer.

Distributed Mission Training

In the mid-1990s, the Armstrong Laboratory initiated an engineering development for a distributed mission training capability. Distributed Mission Training is a system that combines virtual simulators, live aircraft, and constructive simulation using network technology for combat training

(Reasor, 1996). Initially, the virtual component had four simulator cockpits connected via a wide area network to a threat cockpit. Technology areas for the virtual cockpit component that required further development included threat systems with multilevel secure networking, visualization, briefing and debriefing tools, mission application tools, and data reduction and analysis tools. The goal of Distributed Mission Training was to provide low-cost, high-fidelity virtual simulation capabilities that exceeded current training opportunities. The Air Force has implemented Distributed Mission Training Centers (MTCs) at a number of Air Force bases for the F-15, F-16, and Airborne Warning and Control System (AWACS) aircraft. F-15C MTCs are at Eglin Air Force Base, Florida; Langley Air Force Base, Virginia; and Elmendorf Air Force Base, Alaska. F-16 MTCs are at Shaw Air Force Base, South Carolina, and Mountain Home Air Force Base, Idaho. AWACS MTCs are at Tinker Air Force Base, Oklahoma, and Elmendorf (Brower, 2003). These centers provide individual and team training at the local base and can be linked to other training centers for team training. The concept of operations provides that the majority of the training will be at the base level (80%–90%) and the wide area teams of team training (10%–20%). Using these centers, the Air Force conducts on a quarterly basis at Nellis Air Force Base an integrated combat training exercise involving live, virtual, and constructive forces in an operationally realistic environment. Virtual simulations of future operational capability can be used to support the development of new combat tactics. The Air Combat Command has established a road map for additional mission training centers for the remaining F-15, F-16, and AWACS bases as resources permit (Herbert H. Bell, December 13, 2005, personal communication).

Computer-Based Training

The Air Force was the first service to recognize the potential of computer-based training and to begin its development (Fletcher & Rockway, 1986). According to Olsen and Bass (1982), development and deployment of the Semi-Automated Ground Environment was the first use of computers in training by the U.S. military. This program, which became operational in the early 1950s, demonstrated that a computer could use radar signals to track aircraft and to control aircraft interceptions.

Advanced Instructional System

In May 1976, a 53-month, multimillion-dollar contract was awarded to McDonnell–Douglas to develop the Advanced Instruction System (AIS; Rockway & Yasutake, 1974). The AIS had two major objectives: (a) to develop a computer-based, multimedia system to administer and manage individualized technical training and (b) to use AIS as a research test bed to determine the cost-effectiveness of training innovations. The original requirements of AIS called for the development of three courses on precision measuring equipment, inventory management, and weapons mechanics and a computer-managed instruction system to manage 2,100 students per day. In 1979, the system configuration was expanded to four courses, and the system could manage 3,000 students per day (Lintz, Pennell, & Yasutake, 1979). The primary use of AIS was to manage student progress and training resources (i.e., computer-managed instruction) rather than using the computer as a tutor (i.e., computer-assisted instruction, or CAI). Fletcher and Rockway (1986) indicated that the AIS was an important milestone in the development of computer-based training because it first applied computer-managed instruction on a large scale in an operational setting. The evaluation of the AIS indicated that the savings in student training time was 40% (Lintz et al., 1979).

Intelligent Tutoring Systems

In 1991, the Air Force Office of Scientific Research awarded the Armstrong Laboratory a grant for the Training Research for Automated Instruction (TRAIN) Project (Regian, 1994). The objective of TRAIN was to codify pedagogical principles for automated instruction. A computer laboratory was developed that consisted of 30 training delivery microcomputers to support data collection during 50,000 subject hours per year. Results from the TRAIN Project showed that four students can be trained on a single computer station at individual achievement performance levels equated to those obtained by students trained on four separate computer stations. These results have substantial implications for the efficient use of instructional resources.

The TRAIN Project forms the basic research initiative of the Armstrong Laboratory's R & D program concerned with intelligent tutoring systems (ITSs). An ITS is a computer-based, intelligent computer-assisted instruction (ICAI) system that provides instruction adapted in real time to students' capabilities. The instructional effect is comparable to that obtained with individualized instruction (i.e., one-on-one tutoring). ITSs are effective, but the time required to develop an hour of ICAI is substantial because it requires a team effort by cognitive psychologists, computer programmers, course development experts, and subject matter experts.

Intelligent tutoring systems have several components. First, an ITS develops a model of what a student knows and compares the this model to what the student needs to know. Next, the ITS determines which curriculum element or unit of instruction is to be presented to the student and the method by which it will be presented—that is, the teaching strategy. During the next step, the ITS selects or generates a problem and develops a solution to the problem using the domain expert, or it retrieves a prepared solution. Then the ITS compares its solution in real time with the student's solution and analyzes the difference between the two solutions. Finally, the ITS presents feedback to the student. With ITSs, students are involved in interactive practice activities through a simulation of the target task. ITSs also make real-time curriculum decisions and manage student learning activities. Air Force military psychologists have developed R & D prototypes to demonstrate the effectiveness of ITSs, but few have been applied operationally. Regian (1996) indicated that 1% to 2% of Air Force instruction is CAI, and a negligible fraction is ICAI. Consistent data show that the instructional effect of CAI is in the 65th percentile above the mean performance achieved by traditional instruction, and the use of CAI results in a 24% reduction in learning time. For ICAI instruction, Regian indicated that the instructional effect is in the 84th percentile above the mean performance achieved by traditional instruction, and the reduction in learning time is 55%.

In the mid-1970s, the AFHRL used an intelligent tutoring system, Sophisticated Instruction Environment (SOPHIE), to assist learners in developing electronic troubleshooting skills. SOPHIE created a reactive learning environment that responded to inputs from Air Force technicians who had located faults in electronic equipment. The technicians could ask SOPHIE questions in English (e.g., "What are the values of various measurements taken on the equipment?"). SOPHIE had three principal components: a mathematical simulation, a program to understand English, and routines to set up troubleshooting routines and maintain information. SOPHIE was important because it generated additional research concerned with troubleshooting and reactive learning environments (J. S. Brown & Burton, 1975, 1978).

SHERLOCK is a tutor that provides coached practice on electronics troubleshooting tasks (Lesgold, Lajoie, Bunzo, & Eggan, 1992). SHERLOCK contains troubleshooting scenarios for an F-15 avionics test station. An evaluation of the tutor's effectiveness in teaching troubleshooting indicated that the group receiving 20 hours of instruction using SHERLOCK performed significantly better than a control group receiving equivalent on-the-job training. The experimental group also performed at a level equivalent to experienced technicians. Another evaluation of SHERLOCK indicated that the tutor accelerated learning and that the experimental subjects achieved the same outcome performance as a control group (Shute & Regian, 1993).

The Armstrong Laboratory developed the Intelligent Computer-Assisted Training Testbed in the early 1990s (Fleming, 1996). The purpose of the test bed was to produce authoring shells that permitted instructional developers and subject matter experts to develop ITSs. The authoring shells are designed to permit the user to develop, use, and maintain these systems on personal computers. ITS development without authoring shells requires the expertise of subject matter experts, cognitive psychologists, and artificial intelligence programmers. The cost to develop an ITS is significant (estimated at $1 million), and they have a long development time—up to 3 years. In addition, they are difficult to maintain and require expensive hardware to develop and use (Fleming, 1996). The first ITS authoring shell, developed in

1991, was the Microcomputer Intelligence for Technical Training (MITT) Writer. Curriculum experts have used MITT Writer to produce an ITS to train troubleshooting tasks. Using MITT Writer, 1 hour's training was developed in an average of 136 hours (Fleming, 1996).

The Rapid ITS Development Shell (RIDES), developed in 1994 (Regian, 1996), provides a simulation of a complex device to be used during curriculum development. Using RIDES, instruction is automatically generated as the curriculum developer defines the course objectives and interacts with the simulation. After the tutor is complete, the ITS determines how to respond to individual students on the basis of the student's interaction with the simulation and the course objectives. Fleming (1996) indicated that several tutors have been developed using RIDES, and 1 hour of instruction has taken 50 hours to develop. Tutors developed by RIDES include B2 Landing Gear Tutors, the Missile Touch Control Tutor, the Horizontal Situation Indicator Tutor, the Health Tutor, the Pulse Oximeter Tutor, the Tutor of Orbital Elements, the Rapid Execution and Combat Targeting Tutor, and the Tutor on the Respiratory System. Authoring shells developed by the Armstrong Laboratory permit tutors to be developed in any domain at a 30-to-1 reduction in cost (W. Alley, 1997, personal communication). Using authoring shells developed by the Armstrong Laboratory, ITSs have been developed that cost only $50,000, an estimated cost reduction of 95%, with a development time of about 7 months, a reduction of as much as 80% (Fleming, 1996). In addition, the ITS can be delivered on personal computers and is easily updated.

Although the use of ITSs has shown great promise, the Air Force has made only limited application of this technology in Air Force training organizations. In 1977, the Air Force Research Laboratory decided to substantially reduce its investment in ITS, thereby substantially reducing the laboratory's ability to conduct R & D in this area.

CONCLUSION

This chapter has focused on the role that Air Force military psychologists have played in working with engineers, personnel and training managers, and operational commands to perform R & D in the areas of personnel research, human engineering, and training and instruction. Of more importance, military psychologists have worked with operators to help refine requirements for R & D and to facilitate its implementation within the operational Air Force following the completion of the research. It is clear that the work by military psychologists in these areas has made substantial contributions to military effectiveness and to national security. Indeed, many of the efforts in personnel research, human engineering, and training and instruction have served as a force multiplier to make the human in operational systems more effective.

9

THE NAVY PERSONNEL RESEARCH AND DEVELOPMENT CENTER

MARTIN WISKOFF AND EDMUND THOMAS

The organizational roots of the Navy Personnel Research and Development Center (NPRDC) go back to World War II, when the Navy's concerns about psychological screening of recruits, their placement into advanced training, and classification of Navy officers into the most appropriate jobs were addressed by uniformed research psychologists. In 1946, a Personnel Research Division was formed within the Bureau of Naval Personnel to expand the research conducted during the war into the areas of personnel training and definition of Navy occupations.

A field team of military personnel conducting occupational analyses onboard ships in the San Diego, California, area was expanded in 1951 into the Navy Personnel Research Unit with a staffing of 40 civilians and a broader mission that included testing and training research. In 1952, a Navy Personnel Research Field Activity was established in Washington, DC, with 43 civilians and 36 officers to address emerging concerns in personnel classifica-

Much of the information for this chapter was taken from a Navy Personnel Research and Development Center publication (NPRDC-AP-99-4), *Voices From the Past—Command History Post WWII to November 1999*, by E. D. Thomas, T. M. Yellen, and S. J. Polese, 1999. San Diego, CA: Navy Personnel Research and Development Center.

tion, qualifications for advancement in rating, career guidance, and literacy requirements for Navy jobs.

NPRDC officially began operations on July 1, 1973, on a tip of land in San Diego, California, called Point Loma, which the Portuguese explorer Juan Cabrillo discovered and first set foot upon in 1542. NPRDC was created with a mission to serve as the principal Navy organization for managing, coordinating, and conducting research to address national security concerns in the areas of manpower, personnel, training, and human factors engineering. The new organization was established to elevate the status of personnel research from that of two separate laboratories to a center of excellence supporting the Navy and Marine Corps. The manpower authorization was 262 civilians, 7 officers, and 19 enlisted personnel. Initial staffing for NPRDC was provided primarily by personnel at the Naval Personnel and Training Research Laboratory in San Diego, California, that had occupied the Point Loma facility since 1951. About 30 people relocated from the Personnel Research and Development Laboratory in Washington, DC, which had been created in 1952, and several also transferred from the Personnel Research Division of the Bureau of Naval Personnel.

RESPONSE TO NATIONAL SECURITY PRIORITIES

NPRDC was formed during a turbulent period in the nation's history. In January 1973, the Paris Peace Accords brought an end to the costly and controversial war in Vietnam, and drug abuse, racial discord, and strong antiwar sentiments had created problems both inside and outside the military. The image of the military had seriously eroded among the nation's youths, and peace demonstrations had occurred on college campuses. The All Volunteer Force signaled the end to conscription and began on the same day that NPRDC was established. As documented by a Defense Manpower Commission (1976) report, an immediate national security concern was the sustainability of an all-volunteer force. A major issue at the time was the representational policy for the services; the commission concluded, "The Services should recruit and assign personnel without regard to representational factors except for women where unique considerations exist" (p. 10). The prospect of meeting recruiting quotas, given the national sentiment, was not encouraging.

The Navy had made several major changes in its personnel policies as a result of growing antiwar sentiments among sailors. The Chief of Naval Operations, Admiral Elmo Zumwalt, had issued a series of memoranda that recognized the need to accommodate and manage social change while maintaining military readiness. Many of these changes were embodied in an emerging Human Goals Program, and NPRDC was heavily involved in the evaluation of these changes.

The Navy was also about 5 years into an extensive military social experiment called Project 100,000. Initiated by Secretary of Defense Robert S. McNamara in 1966 as a part of President Johnson's War on Poverty, each of the armed services had inducted between 15% and 17% of their recruits from among applicants who did not meet accepted minimum aptitude standards. The services were tasked with training and preparing these individuals for meaningful civilian jobs at the end of their 2-year enlistments (Baskir & Strauss, 1978). NPRDC, along with the other service laboratories, participated in the assessment of the performance of the Project 100,000 personnel.

At the height of the Cold War, and with President Reagan's goal of the 600-ship Navy, there was intense activity to man the force. NPRDC's contribution in the 1980s centered on research to support the selection and assignment of large numbers of personnel, develop manpower tools for managing this large military force, improve the delivery and availability of training, and address social issues as substance abuse and race and gender integration. The end of the Cold War gave impetus to research into downsizing and productivity improvement and into developing a force able to meet such varied missions as peacekeeping, drug interdiction, and counterterrorism.

At its peak in the early 1980s, there were more than 400 personnel working at the NPRDC facility, and the orientation was decidedly multidisciplinary. Of the civilians, about two thirds were scientists and technicians in the fields of psychology, education, mathematics, statistics, operations research, economics, and computer science. Organizationally, there were three laboratories in manpower and personnel, training, and human factors, with major programs within the laboratories. The research had achieved national recognition in several areas, and optimism was high.

A shock came in early 1987, when Secretary of the Navy John Lehman attempted to close NPRDC by the end of the fiscal year. Although his efforts were reversed by his successor James H. Webb, they did lead to a comprehensive review of the entire program by a steering committee of high-ranking officers, which concluded that the NPRDC should continue as a separate shore activity. However, the review resulted in the transfer of the human factors engineering function to the Naval Ocean Systems Command in San Diego, California.

As with most military organizations, the end of the Cold War and cuts in operating budgets affected NPRDC. By October 1994, under severe funding constraints, NPRDC reduced its staff of 228 civilian personnel to 154. One year later, the Navy recommended to the Base Realignment and Closure Commission (BRAC IV) the disestablishment of NPRDC and relocation of its functions (Department of the Navy, 1998). NPRDC's training research mission was transferred to the Naval Air Warfare Center in Orlando, Florida, in early 1998. NPRDC was officially disestablished in November 1999, and its remaining function, manpower and personnel research, was assigned to a new organization, the Navy Personnel Research, Studies and Technology division of the Navy Personnel Command in Millington, Tennessee.

RESEARCH AND DEVELOPMENT ORIENTATION

NPRDC primarily conducted applied research in support of Navy and Marine Corps headquarters and fleet commands with the goal of applying knowledge or technologies to military issues. Basic research funds targeted to discovering new knowledge or technologies represented only a small fraction of the annual operating budget. Instead, NPRDC capitalized on the work sponsored by the Office of Naval Research (ONR) and other basic research organizations for discovery of new technologies. The research in several areas complemented ONR's basic research program in behavioral science. For example, the ONR item response theory program laid the theoretical groundwork for NPRDC's extensive research into computerized adaptive testing (CAT). ONR's cognitive processing research positively influenced NPRDC's long-range program to develop procedures to enhance the learning of complex skills.

This applied focus also led NPRDC to obtain research requirements and funding support from outside the organization; about 40% of operating costs were funded directly by customers and consumers (e.g., Navy manpower resource planners, Navy Recruiting Command, Naval Training Centers, etc.) rather than the NPRDC programmed research budget. These projects capitalized on existing NPRDC strengths to meet specific sponsor needs and requirements. Although many of these projects were short term (2 to 3 years), others extended over many years and were funded through partnerships between NPRDC and other agencies.

Many of NPRDC's research projects required access to subjects and data from operating environments. However, each Navy command's operational schedule had to fulfill a diverse range of mission requirements that left relatively little room for variation. As a result, opportunities for using laboratory paradigms in which experimental and control subjects are treated differently were extremely limited. Therefore, there was a greater emphasis on observational rather than manipulative research and development efforts in the fleet. Within these military restrictions, the Navy and the Marine Corps cooperated in providing NPRDC with sufficient numbers of subjects on whom significant research could be conducted. On the other hand, NPRDC capitalized on access to large military databases and the opportunity to conduct longitudinal research. The military is a relatively closed system, and individuals can be tracked and measured over the course of their service careers and even beyond.

The 26 years of NPRDC's existence also witnessed supportive relationships and joint projects with other government research institutions, particularly the U.S. Army Research Institute and the Air Force Human Resources Laboratory, the other major service behavioral science laboratories. Some of these cooperative endeavors were driven by Department of Defense policy. For example, in May 1974, the Assistant Secretary of Defense (Office of Manpower and Reserve Affairs) mandated replacing individual service test batteries with the Armed Services Vocational Test Battery (ASVAB) for selecting enlistees and placing them into military occupations. This led to the establishment of the joint-service ASVAB Steering Committee and the ASVAB Working Group, organizations that although changed over the years still exist today (ASVAB Working Group, 1980).

Another major Department of Defense initiative in 1974 was the establishment of the Manpower Research and Data Analysis Center, renamed the Defense Manpower Data Center (DMDC) in 1976. Created to collect and maintain accurate, readily available manpower and personnel data, DMDC evolved into a stable source of data and a research resource for the service laboratories (DMDC, 2002).

NPRDC offices were formed to facilitate ongoing support to Navy and Marine Corps headquarters and fleet units. An applications support office was designed to improve interactions between researchers, sponsors, and users. A studies and analysis group was created to carry out analytical studies of a quick-response nature. A Navy Science Assistance Program (NSAP) was established to maintain an active liaison with fleet staffs and to work with NSAP coordinators at other research laboratories to identify near-term fleet problems appropriate to NPRDC's mission and research capabilities. A technology transfer capability was also introduced to respond to many requests from local, state, and national agencies. NPRDC was also a member of the California Consortium of Federal Laboratories, which was chartered to identify military research and development products that could be applied to solving problems in the civilian sector.

MAJOR RESEARCH AREAS

The NPRDC program responded to national security concerns within the framework of four major areas of behavioral science research: manpower, personnel, training, and human factors. A serious consideration driving research in each of these areas is the uniqueness of the environments and platforms within which the Navy operates. Military systems are entirely based on an upward mobility concept; people begin at the first rung of the career ladder, and personnel replacements are obtained only from the entry level. Because each Navy command is also in a constant state of personnel turbulence (i.e., crewmembers and officers rotate in and out of positions on cycles of 2–3 years), replacement strategies must be continually optimized. The Navy ship platform is harsh and demanding and requires high levels of performance that are not easily met by the small number of officer and enlisted personnel manning the ships.

Manpower

Manpower research is concerned with modeling and managing aggregates of people in the workforce to satisfy planned requirements while maximizing the Navy's overall readiness. Programmatic work on force management represented a substantial component of the NPRDC effort over the years. The manpower research program employed professionals with training in operations research, mathematical modeling, economics, and large database design working within the following five domains:

1. *Manpower modeling.* NPRDC and its predecessor laboratory in San Diego, California, were pioneers in developing and implementing mathematical models for manpower planning that helped transform the way the Navy manages its human resources. The first such model, the Advancement Planning Model, was developed in 1965 for use on the Navy's mainframe computers. It calculated the monthly number of enlisted advancements of Vice Petty Officers in some 200 skill communities. The Advancement Planning Model was used for 25 years and was replaced by a PC version in 1990. This work and other related models extended the Navy's capabilities for forecasting manpower supplies, predicting force losses over a horizon of several years, determining and reducing personnel costs, more effectively managing incentives, and optimizing the match between human resources and job requirements.

2. *Distribution modeling.* Research into the assignment and distribution of personnel and positions was a core mission and led to the development of several operational systems, one of which is an online information and decision system for both sailors and those responsible for their assignments. Sailors around the world can gather information about possible job vacancies and even apply for desired jobs online. Also invaluable to the Navy as world issues affect the workforce are decision support systems developed by NPRDC to determine the trade-offs of assignment policy goals and assist in the execution of the assignment process.

3. *Information systems.* As early as 1977, NPRDC developed the Navy's first executive-level information system that consisted of databases and models that Bureau of Naval Personnel top management could query during the course of executive dis-

cussions. A more recent information system used a series of mathematical procedures in support of the Navy alcohol and drug screening efforts, resulting in a random selection and timing algorithm that provided the needed deterrent factor while improving the accuracy and efficiency of testing. A Web site was designed to provide accurate centralized information for drug and alcohol program managers to help the Navy carry out its zero-tolerance policy and for sailors seeking assistance.

4. *Recruiting systems.* Recruiting of more than 50,000 new individuals a year to serve in the Navy is a critical and difficult function. NPRDC supported both the management and personnel sides of this mission. In the late 1970s, as microcomputers became available, NPRDC developed an automated system to support individualized testing, vocational guidance, assignment prediction, and management support. Another major program was directed to measures for selecting recruiters, increasing recruiter productivity, and managing the recruiting process.

5. *Training reservation systems.* The Navy operates training facilities in about 400 different locations, with an estimated 350,000 students attending one or more classes every year. Scheduling training that can last from 1 day to more than 6 months is a major issue, especially in a time of national emergency. NPRDC adapted airline industry booking technology that significantly improved the maintenance of accurate class reservation information, reduced the number of unfilled school seats, improved fleet manning and readiness, and minimized waste of student time awaiting instruction or job transfer.

Personnel

NPRDC's program of personnel research was most closely aligned to the professional area of industrial–organizational psychology. Psychologists, sociologists, programmers, and researchers with behavioral science and quantitative backgrounds were the primary workforce. Research focused on recruiting strategies and applicant screening, selection, and classification; predicting and measuring school and on-the-job performance; assessing attitudes, morale, and job satisfaction; designing and monitoring programs aimed at improving quality of life; and examining and evaluating strategies to improve organizational effectiveness. This research can be clustered into the following seven subject areas:

1. *Officer selection.* The selection of officers was a core NPRDC project whose origins extended back to the 1960s, when im-

proved aptitude tests were developed for officer candidate selection. Starting in the 1970s, seminal work was conducted on developing vocational interest scales on the Strong Vocational Interest Blank for use in selecting the most promising officers to the Naval Academy and the Naval Reserve Officers Training Corps. A selection system consisting of aptitude, interest, and other measures was developed and validated against measures of school and job performance. Using longitudinal designs, NPRDC made continuing recommendations over its lifetime for improvement in selection procedures to enhance their overall usefulness.

2. *Enlisted selection.* Another established NPRDC research program was the development and validation of measures of enlisted applicant aptitudes, skills, and intellectual characteristics. Throughout NPRDC's history, its researchers monitored ongoing developments in testing research to identify promising new instruments, such as dynamic motion, spatial reasoning, personality constructs, and biopsychometrics (i.e., how brain wave measures relate to successful on-the-job performance) that potentially could add to the predictive value of operational test batteries. Tests within the Navy Basic Test Battery and later the ASVAB were constantly validated and used in assigning personnel to Navy schools and jobs.

3. *Computerized adaptive testing.* NPRDC's spearheading of an Office of the Secretary of Defense–supported program to design a computerized adaptive version of the ASVAB represents one of its most significant contributions to the Navy, Department of Defense, and industry (Sands, Waters, & McBride, 1997). CAT–ASVAB was a product of research in cognitive testing, applied computer technologies, and decision sciences and represented a major leap forward in personnel testing. Fully operational in mid-1997, CAT–ASVAB was the first large-scale computerized testing program in the world. The program saves the Department of Defense millions of dollars by testing nearly all military applicants in half the time it would take with paper-and-pencil tests with greater flexibility, security, and standardization of administration.

4. *Enlisted classification.* Beginning in the 1960s, the availability of mainframe computers together with earlier research by the Air Force enabled Navy personnel researchers to develop mathematical models and person–job matching algorithms for optimizing the assignment of recruits to technical schools and first duty stations. Advances in computing over the years facilitated the design of more sophisticated technologies that

considered both the needs of the individual (e.g., the recruit's training preferences) and organizational goals (e.g., assignment of minorities to technical training schools).

5. *Social issues and quality of life.* Work in the 1970s and 1980s provided support for the Navy's Human Resources Management program to assess and manage unit-level problems such as drug and alcohol abuse, race and ethnic relations, and overseas diplomacy. Projects conducted in the late 1980s for the Marine Corps demonstrated relationships between quality of life services, such as child care centers for parents and gymnasium facilities for single personnel, and personnel retention, performance, and readiness. Orientation materials were prepared for families scheduled to be stationed overseas, and a gamelike simulation was developed to involve them in the dynamics of being in a strange country. This product, Bafa Bafa, is now being used as a training tool by international companies sending executives and their families overseas, an excellent example of technology transfer from government to industry.

6. *Personnel surveys.* A strong personnel survey capability was developed in the 1960s that led to NPRDC serving as the central clearinghouse for all Navy surveys by 1975. The NPRDC worked hand-in-hand with Bureau of Naval Personnel managers for over 25 years and pioneered the application of survey technologies to the study of equal opportunity, gender integration, and sexual harassment issues. Statistically reliable sampling strategies were developed that used only about 5% of the enlisted force to gather high-quality information with minimal disruption to the fleet. Years of research and operational use of computer-administered surveys led to the design of a Web-based customer satisfaction survey system, implemented Navywide in 1998, which enabled Family Service Centers and Military Welfare and Recreation facilities to monitor their own clientele.

7. *Productivity enhancement.* A productivity enhancement program to address worker motivation and performance started in 1976 and focused on organizational assessment, diagnosis, intervention, and readiness. The emerging principles from the academic and industrial communities for managing worker incentives using performance-contingent rewards were applied in the design of performance-based incentive systems in six Navy shipyards. The NPRDC also operated an experimental productivity laboratory in the 1980s to examine the effectiveness of various strategies for improving worker quality,

output, and satisfaction. This work led to the development in the 1990s of Navy Total Quality Leadership Centers on both coasts.

Education and Training

Training and education in the Navy is a costly and labor-intensive undertaking. In a typical year, some 300,000 enlisted personnel are involved in formal training. NPRDC's education and training research program addressed a broad spectrum of issues for the Navy and Marine Corps ranging from curriculum development and standardization to foreign language training, leadership development, and several forms of automated instruction. The program was concerned with improving access to and delivery of training, enhancing the quality of education, and increasing overall effectiveness. A close relationship was maintained between the ONR cognitive sciences research program and the cognitive scientists, educational psychologists, subject matter experts, and curriculum development specialists who performed research on the following topics:

1. *Job-Oriented Basic Skills (JOBS) training.* In early research conducted in support of Project 100,000, NPRDC's unique responsibility was the piloting of courses in remedial reading and the development of simple job performance aids (e.g., a conversion chart for weights and measures) to compensate for skill deficiencies of lower-aptitude personnel. In 1978, the Navy implemented the JOBS training program to provide lower-aptitude personnel with job-oriented basic or prerequisite skill training needed to successfully complete Navy technical schools and perform to fleet standards. Evaluation of JOBS student participation and attrition over an extended time period (from 1979 until 1987) demonstrated a high level of success in providing the Navy with an alternative technical manpower source.

2. *Computerized training.* NPRDC pioneered some of the earliest breakthroughs in applying computers to training and training management. A major effort during the mid-1970s examined the use of computer-assisted instruction in the Basic Electricity and Electronics School in San Diego, California, tied into PLATO IV, a mainframe computer–based training system at the University of Illinois, Urbana. A computer-managed instruction system developed for the aviation technical training command in Memphis, Tennessee, in the 1970s was at that time the largest computer-managed instruction program in the world for managing trainee scheduling, test grading, and student feedback.

3. *Computer simulations.* A significant NPRDC research thrust that traces to the early 1980s was the improvement of computer simulations to reduce training costs. Several products were developed in support of electronic equipment maintenance, radar operations, and simulations of a ship steam plant. A 1984 display in the rotunda of the Senate Office Building of these technologies informed Senate staff members of the value of retraining and technology transfer.

4. *Automated instructional design systems.* The Authoring Instructional Materials project developed automated systems for designing, developing, and producing instructional materials for conventional and computer-delivered courses. Implementation of this project in 1990 reduced the time needed for these functions by 20% to 50%. Further work led to a system for a paperless classroom and automated classroom procedures. The system provided classroom instructors with computerized programs for personalization, visual aids, and video and computer graphics displays.

5. *Distance training.* NPRDC established a video teletraining (VTT) research laboratory in 1990 in support of the Navy goal to provide a wide spectrum of technical courses to both active duty and reserve personnel with fewer instructors, fewer classrooms, and fewer travel dollars. A series of studies concerning the conditions under which VTT was most effective, materials to be used, and the interactivity needed between instructors and students led to routine use of VTT in both lecture and laboratory Navy courses. VTT student achievement was found to be similar to that in traditional classrooms, and training costs were cut in half.

6. *Antisubmarine warfare training.* Over 2 decades, NPRDC developed computer-assisted instruction techniques that increased student performance and cut training costs by $10 million to $20 million annually in delivering instruction for the Navy's carrier-based antisubmarine warfare systems. Projects included shipboard minehunting sonar training and examination of an integrated undersea surveillance system. This extensive work led to the development and implementation of a highly sophisticated Interactive Multisensor Analysis Trainer that replaces conventional rote memorization drills with visualization techniques, providing students and operators an understanding of ocean dynamics, acoustics, and tactics. The scientific breakthrough achieved by this trainer is the unique combination of high-quality, physics-based models, databases, and simulation of the ocean environment that

allows students to experience a "what-if" environment in the classroom.

Human Factors Engineering

Human factors engineering was a small component of NPRDC's mission during the first 15 years that examined the interactions between the working environment, hardware, equipment, and the human operator. From a human perspective, the program was aimed at enhancing the quality of performance and improving working life. Many of NPRDC's human factors research projects were conducted in collaboration with the Navy's research and development hardware laboratories as they planned new equipment and systems to meet national security needs for a 20- to 30-year planning horizon. For example, NPRDC researchers provided preliminary and contract design support (combat, hull, and machinery systems) for the introduction into the fleet of the guided missile destroyer (DDG-51) that had reduced detectability and likelihood of being targeted by enemy weapons and sensors. NPRDC researchers also worked directly with ships and commands in addressing current operator problems with (a) hardware, equipment, and platforms; (b) determining how factors such as stress, boredom, or ship's motion affect human performance; and (c) recommending improvements in human–machine interface and operator performance. Two specific examples of successful human engineering NPRDC projects involved mine-hunting sonar and a computer-assisted fault detection system. NPRDC established baseline human engineering requirements and procedures for operating minehunting sonar in a ship environment. Researchers examined ways of maximizing visibility on sonar displays, types of visual cues operators needed to recognize and discriminate mines, and the skills and technical training that were needed to prepare operators for a new system. NPRDC also developed a fault detection system to provide automated monitoring of high-pressure water delivery systems used in fighting fires to reduce the possibility of system damage and an alarm system that alerted personnel immediately when problems were detected. Because they operate at high pressure, these systems occasionally rupture, resulting in flooding, damage, and the loss of firefighting capabilities.

CONCLUSION

NPRDC had a relatively short but highly successful existence as a Navy research center. Research funding and priorities reflected the National Defense Strategy as well as the Navy's own strategic plans and doctrine. However, although Navy and Marine Corps sponsors always considered NPRDC's products and services to be important, management responsibility for NPRDC moved from one headquarters organization to another over its lifetime. Advocacy for portions of the research program was strong, but support for NPRDC as an integral organization was sometimes tepid, and in some cases attitudes toward it were adversarial. Because of diverse sponsorship and research organizations with competing missions, NPRDC was not able to hold together an organization supporting manpower, personnel, training, and human factors components.

However, much of the research strategy was an outgrowth of the knowledge and skills of dedicated NPRDC personnel, who made significant and unique contributions to the body of scientific knowledge and had a major influence on many social issues. Studies that were conducted on such concerns as women's role in the military, zero tolerance for drug use, quality of life, and diversity made significant contributions to the military and to society at large.

The NPRDC manpower research program is an excellent example of an institutional capability that was created in the San Diego, California, laboratory before NPRDC's establishment and fine-tuned during the life of the organization. This program was tailored and totally responsive to emerging Navy and national security concerns over the years and enabled the Navy to be the most advanced of the four services in using scientific manpower planning and decision making.

NPRDC's development of the CAT–ASVAB is a textbook case of how a new technology is identified, investigated, and applied to solve military needs. It is one of the most significant contributions NPRDC made to the Navy, the Department of Defense, and industry. The Department of Labor, the Immigration and Naturalization Service, and the Educational Testing Service used the CAT–ASVAB model in developing their own tests. In addition, the work directly led to programs to implement adaptive testing in other countries, including Germany and Belgium.

The NPRDC training research program was on the cutting edge in applying technology to Navy needs. For example, in the 1980s, NPRDC established a fully equipped portable classroom on wheels that could be stationed alongside ships to deliver specialized training. Another successful product, called *Batman and Robin*, was widely used for training officers in tactical decision making (Federico, Bickel, Ullrich, Bridges, & Van de Wetering, 1989.) This interactive computer simulation was strongly grounded in both theoretical and empirical research, despite its arcade game–like name.

A significant portion of NPRDC research was long term, extending over many years and capitalizing on the continuing development of researcher talents. NPRDC efforts were embedded into a programmatic approach that supported major sponsors over a period of years. NPRDC was able to serve as the institutional memory that many policymakers, customers, and users could draw on for quick responses or more extensive policy analyses. Researchers shared knowledge extensively with their peers from other government laboratories and the research arms of other countries at meetings and conferences such as The Technical Cooperation Program, North Atlantic Treaty Organization panels, and the International Military Testing Association. Two books were edited under sponsorship of the International Military Testing Association to describe advances in the areas of instructional technology (Ellis, 1986) and military personnel measurement (Wiskoff & Rampton, 1989). NPRDC researchers served as faculty members at several local universities and also were active participants in professional organizations such as the American Psychological Association, the American Educational Research Association, and many others. NPRDC received several awards from professional organizations, including a 1999 award from the Military Psychology Division of the American Psychological Association for outstanding contributions in the development and application of CAT. NPRDC researchers also conceived the division's journal, *Military Psychology*.

As a discrete organization, NPRDC existed for only 26 years. However, the Navy's capabilities for performing applied research in manpower, personnel, training, and human factors in fact started much earlier and continue today, albeit in different organizations. NPRDC's efforts were recognized and acknowledged by the Secretary of the Navy in June 1998 with the award of the Meritorious Unit Commendation, one of the Navy's highest awards given to shore-based organizations.

IV

CLINICAL AND COUNSELING PSYCHOLOGY

The recognition and treatment of mental illness evolved over the course of the 20th century. When World War I began, physicians treated mental illness in hospital units; psychologists administered tests but did not provide direct therapeutic services. There was no provision for psychologists in uniform. During World War II, psychologists were added to treatment teams, and in 1944, courses for Army clinical psychologists were begun at Fort Sam Houston, San Antonio, Texas. The Navy offered commissions to clinical psychologists in the Medical Department.

After World War II, American hospitals were faced with the treatment needs of large numbers of veterans with mental illness. The Veterans Administration and the U.S. Public Health Service developed treatment models that included clinical psychologists, the government created the National Institute of Mental Health, and the American Psychological Association helped develop quality university graduate training programs in clinical and counseling psychology. Responsibility for psychotherapy was extended to clinical psychologists, and the independent profession of clinical psychology was established, allowing for the creation of a body of research and scholarship. With the reorganization of the national security organizations in 1947, uniformed psychology grew. Inservice opportunities evolved for internship training programs, postgraduate fellowships, nontraditional positions, and new career opportunities in operational positions.

The discipline of military psychology has always been concerned with adjustment and mental health, important in sustaining military readiness. Military training is realistic and stressful, and psychological screening programs are used to predict individuals who may show difficulty adjusting. Interest in operational and combat stress reactions derived from the concerns about their negative impact on military unit functioning and availability of personnel. With the improvements in tactics and weaponry, more frequent deployments, and longer operations characteristic of 21st-century military operations, stress levels have increased. Mental health teams and forward-deployed psychologists are providing direct assistance, shedding their traditional hospital-based practices. The efforts of military psychologists to maintain the personal resilience and mental health of service members are critical to addressing the nation's security challenges.

10

NAVY CLINICAL PSYCHOLOGY: A DISTINGUISHED PAST AND A VIBRANT FUTURE

MORGAN T. SAMMONS

The history of clinical psychology in the United States is ineluctably intertwined with the military. From the earliest days of World War I, less than 20 years after Lightner Witmer can be considered to have established the discipline of clinical psychology (Routh, 1996), psychologists have been involved in both military research and in direct service provision to the men and women of the military. McGuire (1990) described how military psychology progressed from the assessment of recruits in World War I to a much broader clinical role following the World War II. McGuire also noted the significant contributions of research psychology in preserving the safety of military personnel and, in particular, the role of Navy research psychologists in redesigning the color of life vests from bright yellow, which was poorly visible in choppy water, to the current and more easily spotted orange-red, thereby saving the lives of innumerable military and civilian personnel in need of water rescue.

The opinions expressed by the author are his private views and do not represent the official views or opinions of the U.S. Navy or Department of Defense.

This tradition of research and clinical practice continues to expand. As of October 2004, there were 131 clinical psychologists (including psychologists in training) on active duty in the U.S. Navy. Nonuniformed civil servant psychologists also provide substantial levels of health care in Navy hospitals and clinics around the world. Active duty Navy aerospace experimental psychologists have long provided valuable research and development services to the aviation community. Navy research psychologists provide empirical examination of other programs involving personnel evaluation. At the time of this writing, there were 22 research psychologists and 25 aerospace experimental psychologists on active duty. Although I do not wish to downplay their substantial contributions toward an effectively functioning military, such as the life-saving example given above, they are not the focus of this chapter.

THE EVOLUTION OF NAVY PSYCHOLOGY

McGuire (1990) succinctly outlined the evolutionary nature of military psychology from its roots in personnel screening, from which the entire field of industrial psychology emerged, to the largely clinical role that grew from the necessity of treating the numerous psychological casualties of World War II. Although more than 500 psychologists served in uniform in World War II, in the aftermath of the war the uniformed psychology corps was disbanded. Several psychologists joined the Medical Service Corps on its establishment in 1947, but this number did not grow substantially until the time of the Korean conflict (McGuire, 1990). In the 30 or so years between then and the mid-1980s, Navy clinical psychology gradually expanded its presence in military treatment facilities (MTFs). The first psychology internship approved by the American Psychological Association (APA) opened at the National Naval Medical Center (Bethesda, Maryland) in 1956 and was one of the first internship programs to acquire APA approval anywhere in the country. Gradually, training programs were established at naval hospitals in Portsmouth, Virginia, and in San Diego, California. These produced many classes of interns. Although the training program at Naval Hospital in Portsmouth was disestablished in 2003, the programs at Bethesda and San Diego continue to graduate predoctoral interns.

As the clinical field matured and as clinicians came into their own within the Medical Service Corps, opportunities for Navy-sponsored postdoctoral fellowship training emerged. Because of the demand for skilled assessment of traumatic brain injuries, neuropsychology was the first area in which Navy psychology sent fellows for postdoctoral training. Because Navy clinical psychology is tasked with assessment and service provision to preschoolers and school-age children in Department of Defense Dependents Schools in many parts of the world, Navy psychologists for the past 15 years have been sent for outservice fellowship training in child psychology. Similarly, with the emergence of health psychology as an area of clinical specialization in the late 1980s, Navy psychology has sent a number of postdoctoral fellows for 1-year outservice training in health psychology, generally to Johns Hopkins University in Baltimore, Maryland, or Dartmouth College in Hanover, New Hampshire.

One of the most visible—and controversial—of Navy psychology's postdoctoral training endeavors was the Psychopharmacology Demonstration Program (PDP). The PDP was a triservice fellowship designed to equip military psychologists with the requisite skills to administer, in the context of an ongoing psychotherapeutic relationship, psychotropic medications of various classes. The specifics of the PDP project have been described elsewhere (Dunivin, 2003; Laskow & Grill, 2003; Newman, Phelps, Sammons, Dunivin, & Cullen, 2000; Sammons & Brown, 1997).

In broad terms, the overall importance of the PDP to the field of clinical psychology in general cannot be overstated and is a highly trenchant reminder of the inescapable links that connect the practice of military psychology with the psychological community at large. It is safe to say that the PDP had a larger effect on U.S. clinical psychology than any other demonstration program undertaken at any time by the profession. The PDP was the germinal step in the successful push toward prescriptive authority for psychologists, and without it, the current successes in this endeavor would surely have failed. As of this writing, psychologists have independent prescriptive authority in Louisiana, New Mexico, and the territory of Guam. Legislation has been introduced or is in various stages of the legislative process in another dozen states. More than 30 states have task forces examining the feasibility of introducing prescriptive authority legislation in their state. Several hundred civilian psychologists have been trained in programs modeled in large part on the PDP. Prescriptive authority has become a part of the fabric of U.S. psychology.

Although the PDP ended in 1997, good ideas have a way of reemerging. At Tripler Army Medical Center in Honolulu, Hawaii, the Navy and Army currently are establishing a fellowship program in health psychology that will include a psychopharmacology track. It is anticipated that in the not-too-distant future, psychopharmacology training will be available to military psychologists who seek to specialize in this important area of practice.

WOMEN IN NAVY PSYCHOLOGY

McGuire (1990) noted that in 1988 women constituted approximately 15% of clinical psychologists on active duty in the Navy. Since that time, the number of female psychologists in the Navy has increased but still lags behind female representation among civilian psychologists. In 2004, only

36% of Navy clinical psychologists were women, a substantial improvement over previous Navy figures but lower than in the psychology community at large (about 52% of APA members are women). Women are better represented in the grades of lieutenant and below, with 44% of this group being female. Fewer field-grade officers are women. For psychologists, only 2 out of 21 Navy commanders (9%) and 2 out of 9 (22%) Navy captains are women. So although women have made significant strides in terms of overall representation in Navy psychology in the past 15 years, much room remains for improvement.

PSYCHOLOGY AS AN OPERATIONAL BEHAVIORAL HEALTH SPECIALTY

In the mid-1980s, a transformation began that has significantly reshaped the role of clinical psychology in the Navy more than any other movement in the field. In this transformation, Navy psychology has shed its traditional hospital-based roots and has increasingly become an operationally oriented profession dedicated to meeting the behavioral health needs of underway forces and combatants in the theaters of operations.

This initiative began with the deployment of a few psychologists in support of search and rescue training programs. For about 20 years, psychologists have been on training staffs in operational warfare units, including Survival, Evasion, Resistance, and Escape (SERE) schools. These schools equip military personnel with skills that may be required to avoid capture, successfully seek escape, and resist interrogation should they be captured by enemy forces. SERE psychologists screen participants for this arduous training and support them throughout the course, which involves realistic simulation of experiences of capture. Psychologists also screen and monitor the activities of instructors in these courses to ensure that punitive or brutal behavior by simulated guards, as famously described by Zimbardo (1973) in the Stanford prisoner experiment, does not occur. Each SERE school now has a *billet* (established staff position) for a full-time clinical psychologist, and other psychologists are exposed to this training regimen for shorter periods of time.

Navy clinical psychology also has had a long and valued association with Navy Special Forces. Since the mid-1980s, Navy psychologists have provided services to training programs for Navy Sea–Air–Land (SEAL) teams, evaluating potential candidates and helping trainees withstand the stresses of this lengthy and extraordinarily rigorous curriculum. Navy psychologists also provide consultative services to SEAL teams and leadership and regularly deploy into combat areas with these units.

Close interaction with Special Forces is an area that is expected to grow rapidly in the future. At the time of this writing, Navy psychology is developing a postdoctoral fellowship in the area of operational psychology. This fellowship will be housed at the National Naval Medical Center. The curriculum will involve instruction in profiling of terrorists and defectors, security consultation and screening in the operational environment and security consultation, the psychology of counterintelligence, and the ethics of operational psychology, among other subjects. Training will take place with various agencies and components of the Navy department that have interest and activities in these areas. Thus, Navy psychology's involvement has been seminal in creating a truly operational mental health asset that can not only consult and train, but also directly provide services to fighting men and women on the front lines.

PSYCHOLOGISTS UNDER WAY: THE CARRIER PROJECT ("SPECIALISTS AT SEA") AND BEYOND

In 1996, a demonstration project was established with the goal of providing immediately accessible mental health services to the men and women serving aboard Navy aircraft carriers and the small vessels that make up a carrier battle group. Before that time, no specialty mental health consultation had been readily available aboard these floating cities of some 5,000 personnel, and sailors or marines experiencing suicidal ideation or other psychological difficulties were required to be medically evacuated to the closest shore-based treatment facility. The losses in manpower and mission readiness and the significant costs associated with transporting personnel in need of services unavailable on board were an ongoing and significant concern to carrier commanders and shipboard medical staff.

Lieutenant Helen Napier (now Dr. Helen Holley) in 1996 became the first psychologist to staff a full-time billet aboard an aircraft carrier, the USS *Kitty Hawk*. Lieutenant Napier acted as a pioneer on two fronts: She was the first psychologist assigned to a carrier, and she arrived on board only a few years after the first women had been integrated into ship's companies. I vividly recall an incident in 1990 that illustrates the then-prevailing opinion regarding the presence of women on underway naval vessels: The skipper of another aircraft carrier (since decommissioned) refused to sail with a female surgeon on board—even if it meant sailing without a surgeon!

The demonstration project was judged a success, and in 1998, the Specialists at Sea program formally establishing billets for psychologists aboard the carriers was implemented. By 2002, psychologists had joined the ship's company of all 12 commissioned aircraft carriers. By any measure, the placement of psychologists aboard aircraft carriers has been an unqualified success. Rates of medical evacuation for psychological diagnoses have plummeted on carriers with psychologists, and psychologists have contributed to a dramatic lowering in rates of administrative separation of under way sailors. In 2001, Lieutenant Commander David Jones, serving aboard the USS *En-*

terprise, helped achieve the lowest number of administrative separations of any carrier in the fleet (D. E. Jones & Lee, 2002). On other carriers, similar results accrued. In the last deployment before the placement of a psychologist aboard the carrier USS *John F. Kennedy,* 28 persons were evacuated for psychological reasons. In the deployment following placement of the psychologist, there were no psychological evaluations. The cost of evacuation of one service member has been estimated to be at least $4,400 (Wood, Koffman, & Arita, 2003), but, as Ralph and Sammons (in press) noted, this estimate is almost certainly too conservative, because it does not calculate losses in morale, productivity, training time, and mission readiness and the high cost of training and transportation of a replacement for the unplanned loss.

A further measure of the success of the carrier psychology program can be found in recent initiatives to place psychologists aboard other under way platforms. Expeditionary strike groups are a new concept in naval warfare designed to provide a flexible platform from which to launch marines and other fighting assets, both at sea and during landing missions. They are typically composed of an amphibious assault ship—a large vessel with a substantial medical department and capacity for more than 3,000 troops—as well as other amphibious ships, destroyers, frigates, a submarine, and an air wing (Navy Fact File, 2004). In total, an expeditionary strike group has a complement of about 8,000 ship's company and expeditionary troops, smaller than that of a carrier battle group but substantial enough to require the services of a dedicated mental health professional. At the time of this writing, Navy clinical psychologists have deployed with five expeditionary strike groups, and a request has been made to establish permanent billets for psychologists on the nine Fleet Surgical Teams that support the medical departments of these strike groups. Thus, as is the case with carrier battle groups, immediately available behavioral health consultation will be seen as a significant value-added component to commanders.

THE OPERATIONAL STRESS CONTROL AND READINESS PROJECT

The most recent example of Navy psychology's increasingly operational stance is its involvement in a program designed to provide the men and women of the U.S. Marine Corps with a continuum of behavioral health services throughout the deployment cycle. In the past, direct access to behavioral health care services for marines has been limited. One psychiatrist and one psychiatric technician were attached to the medical company at the three Marine Division sites (Camp LeJeune, North Carolina; Camp Pendleton, California; and with the Third Marine Division in Okinawa, Japan). Other services were provided via the local MTF or, when deployed, by psychologists and psychiatrists temporarily assigned to the unit during the period of deployment. In the past, marines, like other service members, have been extremely reluctant to seek out mental health services. As Hoge et al. (2004) demonstrated, a perception of stigma associated with seeking mental health services was the primary reason why service members deployed to Iraq were reluctant to seek such services. It is likely that this same perception of stigma has kept marines from seeking mental health services in other circumstances as well.

Deployed marines are often tasked with the most dangerous missions in support of national security, and they perform heroically in this role. They also face a unique set of challenges. Rates of substance abuse are higher in the Marine Corps than in other branches of the service. Suicide rates tend to hover around 13 per 100,000 in the Marines—lower than in comparable-age civilian populations but higher than in other branches of the service (the Navy, in contrast, tends to have suicide rates between 10 and 11 per 100,000). Mental health issues are the second most common reason for unplanned separation from the Marines, second only to orthopedic injuries. Accident rates tend to be higher in the Marines than in any other branch of the service.

The Operational Stress Control and Readiness (OSCAR) project was designed to provide a continuum of readily available mental health resources across the deployment cycle by mental health personnel who form part of the unit. The project was initiated on a limited scale in Camp LeJeune in 2000, and in 2003, it was expanded to include all three active Marine divisions. Existing mental health resources have been augmented by either psychologists or psychiatrists who work closely with line commanders and other medical and mental health assets to address the needs of Marines both while deployed and in garrison. The project's goals, as outlined in Ralph and Sammons (in press) and elsewhere are to reduce administrative separations, medical discharges, and other unplanned losses due to mental health problems; to reduce manifestations of unmet mental health needs, such as suicidal behavior and alcohol-related incidents; and to reduce lost manpower and productivity associated with removing a marine from the unit to provide psychological services.

Thus, the OSCAR concept involves some fundamental shifts in the way mental health services are provided. To meet the needs of marines, mental health must become operational. Mental health providers must understand the needs of troops and commanders in the deployed environment and be able to provide an expeditionary mental health force that deploys and redeploys with the unit, thereby allowing an informed perspective on the specific challenges and issues faced by the unit as well as continuity of mental health services across the deployment cycle. Provision of services via embedded mental health personnel will also address the critical issue of credibility. Marines are far more likely to be receptive to mental health consultation if the "doc" has deployed to the field and has experienced the same rigorous demands placed on marines deployed in forward positions. Mental health personnel are also better able to understand the interactions of the marines within the unit and chain of command, where cohesion and a sense of unity are paramount. Thus, the ability of the provider to render expert consultation not only to marines but also to leadership is enhanced, and the issue of stigma in seeking services is reduced. Additionally, the OSCAR project seeks to expand the range of available mental health services through peer counselors. The use of peer counselors can stretch available mental health resources, tends to be easily accepted by colleagues, and can assist in reducing barriers to care, because intervention by a peer counselor is often seen as less stigmatizing than a visit to a mental health professional. Peer counseling has been used successfully to treat young men with sports-related injuries (Wu, 2001), a group with issues similar to those of marines.

Peer counselors in the context of the operational Marines environment tend to be more senior staff noncommissioned officers (SNCOs) who, for whatever reason, are unable to perform the duties in their primary job designator. For example, the personnel and administrative skills of a staff sergeant machine gunner who suffered an orthopedic injury that prevented him from deploying to the field but did not require a medical discharge from the Marines can be trained to act as a peer counselor. He would have an intimate familiarity with the marines in his unit and is likely to have served in loco parentis for many of the younger marines. The formalities required by the enlisted–commissioned officer relationship that may create distance in the psychotherapeutic relationship can also be avoided by using SNCOs. With on-the-job training by mental health professionals, these skills can be expanded on to enable a marine to provide basic peer counseling techniques, preliminary assessment, and cofacilitation skills in group therapy situations. The peer counseling model using trained SNCOs has been successfully implemented at the Second Marine Division in Camp LeJeune, North Carolina. This implementation, along with other aspects of the OSCAR program, has resulted in more rapid return to the unit; a decrease in lost time associated with seeking out specialty mental health services at the MTF; and, according to preliminary data, a reduction in administrative separations (Grefer & Harris, 2003).

Finally, the OSCAR model posits that services will be provided from the perspective of a community mental health and preventive approach. As is often the case in other branches of the military and with mental health service provision in general, the vast majority of mental health services are provided on a consultative basis only. This is an effective mechanism for apportioning the services of often-scarce mental health specialists, but it does little to address unmet needs, and it does not allow a more proactive approach to the identification and amelioration of common behavioral health problems. An office-based, consultative model is unlikely to meet the needs of line commanders, who may require ongoing, real-time consultation to solve behavioral health issues affecting unit morale and readiness. Evidence of the efficacy of the community mental health approach in OSCAR rapidly accrued. At one Marine division, the previous psychiatrist, who had used an office-based consultative model, had a total of 90 patient contacts in an entire year. After implementation of the OSCAR project at that division, the number of patient contacts rapidly rose to 90 in a single month, and those numbers continued to grow as the program expanded.

At present, the precept of continuous mental health service provision under the OSCAR model is being tested in real time. OSCAR-designated psychologists and psychiatrists are deploying with the First Marine Division in Iraq, and their counterparts who remained in the United States will soon deploy to the field to take their places. Lieutenant Commander Gary Hoyt was the first Navy psychologist designated to deploy to Iraq with the Division. He was accompanied by Lieutenant Commander James Reeves, a Navy psychiatrist. The skills of both these providers demonstrated the essential need for a continuum of mental health services for deployed personnel and for personnel returning to their home bases in the United States.

NAVY CLINICAL PSYCHOLOGY AND THE COMBAT OPERATIONAL STRESS RESPONSE

Since its recognition as a clinical entity, the prevention of posttraumatic stress disorder (PTSD) has become a major focus of military mental health. Because PTSD often is associated with long-term impairment in vocational and interpersonal adaptation, and because it afflicted many veterans of the Vietnam conflict, the Department of Veterans Affairs (VA) had traditionally taken the lead in this area, and active duty service members with PTSD were not commonly treated while in the military. In more recent years, however, there has been increasing recognition of the fact that early detection and management of combat operational stress response (COSR) serves as a force multiplier—that is, that early detection and effective treatment can return service members to the field who otherwise would be lost. Thus, effective techniques for assessing and intervening in COSR have become the focus of much research and clinical activity in the services as well as in the VA.

Events since September 11, 2001, have added significant impetus to the treatment of COSR. The aging of the Vietnam cohort resulted in some diminution in the numbers of veterans seeking services for PTSD in both the VA and military systems, but the routine deployment of large numbers of combatants to Iraq and Afghanistan required more aggressive systems to identify and manage active duty members with combat-related stress disorders. In March 2004, the U.S. Army released the report of the Mental Health Advisory Team that identified specific shortcomings in terms of service provision to soldiers deployed in Iraq and Afghanistan (Operation Iraqi Freedom Mental Health Advisory Team, 2004). This report found that if a soldier received mental health services in the field, the chances of return to duty approached 100%. If, however, the soldier required evacuation from the field, the chances of returning to the unit were negligible. The report also found a high degree of inconsistency in how behavioral health providers interpreted and applied the precepts of COSR. This study and a subsequent analysis published in the *New England Journal of Medicine* (Hoge et al., 2004) suggested that the number of returning service members meeting criteria for a combat stress disorder diagnosis might be as much as 15%—approximately the same percentage as Vietnam-era service members diagnosed with PTSD.

Hence, there has been considerable recent impetus toward the development of effective COSR treatments that can be applied both in theater and in fixed MTFs and toward the definition of evidence-based, effective treatments that can be replicated and universally applied. Previously, military mental health providers were routinely, if not systematically, taught the principles of critical incident stress debriefing. In the past several years, however, the efficacy of this technique has been increasingly questioned (e.g., Kenardy, 2000; Wessely, Rose, & Bisson, 1999). The findings of the Mental Health Advisory Team study make this much more than a theoretical debate about the merits of a particular intervention. A doctrinally consistent, systematic approach to the management of COSR will demonstrably improve outcomes and enhance mission readiness.

Some steps toward this goal have been taken. The VA, in conjunction with the Walter Reed Army Institute of Research, has published guidelines for the management of acute and chronic stress (National Center for Post Traumatic Stress Disorder, 2004). These guidelines contain evidence-based treatments that are applicable to the deployed environment, and along with recently published clinical practice guidelines for the management of both acute stress and PTSD (Department of Veterans Affairs, 2004a) in the primary care arena, they should materially assist in devising a consistent, effective set of interventions for these disabling conditions.

CONCLUSION: EXPANSION OF NAVY CLINICAL PSYCHOLOGY BEYOND TRADITIONAL ROLES

Navy clinical psychology has shown a pattern of growth that has progressed from its early roles in personnel selection to a broad range of clinical, operational, and leadership responsibilities. In doing so, Navy psychologists, along with their colleagues from other branches of the armed forces, have played a key part in shaping the future of the profession as a whole. The Navy's predoctoral internship training programs were among the first recognized by APA. Now its postdoctoral fellowships reflect the highest standards of subspecialty training and, in programs such as the operational psychology fellowship and the PDP, are likely to establish standards of training for the community at large.

Navy psychologists' use of a community behavioral health intervention model in providing services to the operational community via the OS-

CAR project and similar undertakings underscores the inefficiencies inherent in providing mental health services from a specialty consultation model. Programs such as the OSCAR project and the Navy's Behavioral Health Integration Project, which seeks to place psychologists directly in primary care clinics so that psychological care is as readily accessible as care by a primary medical provider, will serve as models to expand the availability of much-needed mental health interventions.[1] Navy clinical psychology has played a fundamental role in the global war on terrorism. In addition to direct support of combatants via the OSCAR program and by SEAL psychologists who provide command consultation and behavioral support under extremely arduous conditions, Navy psychologists have been detailed to provide consultation to commands dealing with terrorists and enemy combatants captured in the Afghan and Iraq conflicts. Navy psychologists have worked with senior military and civilian leadership not only to deal with the immediate response to terrorist attacks but also to provide advice on how to strengthen resilience in first responders, military caregivers, and the public at large.

Finally, no discussion of Navy clinical psychology would be complete without a discussion of psychologists' roles as leaders, for every Navy psychologist is expected to be, in addition to a skilled clinician, a consummate leader of military men and women. In the past 6 years, psychologists truly have stepped into the role of military leadership. Before the early 1990s, leadership opportunities for clinical psychologists were relatively uncommon. Captain Robert Herrmann was the first psychologist to head the Medical Service Corps in 1962, and Captain Paul Nelson was the second in 1978 (McGuire, 1990), but these were two remarkable exceptions. Although no Navy clinical psychologist has yet reached flag rank or commanded a Naval hospital, several serve on the executive committee of the medical staff in Navy MTFs, and some serve in executive medicine capacities in the largest treatment centers. At the time of this writing, Captain Freda Vaughan has served as the director for medical services at Naval Medical Center, San Diego, California; and Captain Richard Stoltz, in addition to being specialty leader for clinical psychology between 2001 and 2004, serves as director for women's and ancillary services at the National Naval Medical Center.[2] Psychologists increasingly serve in other decision-making capacities both nationally and locally, and their contribution to leadership, as well as their continued development of programs and curricula that shape the field at large, will represent the enduring legacy of Navy clinical psychology.

---

[1] Space precludes a thorough discussion of the Behavioral Health Integration Project. This project was initiated in 2003 and is closely modeled on an extant project set up via the auspices of Colonel Skip Moe in the Air Force. It is currently operational at five MTFs in the Navy, with plans for significant expansion in the coming years.

[2] At the time of this writing, I am the Specialty Leader for Navy Clinical Psychology and serve as Director for Clinical Support at the U.S. Navy's Bureau of Medicine and Surgery.

11

CLINICAL PSYCHOLOGY IN THE U.S. ARMY: 1946–2004

ROBERT S. NICHOLS

Beginning only a few years after World War II in 1945 and continuing for the rest of the 20th century, the United States did something it had never done before: It maintained large military forces in peacetime. Because its leaders felt that the country was threatened by the power and hostile intentions of the Soviet Union and other communist nations, they decided that the country could defend itself only by having military forces so powerful that no potential enemy could overcome them. This period, called the Cold War, ended only after the Soviet Union disintegrated in 1989. To conduct this "war," several million military personnel were needed, and those in combat units had to be sustained by thousands of support personnel, including large numbers of health care providers.

This chapter explains the history and achievements of the remarkable clinical psychologists who served in the U.S. Army after World War II. It describes how they were recruited and trained; explains the services they provided; and discusses the important professional innovations they made, innovations that have greatly benefited soldiers and their families.

OBTAINING CLINICIANS FOR THE POSTWAR ARMY

Among the skills in demand in the post–World War II period were those of clinical psychologists. At first there were great difficulties in recruiting, training, and assigning an adequate number of clinicians, but half a century later, the Army and the other armed services had developed a highly competent group of clinical psychologists who were playing a key role in supporting military personnel and military activities. By the start of the 21st century, there were nearly 500 uniformed clinicians in the Army, Navy, and Air Force caring for the men and women in uniform and their families and also helping retirees and their families.

There was a severe shortage of mental health personnel in World War II, and any capable worker was desperately needed. Psychologists could do testing, patient administration, and consultation with soldier's units, and some could do therapy. Therefore, they were eagerly sought after. By the end of the war, around 450 Army clinical psychologists were in place (Menninger, 1948; Seidenfeld, 1966). Many had less than doctoral training, and some could function only as psychometricians, but as a group they had shown the value of psychological services. As a result, when the postwar buildup began in 1948, the Army recognized the need for clinical psychologists. However, because the wartime psychologists had left the service and did not want to return, new clinicians were needed. In 1949, there were only 3 clinicians on duty, although 90 were needed (Ginn, 1997, p. 219).

In its first effort to obtain more clinicians during the late 1940s, the Army Medical Department (AMEDD) sent to graduate schools a small number of active duty officers who had some training and experience in psychology but not at the doctoral level. Unfortunately, the attrition rate was high. Fewer than five earned their degrees. Although this program did not meet the total need, it produced several senior-rank psychologists who provided essential leadership for clinical training and service programs. Two of them, Colonel Ernest K. Montague and Colonel Wendell R. Wilkin, later served as psychology consultants to the Army Surgeon General.

Army Senior Psychology Student Program

To cope with the continuing severe shortage of clinicians, a much larger recruitment and training program was suggested in 1948–1949 by the newly assigned psychology consultant, Lieutenant Colonel (later Colonel) Charles S. Gersoni. He was one of the fully trained psychologists who had served during World War II and left the service when the war ended. He agreed to return to active duty and serve as the psychology consultant. He proposed a new training program called the U.S. Army Senior Psychology Student Program (SPSP). That plan was announced in Army Special Regulation 605-60-40, May 16, 1949, and publicized in the *American Psychologist* (U.S. Department of the Army, Office of the Surgeon General, 1949). It offered officers' commissions, in the rank of second lieutenant, to psychology students who had completed 2 years of graduate work. They would enter on active duty for 1 year to continue at their graduate school; serve another year as an intern at one of the large Army hospitals; and then, after finishing their studies, serve for a 3-year "payback" period and be offered a commission in the regular Army.

When the SPSP program started, the draft was still in effect, and soon afterward (June 1950) the Korean War began, so a large number of high-quality students sought acceptance. Most had no prior military service, but some were veterans who had an extra incentive: Their wartime military service made them eligible for higher military pay than the nonveterans.

About 70 psychologists completed the program (Ginn, 1997, p. 219), and they provided valuable payback services during the 1950s. In addition, another unanticipated source of psychologists emerged during the war, when it became possible to "recapture" some World War II psychologists who had accepted commissions in the reserve when they were demobilized at the end of World War II. They were recalled to active duty for the duration of the Korean War. However, there still were not enough psychologists. In March 1953, there were only 6 on duty, whereas hospital commanders said 149 were needed (Ginn, 1997, p. 256), and most left after the war ended.

The SPSP program was terminated in 1954 because it was assumed that those who had been trained would stay on duty, but this turned out not to be true. Some of the veteran psychologists did stay on duty long enough to qualify for their 20-year retirement, but almost all the nonveterans left when they could, so the severe shortage of psychologists continued.

There were three main reasons why the nonveterans left and why direct recruitment rarely succeeded. First, most young psychologists of that era, though not all, were averse to military service. They were liberal, distrustful of authority, and suspicious of the military. Second, there was a severe national shortage of clinicians, so attractive civilian opportunities were plentiful. And third, the new military psychologists faced some unpleasant situations. Their pay was less than they could earn in civilian life, and their entry rank as lieutenant was lower than that of captain given to psychiatrists and other physicians, so they were usually controlled by physicians, and some had trouble shaping their own roles. Still, many of them liked their professional work, the extensive clinical and educational opportunities, and the fact that medical care was free so that a client's ability to pay was not a problem. Nevertheless, the disadvantages outweighed the advantages for most, and they left the Army.

Despite the fact that the first SPSP program had not fully succeeded, it had to be restarted in 1957 (Ginn, 1997, p. 291) because it had proved to be the only way to obtain significant numbers of psychologists. In fact, this type of train-and-retain program has continued in operation to the present with only minor modifications. One change is that they are now used by all three military services, whereas the 1949 program was only for the Army. However, the most important change has been that considerable preference is given to students who have some prior military service and are therefore more likely to remain on duty after their obligated service has ended.

Educational Standards for Army Clinicians

In addition to proposing the student program, Colonel Gersoni made another crucial decision. He insisted that Army psychologists have a doctoral degree. Today this seems to be a rather obvious decision, but at the time, in 1948, it showed both courage and foresight. At that time, American psychologists were still deciding on the desirable level of training for clinicians. Some well-respected civilian psychologists and some Army clinicians had only a master's degree. Indeed, a small number of those master's-level psychologists who performed competently and ethically within their individual level of training were kept until retirement. However, Gersoni believed that the effectiveness and long-range value of Army clinicians would depend on their being highly trained and capable of a much broader range of functions. He also knew that having a doctorate, rather than a master's degree would gain psychologists much more respect from their doctorally trained colleagues in other AMEDD specialties. He understood, of course, that doctoral-level clinicians would be harder to recruit and retain. He also knew that the Navy and Air Force did not then require the doctorate. (In fact, they did not do so until many years later.) Nevertheless, he insisted on the doctorate. Fortunately, at the same time, the Veterans Administration also chose the doctoral standard. When these large and highly visible federal agencies decided to require the doctorate, their decisions helped make that degree the U.S. standard for educating and licensing psychologists.

Organizational Structure for Army Clinicians

It was decided in 1947 (Army–Navy Medical Services Corps Act, 1947) that because clinical psychologists were health care providers, they should belong to the Army Medical Service Corps, a component of AMEDD. Therefore, AMEDD was responsible for deciding how many psychologists were needed; for recruiting, training, and assigning them; and for creating policies governing their work.

The initial estimate was that 90 clinicians would be needed, but that goal soon changed as the Army expanded during the Korean War. However, for the reasons cited earlier, the desired numbers were not achieved during the 1950s and 1960s. There was a constant struggle with the medical department, who did not know how to use psychologists or other nonmedical personnel, for the authority to train more clinicians while assigning the few who were available to the positions where they were most needed.

Assignment options for clinicians changed considerably between 1950 and 2000, becoming both more numerous and more diverse, but for the first 2 decades the limited number of clinicians were usually assigned to the small hospitals at military posts and overseas or to larger Army hospitals such as Walter Reed Army Medical Center in Washington, DC, and Letterman Army

Medical Center in San Francisco, California. Hospitals at U.S. bases and overseas usually had only one psychologist who was assigned to a team headed by a psychiatrist that also included social workers, nurses, and some enlisted assistants. The larger hospitals had several psychologists, still under the psychiatry department, and hospitals that conducted internships had the most psychologists. Overseas hospitals got priority. A few psychologists taught at the Army's medical training center in San Antonio, Texas; one was at the U.S. Military Academy at West Point, New York; and one was at the military prison at Fort Leavenworth, Kansas.

SPECIAL EVENTS IN THE FIRST 25 YEARS OF ARMY MEDICAL DEPARTMENT PSYCHOLOGY

The first event that affected Army psychology practices after adoption of the SPSP was the outbreak of the Korean War on July 25, 1950. Thus, the first SPSP clinicians became heavily involved in caring for the war's psychological casualties once they were returned to the United States. The methods of care were those used in World War II, carried out by teams consisting of a psychiatrist leader and various other mental health professionals, including the new psychologists.

The next change came when the Korean War ended in 1953. Psychologists and other mental health specialists could now turn to the broader AMEDD mission, which was to provide care for all soldiers and their families. The new, broadly trained psychologists identified many other groups that needed more care than they had been getting and helped create or expand programs for them. Psychologists identified and assisted children with special needs, did joint work with the post-dependent school systems, and expanded their liaison with pediatrics. Improved test methods were used to provide care to persons with organic brain damage, both those in uniform and family members. At other posts, many other programs started or expanded and were greatly appreciated. In addition, social work was removed from psychiatric control and freed for its broader activities throughout AMEDD.

The Vietnam War (1965–1973) produced enormous changes. The Army doubled in personnel, and the number of psychologists rose to 139, the largest since World War II (Ginn, 1997, p. 354). Some of the extra psychologists went to new units, and a number of others addressed special needs that emerged during the war, such as programs to stop drug abuse among soldiers in Vietnam before they returned to the United States or to deal with race relations issues in Vietnam and the United States. Some of the new programs were later ended when the soldiers returned from Vietnam and were absorbed into society. Others, such as an organizational effectiveness program, ended several years later when their value remained unproved. Later during the war, promotions speeded up, and army officers reached the rank of captain after 2 years, less time than graduate students spend in graduate school. So bringing psychologists on duty as captains finally became possible.

When the draft ended in 1973, the Army began to reshape itself as an all-volunteer force, and its clinical psychologists helped the Army rebuild itself and recover from Vietnam. In 1972, a psychologist was added to combat divisions (Ginn, 1997, p. 355) following a model created many years earlier, when each division got one psychiatrist and, later, one social worker. Other clinicians were assigned to special drug and alcohol treatment programs and to research, and I became the first clinician on the faculty of the Army War College at Carlisle Barracks, Pennsylvania (1973–1978).

During the late 1970s, 1980s, and 1990s, there was an even greater range of assignments. Clinicians worked with military intelligence agencies, programs for exceptional children, the Special Warfare Center at Fort Bragg, North Carolina, and the National Defense University in Washington, DC. Beginning in the 1990s, there were so many unique assignments and senior psychologists to fill them that it is not feasible to list them all.

In addition, there has always been one especially important assignment: duty as the clinical psychology consultant to the Army Surgeon General. This consultant provides crucial advice about the recruitment, training, assignment, and professional duties of clinicians. For many decades, this was a full-time assignment, but in 1982, it became only a part-time duty for a senior clinician who had another primary assignment. This has reduced the time available for consultant work, but the consultant still plays a vital role in shaping Army clinical psychology.

DUTIES OF CLINICAL PSYCHOLOGISTS

It is beyond the scope of this chapter to describe all the duties clinicians perform, but key aspects of their work deserve mention. The most typical role for Army clinical psychologists during World War II and in the early 1950s was as psychometrician helpers to psychiatrists (Hutt, 1947). However, during the past half century, the duties of psychologists have expanded to include a wide range of diagnostic, therapeutic, consultative, and research activities. This role expansion is similar to the growth that also occurred among civilian clinical psychologists since World War II. Accomplishing this growth in roles during the early postwar decades sometimes required contending with psychiatrists who wanted to continue as the unchallenged leaders and controllers of AMEDD behavioral services. In 1959, the Army issued regulation AR 40-216 outlining the role of Army neuropsychiatry. It emphasized prevention and the importance of close contact with the troops, which was good. However, the regulation clearly placed other mental health professionals in subordinate roles to psychiatry, which was not good.

Army social workers achieved independence from psychiatry by the early 1960s, and clinical psychology also fought for, and slowly gained, more autonomy. Autonomy for clinical psychology was desirable not only because uniformed psychologists were best supervised by peers who understood the discipline's practice, research, and consultation functions, but also because of considerations of career progression and professional status. To advance in their careers and to have their contributions recognized as equal to those of other health care providers, psychologists had to demonstrate leadership capabilities, and the establishment of a separate psychology department afforded such opportunities. However, AR 40-216 remained unchanged. In 1972, I became the first psychologist to be assigned as chief of a Mental Hygiene Consultation Service (MHCS; discussed later in this chapter), a role previously limited to psychiatrists, but this step forward still was considered an exception to policy. Fortunately, as time went on, more psychologists and other nonphysician mental health workers were allowed to serve in broader roles, including leadership positions, and these broader duties are now routine. However, it was not until 1984 that AR 40-216 was rewritten in a way that at least partly recognized the greater autonomy and broader roles of nonphysician professionals.

One person who greatly helped psychologists in developing their programs, in both the Army and the other services, was Patrick DeLeon, a psychologist on the staff of Senator Daniel Inouye. With his background as both lawyer and psychologist, DeLeon recognized that federal psychological programs, including military ones, could serve as national models. Working with Inouye, he helped enact important legislation that greatly assisted military psychology. For example, in the late 1980s, DeLeon worked with Congress to direct the military to make the large psychology services at hospitals where psychologists were trained separately from psychiatry departments and thus free of psychiatric control. Independent departments of psychology are better able to shape their own training and service programs in collaboration with, but not under the control of, psychiatry and other mental health disciplines. The main advantages are more autonomy and freedom to innovate.

DeLeon also played a key role, in the early 1990s, in obtaining authority to operate the Department of Defense Psychopharmacology Demonstration Project, which taught clinical psychologists to safely and effectively prescribe psychotropic medications. DeLeon's support for these and many other psychological advances both outside of and inside the military promoted his election as president of the American Psychological Association (APA) in 2000.

Today, Army clinical psychologists perform a broad range of duties and have autonomy similar to that of their civilian colleagues. They must be licensed in some states and are encouraged to obtain added certification from the American Board of Professional Psychology. This certification has become even more important since a law was passed (again with DeLeon's involvement) giving extra pay to clinicians who have it.

Military clinicians work not only with soldiers but also with their spouses and children as well as with retirees and their families. Hence, they deal with a broad range of clients and a wide variety of clinical problems. Like their civilian colleagues, they do many kinds of evaluations; provide many types of therapy; consult with commanders, supervisors, chaplains, lawyers, and medical, educational, and social service personnel; and also perform research.

CREATING A SIZABLE GROUP OF CAREER CLINICIANS

By the 1970s, the Army was beginning to assemble a sizable group of psychologists, numbering 139 in 1972. Much of the credit for this expansion belongs to Colonel Charles A. Thomas Jr., who was psychology consultant from 1966 to 1974, longer than any other consultant. He made many improvements in the situation of clinicians. Their entry rank was increased to captain, which meant better status among their medical colleagues, greater acceptance as consultants to nonmedical personnel, and higher pay that was more competitive with civilian salaries. Colonel Thomas also realized that it was much easier to make a career-motivated military psychologist out of someone with prior military service than to persuade civilians to become career psychology officers. Therefore, he retrained some career officers as psychologists. Increasing autonomy for psychologists, greatly expanded training opportunities, and a wider variety of interesting assignments also made these careers more attractive.

The result was that by the end of the 1970s and continuing to the present, the supply of career psychologists has greatly improved. There are also a sizable number in the higher ranks of major, lieutenant colonel, and full colonel who fill key managerial and policy-level psychology positions. The decades-long struggle to create a significant group of Army clinical psychologists had come much closer to achieving its goal.

CONTRIBUTIONS OF ARMY CLINICAL PSYCHOLOGISTS
TO MILITARY PSYCHOLOGY

Contributions of Army clinical psychologists to military psychology have included providing high-quality behavioral health care; creating mental hygiene consultation services; providing the therapeutic value of environmental change; training clinicians; developing postdoctoral specialization training; providing realistic, community-oriented training; using paraprofessionals; authorizing payments to clinical psychologists without medical referrals, and establishing the Department of Defense Psychopharmacology Demonstration Project.

Providing High-Quality Behavioral Health Care

The most obvious contribution Army clinicians have made is the excellent care they have given thousands of soldiers and their families. More than 8 million people are now eligible for military medical care, which is available to all active and retired military personnel and their families. This care includes a large amount of behavioral health care, much of it provided by psychologists in the Army and the other military services. The number of clinical psychologists in the military has varied considerably over the years, but at the time of this writing (June 2005), nearly 500 clinicians are on duty, of whom about 120 are in the Army.

The primary mission of these clinicians is to help soldiers, but they also provide a great deal of assistance to soldiers' families. In some areas of the country, such as Washington, DC, where there are extensive military medical facilities, they also care for retirees and their families. They care for a great many children, especially at overseas bases where families accompany soldiers but where civilian medical facilities are not available or suitable. Army clinicians see patients of all ages with all kinds of clinical conditions and have challenging, diversified practices. Another source of variety is that clinicians from any service can see clients from any other service, so it is not unusual for an Army psychologist to treat a Navy child or an Air Force wife.

Understanding the Soldier's Environment

The military environment has many unique characteristics, and the Army has shown that mental health professionals are not fully effective unless they understand the special conditions under which soldiers and military units function. Failure to do so causes clinicians to make inaccurate diagnoses and to use less effective treatment methods. In addition, clinicians doing consultation may give inappropriate advice to commanders and their staffs if they do not understand the rules that govern operation of military units and their staffs. For these reasons, clinicians function best if they are in uniform and located close to the soldiers they serve, sharing the same circumstances and operating under the same rules.

Creating Mental Hygiene Consultation Services

During World War II, Army mental health workers, including psychiatrists and social workers as well as psychologists, began to create a new kind of agency, the MHCSs mentioned earlier in this chapter (Guttmacher, 1966). MHCSs focused on the well-being and effectiveness of the entire military community. In addition to providing normal diagnostic and therapeutic services to soldiers, MHCS staffs played important roles as consultants to commanders and supervisors, advising them on how to reduce the incidence of mental problems and how to assist persons with such problems. By the early 1950s, MHCSs were operating at most posts, and in many ways they were prototypes for the civilian community mental health centers that developed in the 1960s. Many Army psychologists and psychiatrists who returned to civilian life had learned the value of the community approach during their Army service and then applied these methods in postmilitary civilian settings, either by working in community mental health centers or by serving in administrative and policy positions as developers of local, state, and federal mental health programs.

The effectiveness of MHCS principles in a garrison situation has also led the Army to apply the same principles while caring for soldiers in field operations. Mental health specialists have been assigned to combat divisions for many decades, and more recently, they have been assigned to Combat Stress Control Detachments that provide preventive, diagnostic, and treatment services to soldiers in field units that do not belong to divisions. These detachments can also augment divisional services when needed.

Proving the Therapeutic Value of Environmental Change

Army clinicians have discovered that changing a soldier's environment can be an effective way of treating mental disorders. Clinicians have many ways of evaluating the working and living environments of their clients and have easy access to soldiers' long-term personnel and medical records, which stay with soldiers throughout their careers. Thus, they have been able to identify patients who were ineffective or troubled in their current settings but who had performed better in previous assignments and to recommend that these clients be placed in more supportive situations or given more appropriate duties and supervision. Because the Army has a great deal of control over soldiers' environments and can significantly change the ways soldiers are supervised and assigned, clinicians' recommendations for suitable changes in the way soldiers were managed often have led to improvements in soldiers' performance.

This recognition of the importance of soldiers' environments has also led to a strong reliance on a community approach. Clinicians identify units or situations where there are more than the usual numbers or types of soldiers with problems. Then they work with the staff of these units in a supportive way to recommend changes in the unit environment to help prevent or correct such difficulties.

A significant example of helping temporarily disabled soldiers has been the Army's success, starting in World War II, in helping soldiers with "combat fatigue" (similar to what is now called posttraumatic stress disorder) return to duty. Soldiers showing symptoms are treated as near as possible to the combat area, preferably by mental health specialists assigned to the unit.

They are given a few days of rest and good food in a relatively safe area with a chance to talk about their experiences and concerns. They are not evacuated far to the rear, and they are reassured that they are not to blame for their problems. They are told, truthfully, that many good soldiers have similar reactions from which they usually recover. They are led to expect that they can soon return to duty, and in fact, most of them do. This approach has helped psychologists "conserve the fighting strength," which is the main mission of AMEDD, and it has also been of great benefit to individual soldiers by helping them avoid long-term ineffectiveness. (For combat stress reaction treatments literature, see chaps. 1 and 17, this volume, and Mangelsdorff, 1985.) Many civilian psychosocial rehabilitation programs use methods similar to those of the Army with temporarily ineffective soldiers. Patients are placed in supportive environments where they can make effective use of aspects of their abilities that are not greatly affected by their mental problems.

Training Hundreds of Clinicians

Over the course of more than 50 years, the Army has financed high-quality doctoral-level training for between 500 and 600 clinical psychologists who have gone on to play important clinical and administrative roles in the Army or in civilian mental health programs. While in the Army, they studied at APA-accredited graduate programs, took APA-accredited internships at Army hospitals, and gained valuable experience during their years of payback service. Those who later returned to civilian life often took with them the community-oriented philosophy and methods they learned in the Army. It is no coincidence that a great many of the concepts and methods used in civilian community mental health work are similar to military mental health practices.

Developing Postdoctoral Specialization Training

By the early 1960s, the Army realized that the abilities and usefulness of its psychologists could be greatly increased by providing them with postdoctoral training in functions that required more or different expertise than they had gained during their doctoral training. As a result, the Army developed programs to expand the skills of its psychologists. I was the first clinician to benefit from this policy when in 1962 I was sent at Army expense to study community mental health for a year at the Harvard School of Public Health in Cambridge, Massachusetts. Later, starting in the 1970s (Ginn, 1997, p. 355), the Army developed postdoctoral training programs in its own facilities in the specialties of child psychology, health psychology, and neuropsychology. These programs still exist, are APA accredited, and have greatly enhanced the skills of many Army psychologists. In addition, because there is a payback service obligation for those who receive this training, these programs have helped improve career retention.

Providing Realistic, Community-Oriented Training

From their earliest days, Army clinical internships were of high quality and fully accredited. They taught the same diagnostic and therapeutic skills found in civilian training. However, in the 1950s and 1960s, the training was conducted in Army hospitals that were located far from troop installations, so many interns had little contact or familiarity with the environments in which most soldiers lived and worked. As a result, when newly trained clinicians transferred from training sites to troop installations, they often experienced culture shock and tended to stay close to the hospital rather than interact with the rest of the post population. It often took them a long time to become comfortable and effective with troops and the general military community, and some never made the transition.

Because psychologists needed more appropriate training to function better in military communities, I was directed in 1969 to establish the Army's first community-oriented internship at Fort Ord, California, which was then a large training post for incoming soldiers. The interns were encouraged to live on the base, where they could and did have frequent contact with the military community's personnel and activities. They learned the skills of community psychology in addition to the more traditional clinical skills. This internship quickly won APA accreditation and lasted almost 25 years. It ended in 1993 because the Army closed Fort Ord when it was downsizing after the Soviet Union collapsed.

In another move to provide more appropriate community training, the Army in the 1970s closed its internships at Army hospitals that were distant from troops. Instead, new internships were created at hospitals that were either at or close to troop installations. As a result, Army clinicians have become better trained for, and more comfortable with, their roles as community clinical psychologists and also work more effectively when assigned to combat divisions or Combat Stress Control Detachments.

Using Paraprofessionals

For more than 50 years, the Army has increased the efficiency of its mental health officers by training enlisted personnel to assist the officers. These soldiers typically have at least a high school education that is supplemented by training at the AMEDD Center and School. They are given a basic understanding of human behavior and methods of behavior change and then work under the supervision of mental health officers. Typical duties include obtaining biographical data from patients, administering objective psychological tests, obtaining data about soldiers' performance in their units, and doing preliminary screening of soldiers with problems to see if intensive evaluation by a mental health professional is required. Such personnel, if properly selected, trained, and supervised, are a major asset in military mental health programs.

Authorizing Payments to Clinical Psychologists Without Medical Referrals

Because many personnel and families who are entitled to military medical care do not live near military posts, in 1967 the military created the Civilian Health and Medical Program of the Uniformed Services (CHAMPUS) program (now renamed TRICARE), which paid civilian providers to give medical care to family members of active duty and retired military personnel. These services could include mental health care from civilian psychologists, but at first psychologists were paid only if a physician referred the patient. Referral screening became expensive, because there was a fee for the screening. Two people were paid, but the screener provided no real service and delayed actual treatment. The procedure was costly and slow and provided an expensive, unnecessary safeguard.

That policy needed to be changed, and in the early 1970s the Army's clinical psychology consultant, Colonel Charles A. Thomas, was able to work with civilian psychology leaders in the APA and in Congress to change it. These civilian leaders persuaded CHAMPUS managers that they should reimburse civilian psychologists without requiring patients to have a referral from a physician. This pioneering effort subsequently led to similar changes in the Medicare and Medicaid programs, and now these programs also pay psychologists without requiring a medical referral. This policy has also been adopted by many civilian health insurance programs (Wiggins, 2001).

Department of Defense Psychopharmacology Demonstration Project

In 1991, in response to a congressional mandate, the military established a demonstration project during which uniformed clinical psychologists were taught how to prescribe psychoactive medications; these psychologists were then evaluated to see whether they could prescribe the medications safely and effectively. The psychologists who received this training came from all three services (Army, Navy, and Air Force), but a major share of the responsibility for operating and evaluating the demonstration was given to Army psychologists.

This program raised a storm of opposition from psychiatry and some other branches of medicine. Organized medicine (the American Psychiatric Association and the American Medical Association) argued that drug prescribing was unsafe if done without a full medical education. The APA agreed that more training would be needed, and the demonstration project provided that added training as an essential component but not with the lengthy training involved in a full medical degree. Psychiatry and some other branches of medicine opposed the program from the beginning, and they were able to force it to an end in 1997. Psychopharmacology continues to be a vigorous, nationwide issue that is argued among the medical professions.

Fortunately, the objections to the program were not strong enough to end it immediately, and 10 psychologists completed the training. Some of them are still on duty and are proving capable as clinicians, teachers, consultants, and leaders. Contrary to some expectations, they have usually been well accepted by their medical colleagues and are seen as a valuable resource. The final evaluation of those who finished the training (Newman, Phelps, Sammons, Dunivin, & Cullen, 2000) showed that they prescribed medications safely and effectively and could also give valuable consultations about suitable medications to other health professionals, especially nonpsychiatric physicians. The evaluation also showed that the training did not cause these professionals to stop using their other psychological skills. Rather, they used medication as an addition to, not a substitute for, other psychological methods.

Although the military psychopharmacology training program has ended, its success has led to the creation of similar civilian training programs, and the military curriculum has been a model for the recommended content and methods of these civilian programs. The excellent performance of the military graduates has also been used in arguing successfully for state laws permitting properly trained civilian psychologists to prescribe psychotropic medications. Two such laws have been passed, one in New Mexico in 2002 and one in Louisiana in 2004, and many other states are trying to pass similar laws.

CONCLUSION

This chapter has provided only a brief summary of the 5-decade history of the Army's development of a capable and versatile group of career clinical psychologists. I hope, however, that it has presented enough information to enable the reader to understand how that group of clinicians was developed and the contributions they have made to military personnel and military operations.

Many factors helped in developing the Army clinical psychology program. The most important has been the effective services the clinicians themselves have provided. These services have been so highly valued that the Army has spent considerable time and money during the past half century on the education and training of psychologists. The Army has also taken other steps to make psychology careers more attractive, such as increasing the rank, pay, variety of duties, and professional status of psychologists.

AMEDD has been supportive of clinical psychology. Although there have been some struggles with psychiatry, most other members of the medical department have provided greatly appreciated support to psychologists. As for the conflicts with psychiatry, they have not been unlike those between other related disciplines, such as optometrists and ophthalmologists, nurse anesthetists and anesthesiologists, and nonphysicians versus physicians seeking command opportunities in medical departments. As each set of new contenders proved their skills and value, they gradually gained the recognition and status they deserved. That has also been true for psychologists.

Army psychologists have received valuable assistance from the APA, which has accredited Army training activities and provided important advocacy for military psychological programs. Army psychologists have also benefited from the congressional support they received in such matters as establishing independent psychology departments and obtaining greater professional autonomy, extra pay for board certification, and psychopharmacology training. Finally, tribute must be paid to all the career psychologists whose commitment to their patients, their profession, and the Army has enabled the Army clinical psychology program to advance from near death in 1949 to its present state of vigor and effectiveness. These clinicians, ably led by the psychology consultants, had to overcome many obstacles, especially during the first 2 decades, but they persisted and triumphed.

12

AIR FORCE CLINICAL PSYCHOLOGY: HISTORY AND FUTURE TRENDS

KARL O. MOE

The history of clinical psychology in the U.S. Air Force is a success story about meeting the genuine needs of a large organization with diverse, worldwide national security responsibilities and interests. The early years of the Air Force reflected an interest in using the capabilities and skills of the emerging profession of clinical psychology. They also reflected a variety of challenges to the status and profession of psychology. More recent experience has demonstrated the significant success Air Force clinical psychologists have had in using their skills and abilities to address national security challenges.

MILITARY PERSONNEL NEEDS: SELECTION, CLASSIFICATION, AND ASSIGNMENT

From the beginning of the Air Force in 1947, officers have provided psychological services in support of Air Force missions. In fact, before, during, and especially after World War II, the military sought reliable assessment and selection capabilities. In addition, a unique military interest has been the need to deal with combat stress casualties and the mental health problems created by war. Finally, the military has often been interested in predicting and preventing psychological problems. Air Force clinical psychology was developed and has grown in response to these various needs.

The Air Force has one of the longest-standing evaluation programs of basic trainees. A long line of research stems from the Behavioral Analysis Service and its predecessors at the Air Force's only Basic Military Training program, which is located at the Lackland Air Force Base, San Antonio, Texas (Cigrang, Carbone, Todd, & Fielder, 1998; Cigrang, Todd, & Carbone, 2000; Englert, Hunter, & Sweeney, 2003; Staal, Cigrang, & Fielder, 2000). The School of Aerospace Medicine also has a long history of psychology applied to flight, fliers, and flying systems. Psychologists have been involved in the human engineering intended to make aircraft safer, easier to fly, and more effective. They are also involved in efforts to select and train fliers and keep them healthy (Staal, 2004). Air Force psychologists have been involved in programs for fliers in each of the military services, the National Aeronautical and Space Administration, and various special duty assignments.

Human Performance

Much of the human engineering literature and work in the Air Force has been done by nonclinicians. Yet there is an implicit focus on human performance in the practice of clinical psychology and a particular emphasis on adjustment issues. Much of the work on adjustment has been done under the broad and vague rubric of "stress management."

To a degree, expanded clinical roles have been facilitated by postdoctoral training sponsored by the Air Force Institute of Technology. Fellowships for clinical psychologists are now commonly sponsored in neuropsychology, pediatric or child and adolescent psychology, and behavioral health psychology (e.g., chronic pain management, stress reduction, weight and smoking reduction, treatment adherence). Additionally, fellowships in aviation psychology are sponsored periodically to support Air Force needs. A fellowship in organizational health psychology has been sponsored in the past and will be again, if demand grows. Finally, a 2-year forensic fellowship with the Air Force Office of Special Investigations has graduated its first fellow and accepted its second fellow. Like the aviation psychology fellowships, forensic fellowships will be offered periodically as needed to meet Air Force requirements.

Maintaining the Force

Psychologists have played key roles in maintaining the mental and physical health of active duty service members and their families. One of the most dramatic illustrations of the pivotal roles played by psychologists occurred in the early 1970s amid growing concern about drug abuse. More psychologists were recruited into the Air Force to help address this important concern. The Air Force's current drug and alcohol abuse program grew from these initial increases in staffing and has become an effective combination of deterrence and, when needed, rehabilitative treatment.

HISTORICAL THEMES

Certain themes running through the history of Air Force clinical psychology highlight the challenges in meeting national security needs and the roles psychologists have played in meeting these needs. Much of the information in this section regarding the history of Air Force psychology is drawn from an informal history recorded by Colonel Thomas Foley, who served as an associate chief of the Biomedical Sciences Corps in the mid 1970s and was, in effect, the chief clinical psychologist for the Air Force (Foley, 1984a, 1984b, 1985).

Recruitment and Retention

Two predominant themes in the early years of Air Force clinical psychology were the twin problems of recruiting enough clinical psychologists and then retaining them for more than the minimum amount of time. In the 1950s and 1960s, not enough psychologists joined the Air Force; the incentives were not adequate. Initially, there were 30 authorizations for clinical psychologists, but typically not all of these billets were filled. The first authorizations for psychologists were easily filled by officers who had been serving in a variety of career fields such as planning, personnel, education, and research. The Korean conflict, with its increased draft calls and recall of reserves, prompted significant interest among those who wished to serve the nation in volunteering for duty as a psychologist. Filling initial billets may also have been helped by the lower requirements for working as an Air Force clinical psychologist (any master's degree in psychology) compared with requirements for the civilian sector (master's degree or doctorate in clinical psychology).

However, after the first psychologists were assigned, keeping billets filled became harder. Despite this problem, there were no recruiting goals for psychologists during this period, and no clinical psychologists have ever been drafted. Attrition was partially made up for through transfer from other areas, such as the Arctic Aeromedical Laboratory, human factors laboratories, and some walk-in volunteers.

By the early 1970s, some fairly successful recruiting attempts were developed to reduce the shortfall of clinical psychologists. Four training billets per year were added to place active duty officers in civilian graduate schools, and the Air Force made the Health Professions Scholarship Program (HPSP) available to civilian clinical psychology students. The HPSP paid for tuition, books, and fees as well as providing a stipend for graduate students past their 1st year of graduate school. In exchange, the student was commissioned and served as an active duty psychologist after graduate school. In 1978, the number of HPSP authorizations per year for doctoral training in clinical psychology was increased to six, reflecting the continued and increasing demand for clinical psychologists. However, the HPSP created only a year-for-year obligation, and few recipients remained on active duty for more than the normal 3-year commitment; thus, the HPSP helped in recruiting but did not improve retention to any significant degree. Sending active duty officers to school at civilian universities, however, aided both recruiting and retention, because the payback was 3 years for each year of training. By the time the payback was complete, officers were more than halfway to a generous 20-year retirement.

There are a variety of traditional inducements for military service, including fair pay, the likelihood of promotions, and good retirement benefits. For the most part, these benefits have not been sufficient by themselves to recruit or retain clinical psychologists. When the military pay scale has led to lower levels of compensation, recruiting and especially retention have been correspondingly more difficult. Similarly, periods of poor promotions and the perception of eroding retirement benefits have led to dissatisfaction and even worse retention problems. In fact, the negative effect of poor promotions has been dramatic at times. At first, poor promotion rates were ascribed to inadequate performance reports, often written by supervisors who had little military experience but who had higher rank and were often in professional competition with clinical psychologists. When this problem seemed to be resolving, the Air Force implemented a forced rating system that controlled the number of officers who could receive "top-block" (outstanding performance) ratings. This system was not required by Air Force regulation, because medical service officers were exempt, but the Air Force Surgeon General implemented it anyway. The status reports prepared by the chief of the Biomedical Sciences Corps in both 1977 and 1978 recommended relief from this self-imposed requirement, but that did not occur until the entire Air Force discontinued limitations on top ratings. These reports also suggested removal of grade limitations and accelerated promotions for Biomedical Sciences Corps officers. These recommendations similarly were not implemented.

Like recruiting, retention was less of a problem initially. In fact, retention of clinical psychologists was exceptionally good until the late 1960s because the original Air Force clinical psychologists had (with one exception) started with only master's degrees but, with Air Force sponsorship, earned doctorates. The 9-year payback obligation put them well on the way to completing the 20 years needed to retire. However, from 1970 until well into the

1990s, the attrition rate for clinical psychology averaged 25% per year, higher than any other allied health care discipline, with the occasional exception of optometry. Of 80 graduates from the clinical psychology internship program at Wilford Hall Medical Center at Lackland Air Force Base, over the 1st 9 years of the program's existence, only 5 remained on active duty beyond their initial obligation, and only 1 stayed long enough to be promoted to major. For the most part, retention lasted only as long as the payback time for training in an Air Force internship, except for active duty officers who had been sent to graduate school with Air Force sponsorship.

Two reports on pay and promotion (Murray, 1978; Seaquist, 1968), 10 years apart, noted essentially the same findings. Both inquired about the reasons for separation at the completion of initial obligations and for separation via early retirement. Both reports determined that the two leading causes were low pay and limited opportunity for promotion. The retirement pension as a retention incentive had value only after the individual had 20 years of service. Relatively remote retirement benefits simply did not encourage people to stay on active duty beyond their initial obligation or beyond the first opportunity for retirement. Likewise, career progression never seemed to be a reality for clinical psychologists through the late 1970s and thus probably contributed to poor retention figures among those who had already gained substantial clinical experience. As psychologists progressed from single-provider clinics in dispensaries to larger clinics in regional hospitals, there was no real change in job title or responsibilities. Further, up to that point, individuals in single-provider clinics seemed to fare better at promotion boards than those who were supervising larger clinics.

A final point in this discussion has to do with professional status. Higher professional status was available outside the military than inside (Foley, 1985). The roles of private practitioner, civilian consultant, and department or division chief within the Veterans Administration and civil service removed clinical psychologists from a subordinate role and provided professional recognition and varying degrees of autonomy. In the Air Force, in contrast, clinical psychologists were subordinate by regulation to psychiatrists and, in many cases, to any physician assigned as an "acting psychiatrist" (Psychiatry Working Group, 1985). Further, psychologists commonly were trained to deliver at least as broad a range of services as the psychiatrists who were appointed as the chiefs of mental health. This situation was not only personally offensive but also limited professional development.

National Security Needs

Clinical psychologists have helped meet the national security needs of the Air Force, a theme that is closely related to that of recruiting and retention problems. There has always been more demand for clinical psychologists than supply (e.g., during the Korean conflict, and extending to 1955, the total number of authorizations rose to 36; the psychologists who filled those billets were all volunteers, and there were never quite enough to fill all the authorizations). This has left service members vulnerable to psychological problems, representing a significant threat to national security. For example, in the Vietnam War, substance abuse was a significant threat to security because of its deleterious effect on reliability. Psychologists were key players in addressing this problem, and in 1969, clinical psychology authorizations doubled from their Korean War level, to 78.

As the numbers of clinical psychologists increased, their visibility and utility brought greater demand for psychology authorizations by the Air Force's major commands. Again, demand increased faster than the ability to recruit. In 1978, there were 110 authorizations for psychologists, but only 89 clinical psychology billets were filled—a far cry from the 26 psychologists less than 10 years earlier. The projection for fiscal year 1979 was 96 psychologists on active duty with all but four having PhDs.

Professional Standards

Increasing professional standards for Air Force clinical psychologists have mirrored—and sometimes led—those of the civilian community. By the mid-1950s, concern about the lower professional standards required for Air Force practice led to the practice of sending one active duty psychologist per year to a civilian graduate school for a doctorate in clinical psychology. By 1960, nearly one third of the active duty psychologists in the Air Force (i.e., 12) had doctorates. However, by 1960 the number of authorizations had climbed again, to 41. Additionally, the standards for joining the Air Force as a clinical psychologist were raised to the doctoral level. Overall, the quality of psychologists in the Air Force had increased, but the Air Force still did not have all the clinical psychologists it wanted.

Internship Programs

One way of addressing the chronic recruiting problem was to develop a "grow-your-own" approach. In 1970, a clinical psychology internship program was started at the Wilford Hall Medical Center to furnish 4 qualified clinical psychologists per year. The number of internship positions rose to 10 in 1972. Another internship program was added at the Wright–Patterson Medical Center, Wright–Patterson Air Force Base, Ohio, in 1977. This program was to provide five qualified clinical psychologists per year. A third internship was opened in the fall of 1987 at Malcolm Grow Medical Center on Andrews Air Force Base, Maryland. This program received accreditation by the American Psychological Association in its 1st year of existence. Since then, the Air Force has trained between 19 and 24 clinical psychologists a year. In retrospect, it is clear that starting internships was an extremely good decision, because internships have become the most important source of new psychologists for the Air Force. Additionally, in more recent years, internships have been the source of the vast majority of psychologists who have stayed on active duty past any initial commitment. Postdoctoral training has also been an effective carrot for maintaining psychologists on active duty. Internships and postdoctoral training both address retention problems and provide the specialized training needed for some of the roles that Air Force clinical psychologists have taken on.

Professional Growth Opportunities

Specific opportunities available for growth as a psychologist within the Air Force have changed over time. Initially, available opportunities tended to develop basic mental health skills but did not provide experience in leadership and management of mental health or general health care activities. Over the years, psychologists have been asked to apply the science of psychology to increasingly varied problems.

Growth of opportunity for psychologists faced some opposition, however. Initially career progression was generally nonexistent for clinical psychologists. Of 110 authorizations in 1978, all but 6 were assigned to the mental health service in hospitals or clinics. Of the 6 at settings outside of clinics, 1 was at command level in Europe, and another was a biomedical scientist staff officer in a job open to any of the allied health care specialties. Even the associate chief of the Biomedical Sciences Corps for clinical psychology, the ranking psychologist, was assigned primarily as chief of an outpatient mental health clinic. In a 1978 study, promotion, which tends to reflect experience beyond providing mental health assessments and therapy, was ranked as the most unsatisfying aspect of the Air Force for clinical psychologists, both professionally and militarily (Murray, 1978).

In the late 1970s, the clinical psychology leadership in the Air Force felt that the key to maintaining a viable number of psychologists was to offer experience that would allow them to compete successfully in the civilian sector, although they recognized that this strategy also would have problems. The motivational ability of marketable experience can change over time or have unintended, deleterious side effects. In the late 1970s, Air Force experience qualified clinical psychologists for jobs in the civilian market worth $5,000 to $10,000 more annually than military compensation. So although the opportunity to obtain superb experience was an invaluable recruiting tool, its effect on retention tended to be negative. The fact that only one clinical psychologist recruited from the civilian sector remained in the Air Force long enough (7 years) to achieve field-grade rank helped to demonstrate that the Air Force had significant retention problems when it came to clinical psychology. It was clear that opportunities that might help recruit psychologists would not necessarily serve to retain them. Additionally, this experience was primarily in the area of direct patient care—assessment and therapy. Lack of potential for career advancement and career broadening was a clear disincentive for staying on active duty. Colonel Thomas Foley concluded his historical review of clinical psychology in the Air Force with the following statement:

> As of 1978, without significant modifications in policies, regulations, and procedures, the clinical psychology service in the Air Force will probably emerge as a force of 100 or more professionals having a small cadre of senior officers with a dearth of professional experience and a large working force of neophyte professionals with a significant annual turnover rate. A situation not commensurate with effective productivity and certainly not cost-effective. (Foley, 1985, p. 5)

CLOSE OF THE 20TH CENTURY

Many significant changes took place during the 1980s and 1990s. At the turn of the 20th century, there were 215 Air Force clinical psychologists engaged in a wide variety of roles supporting the Air Force's missions and national security. Establishing and maintaining an adequate force of clinical psychologists has always been a national security concern, because psychologists have always served as force maintainers and extenders, but during the years since 1978, roles for psychologists have expanded dramatically. This expansion reflects the wide variety of skills and abilities that psychologists can bring to the Air Force's changing needs.

Since 1990, increasing numbers of psychologists have served as medical squadron commanders, and four have served as hospital commanders. One has served as the deputy command surgeon for the Pacific Air Forces, and another has served as the command surgeon for the Joint Task Force in Southwest Asia. At the end of 1998, the chief of the mental health flights (the Air Force equivalent of a mental health division or department, depending on the size of the organization) at four of the Air Force's eight largest hospitals and medical centers were psychologists. This is a significant improvement on the situation in 1978.

Continuing to reflect the broad usefulness of clinical psychologists' skills and abilities, the number of authorizations for clinical psychologists has continued to climb since 1997. This is most remarkable because the number of people in the Air Force, and especially in the Air Force Medical Service (AFMS), has declined steadily since the end of the Cold War. The central reason for this continued growth in psychology numbers seems to be growing respect and increasing demand for the skills, talents, and abilities that psychologists bring to a wide range of real-world problems.

The success of psychologists at functioning in and out of traditional clinical settings has been reflected in far better promotion statistics in more recent years. In sharp contrast to the first 25 years of Air Force psychology, psychologists have been promoted to higher ranks with great regularity beginning in the late 1980s. During the 5-year period from 1994 to 1998, 30 of 36 captains were promoted to major when they were eligible. Similarly, 27 of 34 majors were promoted to lieutenant colonel, and 5 of 8 lieutenant colonels were promoted to colonel. These rates exceeded the promotion opportunity (for captains, 80% promotion opportunity, 83% promoted; for majors, 70% promotion opportunity, 79% promoted; for lieutenant colonels, 50% promotion opportunity, 63% promoted) and indicate that psychologists were appreciated and were well supported.

NATIONAL SECURITY ROLES

Seaquist (1968) reported that psychologists experienced frustration over not being able to use the skills they had in areas other than diagnosis and treatment; they also desired to consult on organizational, community, and leadership needs and to engage in community outreach and education. In the years since Seaquist's report, psychologists have, in fact, used their expertise in organizations, systems, learning, leadership, and other areas in settings such as the Air Force Survival School, special operations, medical intelligence, the Air Force Safety Center, the Air Force Inspector General's medical team, Major Command staffs, Air Staff, Department of Defense staff positions, the National Security Agency, the Office of Prevention and Health Services Assessment, investigative agencies, and occupational health clinics. Psychologists are on the faculties of the U.S. Air Force Academy and the Uniformed Services University of the Health Sciences as well as professional military education colleges such as the Industrial College of the Armed Forces and the National Defense University. Even when assigned to mental health clinics, psychologists participate in command and organizational consultation, hostage negotiations, stress debriefing, health promotion activities, and community development as well as preparing for combat or disasters.

STRATEGIC HEALTH CARE INITIATIVES

Since the mid-1990s, Air Force health care has been guided by five strategic initiatives designed to support the Air Force's national security missions, and psychologists have contributed substantially to each: supporting medical readiness, managing health care, providing mental health care, enhancing health promotion and disease prevention, and promoting customer satisfaction.

Supporting Medical Readiness

The first initiative has been to support medical readiness. Since the end of the Cold War, the U.S. military has experienced a higher rate of deployments than ever before. The military has been called on to provide support for, or in some cases to manage, a wide variety of contingencies, including the invasion of Haiti; the police action in Panama; the rescue and containment of Haitian and Cuban refugees; interventions in Bosnia, Croatia, and Somalia; Operations Desert Storm and Desert Shield; Operations Southern Watch and Northern Watch; and Operation Desert Fox. To better support the line of the Air Force, the AFMS completely reengineered the medical readiness plans, including the mental health mobility plans, beginning in 1994. Air Force psychologists led much of the reengineering effort in the area of mental health. Mental health mobility teams are now far better trained and equipped to meet the mental health needs that can be expected during a deployment. Current plans are for a mental health rapid response team to deploy virtually any time that medical assets are deployed.

A mental health response now commonly occurs when there is any significant community trauma. The American Red Cross has the primary responsibility for responding to most major civilian disasters within the United States. However, the military recently has provided mental health professionals to staff the Red Cross Service Centers. Considering the substantial amount of training that military mental health providers have in responding to traumatic events, it seems both likely and appropriate that military psychologists will continue to support the Red Cross. The view of the Air Force mental health leaders has been that responding to civil disasters has provided good experience and training for its mental health personnel. The Air Force was involved in the response to the Ohio River Valley flood in March 1997; the Red River Valley flood, which included the vicinity of Grand Forks Air Force Base, North Dakota, in April 1997; and the bombing of the Murrah Federal Building in Oklahoma City, Oklahoma, in April 1995. In each of these cases, the Red Cross and the Federal Emergency Management Agency, along with state and local agencies, provided the organizational structure and leadership, but Air Force psychologists and other mental health personnel provided direct support and expertise.

Air Force efforts to maintain a fit, healthy force as a part of medical readiness have provided a number of important applications to society. Two of these include well-researched and effective smoking prevention programs (Klesges, Haddock, Lando, & Talcott, 1999; Russ, Fonseca, Peterson, Blackman, & Robbins, 2001; Ward et al., 2002) and a corporate approach to suicide risk reduction that has been widely acknowledged as effective, efficient, and achievable (Knox, Litts, Talcott, Feig, & Caine, 2003; Litts, Moe, Roadman, Janke, & Miller, 1999). The Air Force approach to suicide risk reduction was cited as a noteworthy model in the report of the President's New Freedom Commission on Mental Health (U.S. Department of Health and Human Services, 1999).

Managing Health Care

The second strategic health care initiative has been to provide compassionately managed health care to Department of Defense patients. Reports from the National Committee on Vital and Health Statistics (2001) and the National Coalition on Health Care (2004) have emphasized the increasing cost of health care. The cost as a percentage of the gross national product rose from 5.2% in 1960 to 11.1% in 1987. Indeed, Sheridan and Radmacher (1992) suggested that given the current rates of growth of health care costs and gross national product, the cost of health care would represent 100% of the gross national product in 2055. Their point, of course, was that something needed to change. Precisely what needs to change is less clear, but it is the belief of Air Force mental health leaders that an important part of the solution lies in delivering behavioral and mental health care in effective ways to a larger number of patients.

The military medical system has long provided care for active duty personnel and their families and for retirees and their spouses. However, with the cost of medical care rising, an increasingly large proportion of the Department of Defense budget has been going to provide for medical care. The Department of Defense has been pursuing the expectation that a carefully but compassionately managed medical care system can help reduce the cost of care provided in military medical treatment facilities. This strategic initiative has focused the attention of personnel in the AFMS on understanding TRICARE (the military health system health maintenance organization) and finding approaches that provide appropriate care within the TRICARE guidelines. Psychologists are uniquely well qualified to participate in these efforts. An important part of managed care has to do with providing effective care in the most economical way possible. There is increasing interest by practitioners in the use of empirically supported treatments (Barlow, 2000; L. L. Street, Niederehe, & Lebowitz, 2000) and in assessment of outcomes (Burlingame, Lambert, Reisinger, Neff, & Mosier, 1995), and development and use of psychometric assessment of outcomes will be a continuing trend (Maruish, 1999). Psychologists clearly have the expertise to provide psychometrically sound approaches to assessing outcomes. Each of the Air Force's clinical psychology internships emphasizes the use of empirically supported treatments and careful assessment of progress and outcomes. Jobes and Drozd (2004) described one innovative collaborative approach to working with actively suicidal patients that was developed in part with Air Force psychologists. This approach combines the strengths of careful interviewing and psychometric assessment with empirically supported treatment.

Providing Mental Health Care

The third strategic initiative of the AFMS was originally described with the phrase "rebuild the direct care system," which referred to an evaluation of the skills and specialties required to provide for both Air Force readiness needs and peacetime medical care for military beneficiaries. An ongoing evaluation of the specialties needed and the numbers of people needed to staff those specialty services using a sophisticated, data-driven process provided Air Force mental health consultants with the best opportunity to date to state forcefully what most providers already knew: A huge portion (12%–48% of primary care patients) of all medical appointments involve mental health and stress management needs (Barret, Barret, Oxman, & Gerber, 1988; Ormel et al., 1994). The result has been that staffing levels for psychologists and social workers are tied directly to the number of primary care managers, which is, in turn, determined by the number of enrolled patients. Some additional psychology staffing was added to meet readiness needs.

With the growing recognition that 50% to 70% of visits to physicians have a primary or secondary stress-related cause (VandenBos & DeLeon, 1988), there has been increasing interest in placing mental health providers directly in the primary care clinics. A pilot project to demonstrate the use of mental health providers in Air Force clinics found positive results (Moe, Lombard, Lombard, & Wilson, 1997). Training new psychologists to work in primary care clinics was started in the Malcolm Grow Medical Center internship in 1999. By the end of the following year, all three of the clinical psychology internships included training on the delivery of psychological care in primary care settings. This model, which was guided to a large extent by successful work done by Strosahl et al. (1994) in civilian settings, has been well received within the Air Force and consistently generates interest at professional meetings (Isler, Oordt, Hunter, & Rowan, 2005).

The Department of Defense Psychopharmacology Demonstration Project is undoubtedly the highest-profile fellowship training in which Air Force psychologists have participated. This demonstration project trained 10 psychologists, including 3 from the Air Force, to safely and effectively prescribe medication for mental health patients. This led to the graduates having privileges in their military hospitals to prescribe psychotropic and adjunctive medications in their military hospitals. These psychologists have done consistently well, have been promoted on time, and have occupied positions of significant clinical and administrative importance. Perhaps most important, they have been well received by their colleagues, including the psychiatrists with whom they have worked. The Psychopharmacology Demonstration Project has served as the model for state-level efforts to develop laws that will allow appropriately trained psychologists to prescribe medications.

Enhancing Health Promotion and Disease Prevention

The fourth strategic initiative has been promotion of health and prevention of disease. It is a given that the Air Force, the Department of Defense, and the country as a whole cannot continue to pay the escalating costs of health care. Allowing costs to continue rising unchecked is not a viable alternative. Costs can be controlled most rationally and compassionately by building healthy communities. Clinical psychology is one of a number of specialties that have played a significant role in health promotion and disease prevention. In fact, psychologists have often filled the highest-level policy and planning positions in this area.

The Air Force has developed and staffed Health and Wellness Centers at every Air Force base, which exist to provide health promotion and prevention services such as smoking cessation classes, exercise classes, and stress management classes as well as services such as blood pressure and cholesterol screening. Virtually any analysis of what it will take to realize the potential of these centers, specifically, or the Building Healthy Communities initiative in general leads ultimately to the conclusion that the behavior of people—both patients and health care providers—must be modified. Psychology is the specialty that most clearly addresses the study of human behavior, and psychologists are involved in many aspects of prevention services.

The smoking prevention and cessation programs are one of the most prominent examples of health promotion (Klesges et al., 1999; Stein, Haddock, Talcott, & Klesges, 1996). Smoking cessation and prevention models developed in the Air Force are frequently referred to and adopted outside the Air Force (Conway, 1998; Haddock et al., 2001). Air Force psychologists contributed significantly to the *Clinical Practice Guideline on Tobacco Use Cessation* sponsored by the Veterans Administration and the Department of Defense (Veterans Administration/Department of Defense Clinical Practice Guideline Working Group, 2004).

Promoting Customer Satisfaction

The final strategic initiative has been customer satisfaction. The Air Force health care system strives to go beyond patient satisfaction; patients may be satisfied with far less than what is needed to ensure that active duty personnel are fit to meet the demands placed on them. The national security obligations to commanders extend this health care system beyond the usual goals of patient care and patient satisfaction. Psychologists are probably the professionals best prepared to objectively assess patient satisfaction and customer satisfaction. Assessment of satisfaction is an important part of ensuring high customer satisfaction, but psychologists offer more than that. In the primary care initiative, mental health facilities are located with other primary care services, enabling consumers to avoid being stigmatized for seeking care.

CONCLUSION

Air Force clinical psychologists have made significant contributions to all of the AFMS strategic initiatives. In fact, they may well provide the keys to success in the most important areas of health care within the Air Force. The future of Air Force clinical psychology will depend, in large part, on how well psychologists are able to continue responding to the needs of the Air Force. The early data are already in on that issue. The key question for clinical psychology and, in fact, absolutely any health care specialty is how well the specialty supports each of the strategic initiatives. Clinical psychology's answer to this question is remarkably strong and positive.

Clinical psychologists have become an increasingly integral part of the AFMS. Initially, fewer than three dozen clinical psychologists were engaging in fairly circumscribed tasks involving evaluation of airmen for suitability and adaptability. More recently, however, more than 200 clinical psychologists have been involved in a broad range of activities, including clinical practice. However, psychologists are also involved in policymaking, research and development, safety, readiness planning, special operations, teaching, and command. The overall growth in the numbers of psychologists on active duty in the U.S. Air Force at a time of significant reductions in force is the most remarkable testament to the respect that Air Force leaders have for clinical psychology and is the best indication available of the direction in which psychology is headed within the Air Force.

V

APPLIED SOCIAL PSYCHOLOGY

World War II transformed social psychological research into a mature scientific specialization. The most comprehensive product of applied social psychological research following World War II was the four-volume *Studies in Social Psychology in World War II* (Hovland, Lumsdaine, & Sheffield, 1949; Stouffer, Guttman, et al., 1949; Stouffer, Lumsdaine, et al., 1949; Stouffer, Suchman, Devinney, Star, & Williams, 1949), known as the American Soldier series. Many of the research themes of postwar experimental social psychology and sociology can be traced to these volumes.

Applied social psychology research in the military has focused on a variety of themes, including personnel selection and classification, measurement of individual differences, attitudes, human performance, team performance, human factors and ergonomics, leadership and leader development, Reserve and National Guard personnel, minorities, gender issues, work and the workplace, and families. The research is often multidisciplinary and collaborative. The Army Research Institute for the Behavioral and Social Sciences has developed new behavioral and social science approaches, methods, and technologies to help the Department of Defense manage its human capital and meet the challenges of a rapidly changing security environment. Psychology has helped the military be responsive to changing societal images and roles.

13

THE U.S. ARMY RESEARCH INSTITUTE FOR THE BEHAVIORAL AND SOCIAL SCIENCES

PAUL A. GADE, JONATHAN D. KAPLAN, AND NICOLE M. DUDLEY

The U.S. Army Research Institute for the Behavioral and Social Sciences (ARI) has been and continues to be an important national resource for behavioral and social science research in support of the national defense. This chapter begins with a brief history of the organization emphasizing its development since World War II. Although the organization known as ARI was established in 1972, its predecessors with various names have been operational since 1940. For details on the various incarnations in ARI's historical development, we refer the reader to the ARI history by Zeidner and Drucker (1988).

During World War II, the major national security concerns that ARI focused on were selecting people for military service and developing new methods for differentially assigning soldiers to Army jobs. Toward the end of World War II and during the Korean War, research at ARI began to focus on

This chapter was authored or coauthored by an employee of the United States government as part of official duty and is considered to be in the public domain. Any views expressed herein do not necessarily represent the views of the United States government, and the author's participation in the work is not meant to serve as an official endorsement.

other behavioral science issues, especially training, human factors, and ergonomics, in addition to its continuing research on soldier selection, classification, and assignment. Social psychology and sociological issues began to be part of ARI's research mission with the launching of Project 100,000 in 1966. Project 100,000, implemented by Defense Secretary Robert McNamara, was a program devised to provide low-aptitude youths with an opportunity to improve their lives through military training and service. In the 1970s, ARI's mission expanded to include social issues in race relations and the acceptance and inclusion of women in the Army. ARI also began major efforts in leader development research in the 1970s. Toward the end of the 1970s, ARI's mission was further expanded to include issues of recruiting, attrition, and retention brought on by the advent of the all-volunteer force.

In discussing the contributions of ARI to national security, we focus on three broad concerns affecting the U.S. Army: (a) selecting, assigning, and recruiting soldiers; (b) training soldiers and developing leaders; and (c) enhancing soldier performance. Within this broad framework, we discuss some of the major research projects that ARI has undertaken through the years and how these have had an impact on the Army and on psychology and other behavioral sciences. A word of caution before we begin: The three categories in this framework are not necessarily orthogonal, and our discussions within them will be necessarily brief. Furthermore, because of limited space, our discussion of topics is selective and not comprehensive. We apologize in advance for the many significant ARI accomplishments we have omitted; however, we have striven to discuss those topic areas that we felt had the biggest impact on the Army, on other military services, and on the behavioral sciences.

SELECTING, ASSIGNING, AND RECRUITING SOLDIERS

Selecting soldiers and assigning them to Army jobs has been a major research theme since ARI's inception. With the advent of the all-volunteer force, recruiting has become a continuing research theme as well.

Selection and Assignment

When World War II started, the military services desperately needed good classification procedures. As Street points out in chapter 2 of this volume, ARI's early ancestor, the Personnel Research Section, developed the Army General Classification Test (AGCT) in response to this need. The AGCT was the testing workhorse of World War II, as the Army Alpha and Beta tests had been in World War I.

In the early 1940s, specific mental tests, such as the general Mechanical Aptitude Test and the Clerical Speed, Radio Learning, and Automotive

Information tests, were often used to supplement the AGCT to assist in classification (Zeidner & Drucker, 1987). By 1947, 10 of these specific aptitude tests, which later formed the Army Classification Battery, had been used. The organization of the specific aptitude tests into the Army Aptitude Area System for differential classification was a major innovation for the military personnel system. This multiple aptitude area system markedly increased differential classification precision and efficiency over that provided by the AGCT during World War II (Zeidner & Drucker, 1987).

With the passage of the Selective Service Act in 1948 (see Uhlaner, 1967a), Congress mandated that the Department of Defense develop a selection and classification test to be used by all of the services. Between 1948 and 1950, with substantial contributions from the Navy, Marines, and Air Force, ARI developed the Armed Forces Qualification Test (AFQT), modeled after the AGCT. The AFQT was the first selection instrument to be used for the uniform mental screening of recruits and inductees across the services. In addition to determining the mental qualifications of recruits during the Korean and Vietnam Wars, the AFQT was used to help achieve an equitable distribution of abilities across the services (Maier, 1993).

After the end of the Vietnam War in 1973, the Army transitioned from a drafted Army to the all-volunteer force (Shields, Hanser, & Campbell, 2001), and using the joint AFQT became optional for the services. Each of the services used its own batteries for selection and classification between 1973 and 1976, with the Army using a version of the Army Classification Battery (Maier, 1993).

In 1976, all services began using the Armed Services Vocational Aptitude Battery (ASVAB) in lieu of their own classification batteries (Maier, 1993; Walker & Rumsey, 2001; Zook, 1996). Updated several times, the ASVAB still serves as an essential military screening and classification tool, and the AFQT score is still used as a general screening tool by the Army and the other services (Zook, 1996).

In the wake of the ASVAB misnorming in 1976 and congressional skepticism about the validity of entry test scores in predicting future performance in the military (Shields et al., 2001), the congressional mandate to show that ASVAB was a valid predictor of job performance resulted in the Army's Project A (see chap. 7, this volume). Project A went well beyond the mandate to validate the ASVAB and included research to validate and expand Army personnel selection and classification techniques. This expansion was made possible by Major General Maxwell Thurman, then the head of the U.S. Army Recruiting Command (USAREC), who pushed for a broader concept of soldier quality. Together with Joyce Shields, then head of the Manpower and Personnel Resource Laboratory at ARI, General Thurman pressed the concept of the whole person evaluation, which incorporated all the diverse characteristics that could influence performance in addition to mental abilities, including psychomotor, spatial, interests, and temperament characteristics. Project A (1982–1989) was to require the measurement of more than 50,000 soldiers in 21 military occupational specialties and, along with its follow-up project, Career Force (1989–1995), became one of the most influential projects in the history of selection, classification, and performance research.

Project A and Career Force provided key answers to the question, "What exactly is job performance?" Intensive analysis of the huge soldier sample yielded five core common dimensions of performance. Two were proficiency dimensions—core technical proficiency and general soldiering proficiency (termed "can do" dimensions)—and three were motivational dimensions— effort and leadership, personal discipline, and physical fitness and bearing (termed "will do" dimensions). Conceptualizing performance this way led to the task versus contextual performance distinction; these components are still seen as key dimensions of job performance (Borman & Motowidlo, 1993). With Project A, the "classic prediction model was born" (Shields et al., 2001, p. 21), and it continues to serve as the dominant prediction model in personnel research in both the military and civilian worlds (J. P. Campbell, 1990).

Recruiting

The all-volunteer force fast became known as the "all-recruited force" in the late 1970s and early 1980s. To meet the recruiting challenge, the Army needed to find out what influenced young Americans' propensity to enlist. ARI first began recruiting research in 1980 with the arrival of General Thurman as the commanding general of USAREC. ARI's recruiting research fell into three broad categories: market analysis and segmentation, recruiter selection and training, and sales aids. The highlights of work in each of these areas are described in the subsections that follow.

Market Analysis and Segmentation

One of the first ARI research efforts for USAREC was the 1981 New Recruit Survey, which was administered at reception stations (now called *battalions*) and was designed to segment new recruits by the reasons they had enlisted. Results from the survey were used to corroborate the theory of a dual recruiting market—those recruits interested in a college education versus those interested in job training—and to structure advertising and enlistment incentives accordingly. ARI continued to refine and administer the New Recruit Survey until 1987, when USAREC made it an annual operational survey.

Perhaps the most important ARI survey of this sort was the 1984 Army Experience Survey to determine what veteran one-term soldiers thought about their Army experience. Commissioned by the Secretary of the Army, John Marsh, and reported to President Reagan, results from this survey showed that veterans overwhelmingly found their Army experience to be a positive, worthwhile life experience. Other ARI research efforts identified the key influencers and the roles they played in the enlistment decision processes of new recruits. Results showed that when making their enlistment decision, potential recruits looked to their parents for emotional support but to teachers and high school counselors for information. ARI research also showed that interviewers administering the Department of Defense Youth Attitude Tracking Survey could generate an accurate predictor of AFQT category using a telephone version of the verbal component of the Computerized Adaptive Screening Test (CAST; Legree, Fischl, Gade, & Wilson, 1998). This interview provided recruiting advertisers with an accurate way to assess the effectiveness of recruiting advising targeted to high-AFQT high school prospects.

Recruiter Selection, Training, and Utilization

Recruiter research in the 1970s and 1980s examined the use of the *realistic job preview*, a recruitment procedure organizations use to present realistic work information, both favorable and unfavorable, to potential employees (Meglino, Ravlin, & DeNisi, 2001). The purpose of the preview is to give potential recruits a taste of what their work requirements are likely to be, thereby allowing them to self-select out of the job or self-adjust to the challenges they are likely to face. ARI also created the Recruiter Development Center (RDC) to help new recruiters adjust to the job and to maximize training benefits in the Army Recruiter Course. As it turned out, the RDC was probably the first use of an assessment center as a training diagnostic. Originally, the RDC was to be an assessment center for selecting recruiters (Borman, 1982; Borman & Fischl, 1980). However, the RDC as an assessment center proved impractical when it became clear that soldiers did not volunteer but rather were selected for recruiter duty. ARI was still able to make use of the RDC by turning it into a realistic job preview and training diagnostic system for the U.S. Army Recruiter School (Borman, Rosse, & Rose, 1982). For this application, ARI scientists received one of the Army's prestigious Army Research and Development awards in 1983. In addition to the research already cited, see Penney, Horgen, and Borman (1999) for more information on ARI recruiter research.

Sales Aids

The development and implementation of the CAST as part of Army's Joint Optical Information Network semiautomated recruiting system was one of ARI's major accomplishments in this area. This test, which takes about 5 to 10 minutes to administer, provides recruiters with a highly accurate predicted AFQT score range so they can determine the prospective recruit's likelihood of earning a passing score on the ASVAB early in the recruiting interview at the recruiting station. The test saves recruiters time by allowing them to focus on the most desirable potential recruits, and recruiters stated that it was symbolic of the new, high-tech Army they were trying to sell. CAST was an important precursor of the computerized adaptive version of the ASVAB, providing the first test of the large-scale application of computerized adaptive testing (Sands, Gade, & Knapp, 1997).

TRAINING SOLDIERS

ARI expanded its research mission into training research when the Motivation and Training Research Laboratory, the U.S. Army Manpower Research and Development Center, and the Behavioral Systems and Research laboratory merged to form what is now known as ARI. Below is a sampling of the major training research issues that ARI has undertaken since that merger.

Computer-Based Training

ARI has been a leader in the use of computers in training, starting with computer-aided instruction, through computer-based training, to intelligent tutors. Arguably the most advanced of these systems was the Military Language Tutor (MILT). MILT began as an ARI research program and test bed to determine the effectiveness of including a natural language processing engine in a language tutor for Arabic, Spanish, and English.

Together with a commercial version of MILT (Multimedia Instructional Tutor Authoring System), MILT is in use at the U.S. Military Academy to both author and deliver foreign language instruction. In 2003, the U.S. Military Academy began using MILT to create and deliver Arabic lessons for the U.S. Army Intelligence School and the Special Operations Forces Language Office. Holland, Kaplan, and Sabol (1999) and J. D. Kaplan and Holland (1995) discussed the development of MILT in more detail.

Distance Learning

ARI has been a pioneer in the use of distance learning techniques for Army training. A prime example of this is Critical Thinking Skills Training. Like technical skills, critical thinking skills are developed over time with appropriate training, practice, and experience. Recently, ARI has successfully transitioned this training from the classroom to open Web architecture so that soldiers can benefit from training and practice in a self-paced program that is always available when they are free to participate (see Katz & Grubb, 2003, and Reidel, 2003, for details on this and other ARI research efforts).

Realistic Battlefield Training

How to practice combat in an environment resembling realistic conditions without endangering trainees poses difficult problems. More than 30 years ago, ARI developed the first practical, realistic field training system for Army tactical units and has continued this training research program to the present. Begun as a simple method for assessing simulated casualties during collective training, the Squad Combat Operations Exercise Simulation (SCOPES) became an essential feature of the Tactical Engagement Simulation (TES; Gorman, 1992). In SCOPES, telescopes are attached to rifles, allowing soldiers to claim hits on opposing soldiers by identifying a number painted on their helmets. In this way TES exercises achieved a close parallel to combat.

SCOPES was extended to mounted forces in Realistic Training (REALTRAIN; Gorman, 1992). REALTRAIN used scopes mounted on tanks and antitank weapons to allow observers to assess target hits, and the effects of simulated mortar and artillery fire were soon added. Although REALTRAIN gave soldiers a much more accurate simulation of actual combat, it was labor intensive, requiring many people to do target assessment. The Multiple Integrated Laser Engagement System (MILES) was the answer to this problem. MILES provided a realistic battlefield training environment for soldiers by simulating direct fire in force-on-force training using eye-safe laser "bullets" from lasers mounted on rifles and guns. Each individual and vehicle in the training exercise had a detection system to sense laser hits and perform casualty assessments. MILES training has proved to dramatically increase the combat readiness and fighting effectiveness of military forces (U.S. Army Research Institute for the Behavioral and Social Sciences, 1995). To this day, MILES is the key training technology of the National Training Center, the Army's premier battle training location.

An important part of this training is the after action review (AAR). The AAR is the Army's technique for feeding back important information following collective training. This review process was developed by ARI in the mid-1970s as part of its TES training program and was influenced by Marshall's World War II oral history techniques (Marshall, 2000). ARI modified the AAR approach in its development of the MILES training system for the Army's National Training Center. Further ARI research led to an improved generation of AAR techniques as part of the Simulation Networking (SIMNET) computer-networked simulation system in the mid-1980s (see Meliza & Tan, 1996, for an example of ARI training research using SIMNET). The AAR technique is now standard in the Army and is widely used internationally (Morrison & Meliza, 1999).

LEADERSHIP AND LEADER DEVELOPMENT

In addition to its work on selection and classification and the conceptualization and measurement of individual and team performance, ARI has conducted and funded leadership research since the early 1970s. Owen

Jacobs, then chief of ARI's Leadership and Motivation Technical Area, together with his colleague Elliot Jaques, conducted an extensive leadership research program based on Stratified Systems Theory (SST; Jacobs & Jaques, 1987). SST specifies "how leader performance requirements change at different organizational levels" and thus "how leader attributes, particularly conceptual capacities, change in corresponding ways" (Zaccaro & Horn, 2003, p. 795).

ARI research based on SST concepts developed both qualitative and quantitative assessment instruments designed to measure cognitive attributes associated with effective executive leadership (Zaccaro & Horn, 2003, p. 798). In particular, SST has informed numerous training products and tools that have been used in military leader development programs and schools such as the Army War College, the National War College, and the Industrial College of the Armed Forces. Furthermore, SST conceptual principles, as elaborated by ARI research, have been adopted as part of the Army's training doctrine (Zaccaro & Horn, 2003). An ARI-sponsored research effort by Zaccaro (2001) provided an excellent overview and state-of-the-art look at executive leadership.

Recently, ARI has turned to Sternberg's theory of successful intelligence to provide a fresh, if somewhat more controversial, approach for gaining insights into leadership processes and leader development in particular (Sternberg et al., 2000). In this pursuit, Sternberg and his associates developed and validated the Tacit Knowledge of Military Leaders Inventory to measure leadership tacit knowledge (i.e., practical, largely unarticulated, procedural knowledge about leadership). They have shown this test to be a measure of practical intelligence that is different from psychometric g, experience, and personality but that, like SST, differs by the individual's level of organizational responsibility. In collaboration with Yale University, ARI researchers have demonstrated early success using this encapsulated tacit knowledge in training programs to accelerate the leader development processes.

ENHANCING HUMAN PERFORMANCE

The Army continuously seeks new ways to improve soldier performance through a variety of means, including new training technologies, selection and assignment procedures, engineering human—machine interactions, and even social programs. This section describes some of the key issues the Army has dealt with since World War II and the research ARI has conducted to address them.

New Age Techniques

The dictionary defines *new age* as "of or relating to a complex of spiritual and consciousness-raising movements originating in the 1980s and covering a range of themes from a belief in spiritualism and reincarnation to advocacy of holistic approaches to health and ecology" (Houghton Mifflin, 2000). In the late 1970s and into the 1980s, a variety of popular and new age human performance enhancement and assessment techniques that were called *human technologies*, such as neurolinguistic programming, parapsychology techniques, and super learning programs, captivated the attention of a group of influential Army officers. In 1984, General Thurman and other concerned Army generals turned to ARI for guidance in evaluating the scientific bases of these human technologies. Because of the often pseudoscientific nature of these technologies, the task was broad and rife with potential credibility issues. As a result, ARI contracted the National Research Council to help with these evaluations. Between 1985 and 1999, the National Research Council Committee on Techniques for the Enhancement of Human Performance published the results of its investigations in five books. Space does not permit a full description of the committee's findings; E. Salas, DeRouin, and Gade (in press) provided a good summary of the most important of these findings.

These results had far-reaching implications for the military services, for the general population, and for psychology in demystifying many popular, but unsubstantiated, programs and practices for enhancing human performance. They saved the Army millions of dollars that might have been wasted implementing practices that would have had little or no positive impact on performance. These studies also provided psychology and the military with solid recommendations for closing the gap between appealing but unsubstantiated organizational practices and theory-based research (Swets & Bjork, 1990). They also indicated how to deal effectively with similar situations in the future.

Human Factors and Ergonomics

Manpower, personnel, training, and human engineering integration (MANPRINT) is the behavioral sciences' successful response to decades of ineffective attempts to influence the development of manned systems in the Army. The concept and method of implementing MANPRINT originated at ARI in reaction to the comparability analysis approach of the Navy's Hardware/Manpower Integration (HARDMAN) I and ARI's HARDMAN II (see J. Kaplan, 1985; J. Kaplan & Hartel, 1988). Comparability analysis assumes that predecessor systems' manpower, personnel, and training by definition predict equivalent requirements for new systems. MANPRINT, in the form of HARDMAN III, was based on the concept of using simulation modeling that incorporates manpower, personnel, training, and human factors data to predict manned system performance. The object of this approach is to identify the levels of these four MANPRINT component areas that result in successful manned system performance, as defined by speed and accuracy.

ARI researchers believed that the MANPRINT concept would become particularly meaningful if it could be implemented as a fully integrated suite of simulation modeling tools that ran on a personal computer and was designed for use by analysts in these fields rather than by professional modelers. These characteristics would both reduce the cost and time of using such tools and remove the problem of specialized modelers not fully understanding the models they were creating.

HARDMAN III's software and initial data and models were completed through close cooperation between the contractors, Micro-Analysis & Design and Dynamics Research Corporation, and ARI personnel and delivered to ARI in 1992. Key data for its personnel prediction module were personnel performance data that were analyzed from the objective portion of ARI's Project A. This enabled HARDMAN III to predict performance based on ASVAB scores or AFQT category. In that same year, the Systems Laboratory of ARI, developers of HARDMAN III, was separated from ARI and became part of the Human Research and Engineering Directorate of the U.S. Army Research Laboratory. Its models were verified and validated in 1994 (see Allender et al., 1994).

A new version of HARDMAN III, the Improved Performance Research Integration Tool (IMPRINT), a Microsoft Windows application, was developed by the Human Research and Engineering Directorate in accordance with the original ARI plans. Over time, the directorate developed new versions with greater enhancements, including the ability to predict goal-oriented behavior and to model at the cognitive level by adding connections to the ACT–R theory of cognitive architecture (alternatively spelled out as Atomic Components of Thought—Rational or Adaptive Control of Thought—Rational). The current version of IMPRINT is in use at 189 locations: 75 Army, 13 Navy, 8 Air Force, 2 government non–Department of Defense locations, 10 universities, and 81 contractors (Lockett, 2000).

The Army selected IMPRINT as the modeling implementation for the Multi-University Research Initiative basic research program at Central Florida University in Orlando. This research program is attempting to understand and predict the effects of various sorts of stress and stress alleviators on human performance. HARDMAN III and its IMPRINT successor have been used to predict the following: task allocation and maintenance, level of automation, mental workload and function allocation, task allocation, maintenance manpower, level of automation, decision making, goal-oriented behavior, workload and information demands, performance degradation, job restructuring and consolidation, and information flow (Lockett, 2000).

Minorities and Women

There was a great deal of tension in race relations in the Army during the 1960s. ARI survey research helped identify the perceived sources of that tension and provided information to the Army leadership on how effective the Army's race relations programs were in reducing those tensions. ARI conducted large-scale surveys in 1972 and 1974 to see what changes in Black and White perceptions of race relations had taken place as a result of equal opportunity interventions administered during the time between surveys. Although they found many positive changes in attitudes and opinions, race relations were still perceived as a problem by both Black and White soldiers. J. A. Thomas (1988) provided an excellent account of the details and impact of that research.

Issues concerning the number of women in the Army and their utilization arose in the mid-1970s. The Army asked ARI to develop a "Test of Women Content in Units" (U.S. Army Research Institute for the Behavioral and Social Sciences, 1977). The results of this test showed that when women constituted as much as 35% of a unit, the highest percentage in the experiment, there was no significant effect on performance. Gender-integrated training became a related issue during the same time period. Research then and again in the 1990s showed that women benefited from training with men and that men did as well as they usually did. *Women in the U.S. Army: An Annotated Bibliography* (U.S. Army Research Institute for the Behavioral and Social Sciences, 2002) provided an excellent summary of this research.

Army Family Research Program

Beginning in 1989, ARI conducted the Army Family Research Program (AFRP) in response to requirements from the chief of staff of the Army, John Wickham Jr., and subsequent Army Family Action Plans. The AFRP, sponsored by the U.S. Army Community and Family Support Center, explored the demographic characteristics of Army families and assessed the impact of family on soldier readiness and retention. The results of the AFRP research had far-reaching effects on Army family policies and programs and on similar research conducted for the other services by the Department of Defense. M. W. Segal and Harris's (1993) *What We Know About Army Families* is a highly useful, readable summary of the program findings (for additional details, see also chap. 16, this volume).

Special Forces

In 1991, ARI developed a comprehensive needs assessment for the John F. Kennedy Special Warfare Center and School and for the U.S. Army Special Operations Command that detailed the behavioral science research requirements for Special Forces and was the guiding document for the ARI's Special Forces research program during the subsequent 7 years (Brooks, 1992). L. Morgan Banks, in chapter 6 of this book, details the research this generated and the impact it had on Army Special Forces. As the U.S. Army transforms to smaller units of operation and expands its missions to include those traditionally in the purview of Special Forces, such as stability operations, ARI's Special Forces behavioral research will become more relevant and useful for the conventional Army.

CONCLUSION

The Army is undergoing a major transformation in how it organizes itself, equips its soldiers, and conducts operations in the aftermath of the terrorist attacks of September 11, 2001. The transformation of the human component of these changes is essential to the Army's overall success in meeting the challenges of a rapidly changing world. ARI is charged with the responsibility of applying new behavioral and social science approaches, methods, and technologies in its research agendas to address the human issues in this transformation. For example, to meet the Army's need for enhanced team performance, cognitive psychology research programs are under way at ARI to identify and understand the input–output cues that lead to effective collective skill development. ARI researchers are exploring the cognitive factors that facilitate and impair team formation and contribute to a sense of trust among team members. Identifying principles for developing shared mental models that influence soldiers' understanding of their commanders' intent and team performance is part of this research as well.

Recruiting and retention continue to be critical issues for the all-volunteer Army. To further address these issues, current ARI research is identifying factors that affect enlistment decision making, including demographics and motivation, and modeling how these decisions develop. Research is currently under way to identify the factors that influence retention decision making, productive behavior, and good citizenship and to model how these behaviors are acquired. Understanding the role of mediators such as personal motivation, job satisfaction, organizational commitment, values, and ethics in recruiting and retention processes is a critical part of this research. This research will enable the Army to more effectively tailor recruiting and retention practices to its changing needs.

Training in complex situations is another critical element in the Army's transformation. Today's soldiers must deal with increased cognitive demands resulting from the technology requirements of digital, semiautomated, and robotic systems. ARI's training research seeks to reduce the effects of information overload through training; to determine how individuals assign meaning and relevance to large amounts of rapidly received, ambiguous data; and to determine how to improve this ability through training. Understanding and modeling the role of feedback and feedback systems in the acquisition, retention, and transfer of individual and collective training and in motivating learned task performance are important aspects of ARI's training research. ARI research is also investigating methods for compressing training time that maximize retention and transfer of training.

In other critical research, ARI is beginning to address the adaptive value of human emotions in calibrating psychological systems in its research programs, as well as self-control and self-awareness, both of which are critical to the success of any military operation. Psychological science does not understand well how emotions, as positive and negative evaluative processes, operate together or in opposition to influence actions and cognitions. Researchers need better measures of affective processes, such as functional neuroimaging, measures of changes in brain chemistry, and more traditional psychological measures. Understanding how emotions can help people calibrate their behaviors and thoughts to achieve internal stability in difficult situations is critical to achieving effective performance on the battlefield and in other difficult military operations.

This has been but a brief overview highlighting a few of the contributions of ARI to national security through its research programs in the past, present, and future. Through its research, ARI continues to be a critical contributor to the national defense by helping the U.S. Army manage its precious human capital effectively.

14

THE RACIAL INTEGRATION OF THE U.S. ARMED FORCES

ALAN GROPMAN

Uniformed military leaders racially integrated the U.S. armed forces in the late 1940s and early 1950s. Their motivation was mainly to make the Air Force, Army, Marine Corps, and Navy more efficient and more effective as fighting forces. The generals and admirals were not acting out of a sense of political correctness, nor were they acting ethically in the spirit of the second paragraph of the Declaration of Independence (that all men were endowed by God with inalienable rights and created equal). Rather, the four combat services altered their policies and practices on race for pragmatic reasons—their ethos was to win wars more effectively and efficiently. Civilian leaders in the Department of Defense played roles secondary to those of the uniformed senior officers (although the secretary and under secretary of the Air Force were definitely sympathetic to racial integration, as was the secretary of defense). The chiefs of staff, commandants and chief of naval operations and their deputy chiefs for personnel reached their decisions on the basis of the performance of Blacks in the Navy and Air Force in World War II and in the Army and Marine Corps in the Korean War, relying practically not at all on input from social science researchers. Military leaders recognized racial segregation as an expensive burden—two sets of everything were required—and, because of the combat and technical service performance of Black service members, as unnecessary.

This chapter focuses principally on the Army and the Air Force, because there were many times more Black officers in those services than in the Navy and Marine Corps, and the enlisted men worked in many more specialties connected to combat missions. It covers the efforts to end segregation in the 1940s and 1950s; discussion of subsequent efforts to promote equal opportunities among military personnel is beyond the scope of this chapter.

OVERVIEW

The military leadership in the mid-20th century essentially had three personnel objectives, and racial integration satisfied all three purposes (Gropman, 1998).

1. *Recruit and retain enough troops to fill the ranks:* White persons were less eager to join the military than Black individuals (still true today; see Gropman, 1998, pp. 52–53), and segregation would deny the services of the fullest use of Black soldiers.
2. *Employ and keep people with adequate skills:* World War II had demonstrated to a few but also key leaders in each service (mainly in the Army Air Force and to a lesser extent Army ground forces) that Black soldiers, given the same training as White soldiers, could perform as well as the latter.
3. *Enlist and retain people who were motivated to serve:* Racial integration would make the services even more attractive to Black persons, leading to greater retention, which would improve the quality of the force (through greater experience) and lower the costs of training new recruits.

Although *integration* is the term most often used to describe the personnel reforms, what occurred in the late 1940s and early 1950s is better described as *desegregation*—the ending of quotas for Blacks and the termination of separate living, dining, and recreational facilities. Blacks began to supervise Whites, but until well into the 1960s (depending on the specific service—more Blacks supervised Whites in the Army and Air Force than in the Navy and Marine Corps), not many Blacks were put in charge of significant numbers of Whites, and there were few Black officers. The highest percentage of Black officers in 1949, when the Air Force desegregated, was in the Air Force, and constituted less than 1% of the Air Force officer corps. Off the military post, there was practically no interracial socializing for decades following official integration, hence the distinction between *desegregation* and *integration*; the Air Force in 1949 was integrated only during the workday, and bonding did not occur off the job. What was unequivocally real following integration, however, was equal opportunity for training and service: One's test scores, once racial segregation was abandoned, determined one's specialty within the services. Promotion to supervisory ranks, however, came slowly and blossomed only during the postdraft era beginning in 1973 (Gropman, 1998).

Military leaders, however, were not concerned with improving the Black race, nor were they eager to advance the condition of Black soldiers in a segregated and racist United States. All services paid little or no attention to the conditions of Black people in civilian communities, where life could be miserable and often dangerous. Black service members everywhere, and especially in the South, had difficulty finding adequate housing, employment opportunities for spouses, and decent education for children until well into the 1960s. Conditions in the rural North were often worse. The armed services racially integrated to make the military a better combat force, but most leaders in all four combat arms were nearly oblivious to the needs of Black service members and their families outside the base fence (Gropman, 1998).

HISTORY PRIOR TO INTEGRATION IN THE MILITARY

Military leadership indifference to Blacks' problems in civilian communities was as old as the service of Black Americans to their country, which began in the 17th century.

Early History

Military service of American Blacks is more than 300 years old. It began in the militia of the British American colonies, and Black soldiers served prominently in the new country's Continental Army beginning in the first battles in 1775 and have fought with honor in every war since (Coffman, 2003; Cornish, 1956; Dobak & Phillips, 2001; Franklin, 1997; Leckie, 1967; MacGregor, 1981; McPherson, 1988, 1991; Nalty, 1986; Quarles, 1953, 1961). The military leadership's position on race has evolved from its racist beginnings in the 17th and 18th centuries, when there were attempts at general exclusion from colonial military service and the Continental Army, to full acceptance in the last 2 decades of the 20th century.

Despite heroic service in many wars, American racism during the first quarter of the 20th century all but obliterated the memory of earlier Black achievement in the military. The Army turned the four historic Black combat regiments (9th and 10th Cavalry Regiments and 24th and 25th Infantry Regiments, all formed in the late 1860s and all of which continued into the 1950s) into service units; the Navy completely barred Blacks from enlisting in 1919, and when it permitted them to enlist in the 1930s, it was only as servants. The Army Air Corps (what Army Air Forces was called 1907–1941) took no

Blacks in any capacity, and the Marine Corps was similarly exclusive (MacGregor, 1981; Nalty, 1986; see also Franklin, 1997; Quarles, 1961).[1]

Following World War I, the Army was concerned that it had not used Black soldiers properly during the war; Blacks had made up only two divisions, or significantly less than 10% of the fighting force. At the request of the Army chief of staff, the Army War College devoted a whole academic year and its entire faculty and student body in 1924 and 1925 to a study entitled "Uses of Negro Manpower in War" (Ely, 1925). This study is an example of vicious racism. It reported that Blacks were a "mentally inferior subspecies" of the human population and "very low in the scale of evolution" and that Blacks had "smaller craniums" and significantly lighter brains than Whites—"35 ounces contrasted with 45 ounces for whites" (Ely, 1925). Blacks, the report declared, were moreover inherently cowardly and immoral and often raped White women. The report also asserted that Blacks themselves had no confidence in Black officers. Despite all of this erroneous and undocumented data, written by future Army senior leaders, the report concluded that Blacks must serve in the military, and in combat, because the United States was their country, too, and if Blacks did not serve, the burden of combat would fall on Whites, which was not good for the country. The report emphasized that if Blacks served, it must always be under the command of Whites.

This report failed to deal with such facts as the fighting record of the historic four regiments and why they were deliberately not used in World War I combat, as well as why and how the most senior Black officer, Colonel Charles Young, was intentionally and fraudulently discharged to prevent him from commanding troops. The absence of Black members on draft boards, moreover, meant that a higher percentage of Blacks were drafted than Whites, and virtually all of the millions deferred from service for dependency or other reasons where White.

Students and faculty at the Army War College restudied the use of Blacks in the Army nine more times between 1925 and 1939, and I read one study written in 1944 by an Army service unit that cited the same pseudoscientific rubbish as the 1925 study.[2] Yet 7 years later, in 1951, the Army leaders of World War II began to integrate the service during the Korean War, 12 years after the last Army War College report. Charles Darwin would undoubtedly be astonished at the revolutionary change in human evolution in just a dozen years (Lee, 1966).[3]

---

[1]For the War of 1812 and the Civil War, see Nalty (1986) and McPherson (1988). For the Indian Wars and the Spanish American War, see Nalty (1986), Leckie (1967), Coffman (2003), and Dobak and Phillips (2001). For World War I and the interwar period, see Franklin (1997) and Nalty (1986).
[2]The 1944 document was destroyed in a fire, so a citation is not available.
[3]Racist nonsense regarding size of craniums and weight of brains is undocumented, and all of this report, by senior leaders who were expected to become even more senior, is never supported with scientifically based material. Perhaps the most damaging statement in the report: "In physical courage it must be admitted that the American Negro falls well back of the white man and possibly all other races" and that "[the Negro possesses] physical, mental, moral and other psychological characteristics"

World War II

Black soldiers who did serve in World War II were allowed to do so largely as a result of political pressure. They were inserted in the Army Air Forces (later U.S. Air Force), for example, by President Franklin D. Roosevelt because he had made a campaign promise to do so. Black personnel worked mostly as laborers and service troops in the Army, and only two of the 89 divisions formed were Black (both with White commanders and senior officers). There were many other small combat units—for example, the 761st Tank Regiment—but all of them had White commanders, except the Tuskegee Airmen units that actually flew in combat. The Marine Corps, the last service to enlist Blacks, did not do so until the war was almost half over. The Navy kept Black personnel in menial roles until forced by the civilian secretariat to open opportunities, but this did not occur until 1945, and the number of Blacks serving as other than stewards was tiny. The number of Black officers in a Navy of several million members did not exceed 100 (Franklin, 1997; MacGregor, 1981; Nalty, 1986).

To take one illustrious example of Blacks in combat, the Tuskegee Airmen, Black fighter pilots in the Army Air Forces, were not permitted into combat until long after the unit had been fully trained and certified for battle, and once in the fight they were almost barred from further combat out of prejudice. In time, the 99th and the larger 332nd Fighter Groups it joined performed outstandingly, scoring at least one major unique achievement: On bomber escort missions, the 332nd Fighter Group never lost a friendly bomber to an enemy fighter, and no other fighter unit (all of which were White) that served as long in the war as the Tuskegee Airmen could make that claim. It was the accomplishments of the Tuskegee Airmen, essentially, that drove the U.S. Air Force to integrate racially ahead of the other services (Gropman, 1998; Nalty, 1986; Sandler, 1992).

Following World War II

President Harry S. Truman is often given most of the credit for racially integrating the military because he promulgated Executive Order 9981 of July 26, 1948 (Establishing the President's Committee on Equality of Treatment and Opportunity in the Armed Services, 1948), but this acclaim is based on superficial historical analysis. He was a lifelong social segregationist who always used the word *nigger* when he spoke privately of Blacks (Berman, 1970; Bernstein, 1970; Gropman, 1998; MacGregor, 1981).[4] Truman had

---

that "made it impossible for him to associate socially with any except the lowest class of whites." There was a sole exception to this rule, however: "Negro concubines who have sometimes attracted men who, except for this association, were considered high class."

[4]Truman was advised by Clark Clifford (see confidential memo to Truman from Clifford in the Clifford papers at the Truman Library, Independence, Missouri, November 1947) to make appeal to Blacks to win reelection (Gropman, 1998; see also Berman, 1970; Bernstein, 1970). Truman abhorred

been reluctant to issue Executive Order 9981, but he believed that such a directive was required by the exigencies of the 1948 presidential election campaign; his closest advisers warned him that he would lose the election to New York Governor Thomas Dewey, a civil rights advocate, if he did not make gestures to win the Black vote. This executive order does not include the words *integration* or even *desegregation* and calls only for "equality of treatment and opportunity" regardless of "race, color, religion, or national origin." Armed forces racial integration and the significant equal opportunity and nondiscrimination apparatuses that followed it were motivated not by a president running for reelection to hold the Black constituency but rather by recognition by a few key leaders in the armed forces that there was no scientific basis for segregation (e.g., racial inferiority) and that segregation was an intolerably expensive barrier to effectiveness.

Once Truman had issued the executive order in July, he took no further action on behalf of Black military people until about the time of the inauguration in January, when he was already safely reelected, and then his measures were half-hearted and ineffective. He established a committee to enforce "equality of treatment and opportunity" but dissolved it in April 1950, long before the Army, Marine Corps, and Navy were racially desegregated. The Air Force began to integrate in May 1949, almost a year after Truman's executive order, but it had announced plans to integrate 2 months before the order. The Air Force leadership's deliberations were not influenced by Truman's views or activities on civil rights issues. The Army and Marine Corps did not integrate until 1951, during the Korean War. The Navy had token integration beginning in 1945 (directed by the civilian secretary of the Navy but not supported by the senior admirals), but it did not pass beyond tokenism until after all the other services had completely integrated (Gropman, 1998; MacGregor, 1981). Five years after the integration directive, more than 99% of Blacks in the Navy served in segregated units, and the Mess Corps (e.g., stewards, cooks) stayed all Black until the end of the 1950s (MacGregor, 1981).

The Air Force decision in April 1948 to integrate racially was based on the need for efficiency and effectiveness. Blacks would serve in any specialty for which they qualified, all racial quotas would be abandoned, and units (and their facilities) would no longer be racially divided. The service began the process in May 1949, and by the end of that year, more than 75% of

lynching and sponsored antilynching legislation, which may in fact be a singular reason he became President in 1945, because Franklin Roosevelt's first choice for Vice President was James F. (Jimmy) Byrnes, who was unacceptable to Blacks (see Robertson, 1994). Truman believed Blacks should vote if they met the same standards as Whites and opposed subterfuges that kept the ballot from Blacks—but he was still a segregationist. Ample proof of this is to be found in the numerous interviews he had with Merle Miller (see M. Miller, 1973). Miller stated, "Privately Mr. Truman always said 'nigger,' at least he *always* [italics added] did when I talked to him" (McCullough, 1992). Truman, however, did more for Blacks than any president between Abraham Lincoln and Lyndon Johnson, but he did not integrate the armed services. The Army Secretary was appalled at the Air Force announcement in April 1948 that the Air Force intended to integrate racially (see Gropman, 1998; MacGregor, 1981).

Black airmen and officers in the Air Force were serving in integrated units (Gropman, 1998).

The prime mover in the Air Force was its deputy chief of staff for personnel, Lieutenant General Idwal Edwards. He had been thoroughly involved in racial matters during World War II because he served on the War Department's McCloy Committee, which was charged with monitoring the use of Blacks in order to get the most from the Black service population. Edwards came to believe that segregation was an unnecessary and expensive burden. He saw maintenance of separate living and dining quarters and recreational and training facilities as a needless expense, and he knew that segregation provoked racial friction. When the Air Force became independent in September 1947, he called for a study on the relevance of racial segregation and the reasons for it. The study director, Lieutenant Colonel Jack Marr, considered the combat accomplishments of the Tuskegee Airmen during World War II and found that these units performed as well as White outfits. The combat Tuskegee Airmen were entirely Black, with no Whites anywhere in the chain of command. No White sergeants supervised mechanics or avionics specialists, and no White captains or colonels supervised or led the aviators on their hundreds of combat missions. Marr also studied the output of technical schools and the performance of Black enlisted men in Army Air Forces service specialties. Marr proved to himself and to General Edwards that Blacks with the same aptitude as Whites, if given the same training as Whites, performed as well as Whites. Segregation, he concluded, because it was not based on inferior performance, was therefore based on prejudice.

Edwards took these findings to the Air Force chief of staff and convinced him that the Air Force would be more efficient and effective if it were racially integrated. There would be no more need for two sets of mess halls, barracks, swimming pools, gymnasiums, and so forth. More important, there were talented Blacks who wanted to join the Air Force, but they would be underutilized in a segregated system. Finally, there were personnel shortages in some specialties at the one Black air base (Lockbourne Air Force Base, Ohio), and at the same time, there were personnel surpluses at this base that were fully qualified to fill billets at other bases, but because all other bases were White, no shortages could be filled by Blacks, nor could White bases with needed surplus people send them to Lockbourne. Edwards convinced the Air Force leadership to integrate, and in April 1948 Air Force Undersecretary Eugene Zuckert and Chief of Staff Carl Spaatz announced that the Air Force would integrate racially. Racial segregation in the Air Force had been based on Army regulations that were carried into the Air Force independence era. These regulations dated back to the interwar and World War II periods, and the Air Force decided to abandon these policies. The process began in 1949 and progressed smoothly (Gropman, 1998).

As in the case of the Air Force, the Army had in a key position a person who believed that racial integration would improve the service—General James Lawton Collins Jr., Army chief of staff during the late 1940s and early 1950s and, most importantly for the purposes of this chapter, during the Korean War. During World War II, Collins's First Army had been involved in a racial integration experiment. Because the Army in Europe at the time of the Battle of the Bulge (late 1944 and early 1945) was desperately short of riflemen, it offered a limited amount of Black service troops (e.g., cooks, auto mechanics, truck drivers, clerks) the opportunity to fight alongside Whites if they volunteered to become riflemen. The quota was soon oversubscribed, with many Blacks eschewing soft behind-the-lines work for combat. About 2,500 soldiers ended up fighting as integrated warriors (in an Army of more than 6 million). The Black soldiers fought well and were widely accepted by their White colleagues.

General Dwight D. Eisenhower's Army personnel office studied the reactions of the Whites in integrated units during World War II and received reports from practically the entire chain of command in the First Army. From the colonels and generals to the sergeants and privates, the Whites who had served in integrated units favored racial integration. Although astonishingly no Blacks were polled, a large slice of the Whites who had served with Blacks were formally surveyed. Shortly after the war, the Gillem Board (comprising three Army general officers) was to make much of this tiny sample and the formal after-action reports of this integration experience as Lieutenant General Alvan Gillem and his two general officer partners studied yet again the uses of Blacks in the Army. The First Army experience was a major factor in the Gillem Board's decision to recommend Army racial integration by the time of the next war (Lee, 1966; MacGregor, 1981).

The Gillem Board had been tasked in October 1945 by Army Chief of Staff General George Marshall

> to prepare a policy for the use of authorized Negro manpower potential during the postwar period including the complete development of the means required to derive the maximum efficiency from the full authorized manpower of the nation in the event of a national emergency. (quoted in MacGregor, 1981, p. 153)

This directive was similar to the order the Army chief of staff gave to the Army War College in 1924, indicating that the senior Army leadership was still searching for a personnel strategy for Blacks. The Gillem Board interviewed numerous people, studied for months, and concluded that racial segregation was inefficient and unnecessary and had to be abandoned. All of the uniformed people interviewed for the study, with three exceptions, called for continued segregation, but the vast majority of the social scientists who testified (though certainly not all) called for racial integration, although none offered hard scientific data, including psychological information. The historians and other civilian specialists argued that prejudice was the basis for racial segregation and that it was a terrible foundation on which to build a personnel policy.

The three officers who called for integration were Colonel Noel Parrish, White commander of Tuskegee Army Air Field and commander of the flying program for Black pilots during World War II; Brigadier General Benjamin O. Davis Sr., the first Black general in U.S. history; and Colonel Benjamin O. Davis Jr., commander of the Tuskegee Airmen units that fought in combat in World War II. General Dwight D. Eisenhower and all the senior Army Air Forces generals—despite the admirable record of the Tuskegee Airmen— recommended continued segregation, but the Gillem Board argued against that tide and proposed that any future war had to be fought with an integrated Army. I believe that the members of the Gillem Board were inclined from the outset to reform racial policies and therefore gave heavier weight to the historians and other civilian social scientists than to the uniformed military in writing their report. The Gillem report was published in an Army Circular, making clear the need for eventual racial integration (Gropman, 1998; MacGregor, 1981).

It is clear in reading the Gillem report that Lieutenant General Gillem and his partners focused heavily on the poll taken in 1945 after the 2,500 service specialist Blacks were given the opportunity to fight in the European theater alongside Whites. The results indicated to them that the overwhelming majority of Whites, whatever they may have believed in the past, accepted Blacks in their ranks (and foxholes) and believed that Black fighters were equal to White fighters. Racial integration, Gillem believed, would therefore promote itself. The Army did not trumpet the results of this poll for a number of reasons (the sample size was small, but residual race prejudice was surely the main motivation), but the Gillem board made this poll a major factor in its proposal to integrate racially by the time of the next war, even though the Army's leadership at large did not accept the Board's advice regarding integration (Gropman, 1998; MacGregor, 1981).

The Black troops who had fought in integrated units had served in the First Army under the command of Lieutenant General J. Lawton Collins Jr. The solid performance of the Black troops and the result of the poll convinced Collins that integration was a necessary reform. He did not have the power then to effect a reform of this magnitude, but he was chief of staff during the Korean War and was unique among four-star generals in the Army at that point: He believed that racial integration would make the Army more effective (MacGregor, 1981).

Korean War

The Korean War theater became a laboratory for analyzing race in the U.S. military. Almost from the start, in June 1950, when North Korea invaded South Korea, there were in the field all-Black and all-White units but also racially integrated organizations put together in the field and made up of remnants of broken platoons, companies, and regiments. When field commanders formed these units, they paid no attention to race, because the exigencies of battle would not permit them to do so. Commanders on the ground in Korea wanted to integrate racially early on, but General Douglas MacArthur would not consider it. After the Chinese came into the war that autumn, however, there were even more integrated units in the field.

Lieutenant General Mathew Ridgway was commander of the Eighth Army and had argued for racial integration when he held that important command. When Ridgway replaced MacArthur as theater commander, his desires governed, and because the chief of staff advocated integration, it was accomplished. Ridgway believed he could fight better integrated than segregated, and that was his motivation. Collins asked Ridgway to send him a message asking permission to integrate to make the resolution formal, and Collins then granted it (so much for Truman's executive order 3 years earlier!). By the time this occurred, the Air Force had completed the racial integration process—there were no longer any all-Black units.

The Army thus began racially integrating during the Korean War and, more significantly, in the combat zone. Many Army leaders in Korea believed their units would be more successful integrated than segregated and advocated integration. The rest of the Army followed suit, and by 1954 the entire Army was integrated (MacGregor, 1981). To gain support for this move with recalcitrant general officers, Collins and Ridgway called on Army researchers and outside research contractors to conduct psychological field surveys to ascertain the attitude of the troops toward racial integration. Ridgway and Collins thought they needed this ammunition, because in the spring of 1951 Lieutenant General Stephen Chamberlin and Vice Chief of Staff General Wade Haislip were vocally arguing against integration. In fact, Chamberlin chaired a board that asserted that racial integration was not in the interest of the Army or the country and that the Army was not ready for such a radical change. In May 1951, Haislip told Secretary of the Army Frank Pace Jr., that "no action should be taken which would lead to the immediate elimination of segregated units" (quoted in MacGregor, 1981, pp. 229–230). Haislip and Chamberlin were supported by Lieutenant General Edwin M. Almond, a combat commander in Korea, commander of the Black 92nd Division in Italy during World War II, and a lifelong segregationist who continued his antipathy to Black troops serving alongside Whites for the rest of his life, well into the 1970s.

To overcome such powerful objections, Project CLEAR (not an acronym, but a code name for the project) was created to survey attitudes of warriors in combat in Korea toward racial integration. The Army's Operations Research Office conducted a similar study at nearly the same time, both using surveying techniques and intense examination of segregated and integrated units in the field. Both of these studies were conducted after Ridgway had sought permission to integrate his forces and Collins had agreed, and both endorsed the similar findings of studies conducted in Europe in 1945: Soldiers who fought in integrated units favored integration. Collins used these studies to justify his decision to integrate racially (MacGregor, 1981).

The Marine Corps, which also had integrated units fighting well on the battlefield in Korea, began to integrate in 1951 and completed the process with equal celerity (MacGregor, 1981). The Navy began racial integration at about the same time but kept the stewards corps all Black until the end of the decade (MacGregor, 1981).

CONCLUSION

The path to racial integration (and beyond it to equal opportunity) was not always smooth. During World War II and afterward, until the Air Force integrated in 1949 and the other services in the early 1950s, there were numerous race riots in military settings. Subsequently, for almost 20 years, until the late 1960s, racial altercations ceased. The fact of integration had brought racial internal peace. But because residual bias brought unequal opportunity, racial friction and sometimes riots occurred in all services in the late 1960s and 1970s. The armed services recognized the necessity of moving beyond integration to equal opportunity and implemented programs and regulations that saw to it that people of any race could self-actualize.

The military racially integrated rapidly and completely in the late 1940s and 1950s, because the core value of the military is cohesion. No military unit can succeed in combat divided—without unity, a military organization will fail in battle. This was the cardinal principle that ensured, in time, the completeness of military racial integration. Equal opportunity, nondiscrimination policies, and a welcoming attitude toward Black soldiers, however, are fairly recent social phenomena. These efforts occurred after the race riots among service members during the Vietnam War, and policies underscoring them followed the 1971 race riot at Travis Air Force Base, California (Gropman, 1998).

Expediency has governed racial practices: When defense needs for manpower were urgent, the United States accepted—and even recruited—Blacks, and when there were no dire needs, Blacks were excluded. Whites currently do not volunteer to serve in proportion to their percentage of the population, and to fill the ranks, Blacks are enthusiastically recruited, eagerly enlisted, and ardently reenlisted. At the time this chapter was written, the Army enlisted force was about 30% Black, whereas Blacks at military age formed about 13% of the U.S. population. Black soldiers have a much higher reenlistment rate than White soldiers; until recently, about 20% of the enlisted recruits in the all volunteer Army, for example, were Black, but because the Black reenlistment rate has been 150% of the White rate, the total Black component is significantly higher than the enlistment rate. Because of the high quality of Black career enlisted personnel, about 39% of the highest ranking enlisted personnel are Black. Every combat military service has a higher percentage of Blacks in its force than the Black portion of the population.

To arrive at this state, the armed forces have had to provide an equal opportunity environment and a nondiscrimination atmosphere. All the services currently perform social science research on racial issues and intelligently and sophisticatedly poll those who enlist, those who reenlist, and those who leave the military. The Department of Defense provides quality education and superior research on equal opportunity practices and nondiscrimination activities at the Defense Equal Opportunity Management Institute at Patrick Air Force Base, Florida. This chapter, however, has described how the services approached and accomplished racial integration (or desegregation) 50 years ago, and not the military today, and at that time social science research played almost no role (Moskos & Butler, 1996).

Race riots have been absent from the military for more than 30 years. This is not to say that racial friction does not occur but rather that there is a proven apparatus for dealing with it. The foundation of that apparatus is social science, especially psychology and sociology, and it was created in the early 1970s and has evolved considerably. During the 1940s and early 1950s, when the services racially integrated, military leaders rarely used the social sciences, and in the few cases they did they used the research to ratify decisions already taken. Racial friction is a luxury today's military cannot afford, and the armed services of the 21st century depend heavily on sociology and psychology to ensure racial harmony. But that is another story.

15

PSYCHOLOGICAL RESEARCH WITH MILITARY WOMEN

JANICE D. YODER AND LOREN NAIDOO

The needs and resources of the U.S. military have helped shape and have been shaped by the contributions of psychologists, and both the military and psychology have operated in the broader context of changing societal trends (e.g., Driskell & Olmstead, 1989). Nowhere is this clearer than in research with women and on gender (see, e.g., Goldstein, 2001). Women's roles in the U.S. military have changed remarkably, from formal exclusion to a position at the center of many contemporary military, and more sweepingly social, controversies (D. Simpson, 1998). This chapter briefly describes how women's participation in the military has changed and considers how the military, psychology, and current societal issues central to women's lives and experiences are interconnected. Specifically, we will explore how military policy and psychological research inform and are in turn informed by societywide debates about women in nontraditional roles, work and family issues, gender integration of the workplace, gender discrimination, and women's health care.

This chapter draws heavily on the entry "Military Women," by J. D. Yoder, in *Encyclopedia of Women and Gender* (Vol. 2, pp. 771–782) by J. Worell (Ed.), 2001, New York: Academic Press.

NATIONAL SECURITY AND THE ROLES OF WOMEN

Women's contributions to national security began informally with nursing in 1775 in the American War for Independence (Hoiberg, 1991; Sherrow, 1996) and were formalized in World War I with the creation of the Army (1901) and Navy (1908) Nurse Corps. Beyond nursing, historians have documented the more informal participation of women in active combat, in artillery units as disguised enlisted "men," in militia units, in frontier warfare during the early years of U.S. history, as spies and scouts during the Civil War (Hoiberg, 1991), and as pilots in World War II (Holm, 1992). Dienstfrey (1988) estimated that 7.4% of women veterans from World War II and the Korean and Vietnam conflicts were exposed to combat; 73.5% of these were nurses.

A watershed for women's expanding participation in the military was the implementation of the all-volunteer force in 1973 at the end of the Vietnam conflict (Hoiberg, 1991). In 1967, the 2% ceiling on women's enlistment was lifted, and in 1974, all occupational specialties were opened to women except those directly related to combat. Beginning in 1975, pregnant women had the option to remain on active duty during and after a pregnancy. The service academies were opened to women in 1976; the women's corps was integrated into the regular organizations with men in 1978; and effective in 1980, officer promotion lists were integrated. A fivefold increase in the number of active duty women occurred between 1973 and the end of the 1980s, expanding women's participation to 10.8% of the military and ranking the United States first in the world for its representation of women. In 1990, intensive news reporting gave Americans their first highly visible glimpse of women being deployed to a war zone in the Persian Gulf. These women made up 7.2% of personnel deployed to the Gulf (40,782 women) and were shown performing a wide array of military jobs, carrying arms, and being exposed to the dangers of combat (Holm, 1992). Thirteen women died, and 2 became prisoners of war (Manning & Wright, 2000).

These events renewed debate about the combat exclusion policy (V. L. Friedl, 1996; Sherrow, 1996). In 1988, the Department of Defense's Risk Rule was revamped with the goals of narrowing and standardizing each service's interpretation of combat. In 1992, service in combat aircraft was opened to women. In 1993, the Navy developed a legislative proposal to permit the assignment of women to combatant ships, and in 1994, the USS *Eisenhower* became the first combatant ship to carry an integrated crew (Thompson, 2000). Most important, in 1994, the Risk Rule was rescinded and was replaced by a directive from the U.S. Secretary of Defense that excluded only assignments to below the administratively broad brigade level with the primary mission of ground combat. Other permissible exclusions involved units and positions required to physically collocate with direct ground combat units, prohibitive costs of providing living space for women, units engaged in special operations missions, or job-related physical requirements that exclude the vast majority of women. These changes opened up 32,699 new positions to women, mostly in the Marine Corps (V. L. Friedl, 1996). All units and positions in the Army became accessible to women except direct ground support and support units physically collocated with them. Overall, women were eligible to serve in more than 80% of military jobs (General Accounting Office [GAO], 1999a). In sum, the United States inched toward elimination of all combat exclusion, coming closer to joining the ranks of countries without such restrictions (Canada, Denmark, Norway, and Belgium).

As of January 31, 2000, women composed 14.4% of the U.S. military. Women served as 14.2% of all officers and 14.4% of all enlisted personnel, but they were severely underrepresented in the Marine Corps (5.9%; Department of Defense, 2000). Table 15.1 provides the numbers and percentages of officers and enlisted personnel who were women and minority women in January 2000. Slightly over half of military women were White (53.7%), 31.9% were African American, and 7.6% were Hispanic. The United States retained its top ranking with the highest female representation in the world; followed by Canada and Israel, each with 11%; and by the United Kingdom, with 6% (B. Mitchell, 1998).

CONTRIBUTIONS AND LESSONS FROM THE BEHAVIORAL SCIENCES

Driskell and Olmstead (1989) concluded that throughout psychology's history, the U.S. military has served as an applied testing ground for psychological processes and, in turn, has been the impetus for research and innovation in the behavioral sciences. A parallel and reciprocal relationship can be found between the U.S. military and the contemporary women's movement (see V. Taylor & Whittier, 1993). Women's inclusion in the all-volunteer force of the 1970s was at the forefront of women's expanding labor force participation during that same period (Reskin & Padavic, 1994). The gradual whittling away of restrictions on women's participation in combat both reflected and reinforced women's overall penetration into male-dominated occupations (Reskin & Roos, 1990). The Tailhook convention incident in 1991 shared center stage with the Clarence Thomas Supreme Court confirmation hearings in the public debate about defining and redressing sexual harassment (Lancaster, 1999). Since the 1970s, controversies surrounding pregnancy and the employment of parents, especially mothers; homosexuality and the workplace; glass ceilings and promotions; and physical abilities testing were all put to the test in the military with input from psychologists. In the remainder of this chapter, we will explore popular attitudes about military women, work and family issues, gender integration of male-dominated occupations, workplace discrimination, and health care.

TABLE 15.1
Active Duty Military Personnel as of January 31, 2000

Personnel category	Total	Army	Navy	Marine Corps	Air Force
Total personnel (n)	1,355,523	468,718	363,342	171,612	351,851
Women (n)	194,605	70,269	48,989	10,167	65,180
Women (%)	14.4	15.0	13.5	5.9	18.5
Minority women (%)	46.3	57.5	45.2	43.1	35.6
Total officers (n)	216,533	76,928	52,993	17,879	68,733
Women (n)	30,656	10,502	7,647	905	11,602
Women (%)	14.2	13.7	14.4	5.1	16.9
Minority women (%)	25.9	33.4	22.3	24.3	21.8
Total enlisted (n)	1,138,990	391,790	310,349	153,733	283,118
Women (n)	163,949	59,767	41,342	9,262	53,578
Women (%)	14.4	15.3	13.3	6.0	18.9
Minority women (%)	50.1	61.7	49.4	44.9	38.6

Note. Data provided by the Office of the Assistant Secretary of Defense (Department of Defense, 2000).

Attitudes About Military Women

According to surveys by Hurrell and Lukens (1994) and Wilcox (1992), although the majority of the general public has supported all roles except ground combat (35%) for military women, patterns within roles closely paralleled gender stereotypes. Public support was near universal for military women in traditional roles: typist (97% of respondents supported it) and combat nurse (93%). Support dropped somewhat for women in nontraditional roles (mechanic, 83%, and air transport, 73%) and declined even more for high-prestige (base commander, 58%) and combat roles (fighter pilot, 62%; missile gunner, 59%; fighting ship, 57%; ground combat, 35%). Most Americans felt that the inclusion of women had raised the effectiveness of the military (22.4%) or made no difference (68.7%; Wilcox, 1992). Indeed, the majority of military women and men attested to their personal readiness, physical and training preparedness, and willingness to deploy to a war zone (GAO, 1999a).

Thus, attitudes about military women reflect patterns common to gender stereotypes in that the more nontraditional the role, the higher the level of dissent (Kite, 2001). However, like gender attitudes in general (e.g., see Misra & Panigrahi, 1995), overall support for military women engaged in a wide array of roles is remarkably favorable.

Work and Family Issues

Much of psychological research focused on work and family issues for military women has concentrated on the unique issue of access of married women and mothers to participation in military service. The Army Nurse Corps in 1901 expressly prohibited nurses from marrying or being mothers

(V. L. Friedl, 1996); some barriers to marriage fell during World War II, and marriage became a basis for voluntary discharge through the late 1970s (P. J. Thomas & Thomas, 1993). Today, marriage is not a legitimate basis for separation in any of the services. In 1973, only 18% of enlisted women were married (compared with 52% of enlisted men), but this figure jumped to 47% by 1991 (56% for men), virtually closing the gendered marriage gap (Office of the Assistant Secretary of Defense, cited in P. J. Thomas & Thomas, 1993).

Questions raised about the compatibility of military service with pregnancy and motherhood have been more difficult to resolve. Lawsuits through the early 1970s ultimately led to a 1975 Department of Defense directive discontinuing involuntary discharges and instituting an optional separation policy initiated by the woman (P. J. Thomas & Thomas, 1993). Since then, military policy has fluctuated between guaranteed separation on request and separation only if conditions warrant and obligations are met. At present, both the Army and Air Force give discharge authority to the installation where the solider is assigned. All services involuntarily terminate women found to be pregnant during basic training.

Although mothers represent about 5% of the total active duty force, they encompass a significant percentage of military women (P. J. Thomas & Thomas, 1993). Large-scale surveys of Navy personnel in 1993 revealed that 34% of Navy women were mothers whose children lived with them (mean number of children = 1.6; P. J. Thomas & Thomas, 1993). These mothers were about equally divided among the categories of single (32%), married to a civilian spouse (32%), and married to a military spouse (37%), in contrast to military fathers, who largely were married to a civilian spouse (91%)— only 6% were single, and only 3% were in a dual military marriage. These patterns make dual military assignments and parental arrangements important issues for a disproportionate number of military women and a proportionally small but numerically sizable group of men.

Women-friendly policies and provisions are developing slowly and being expanded to encompass men. A landmark 1973 Supreme Court decision decreed that military women were entitled to the same benefits for their dependents as military men (P. J. Thomas & Thomas, 1993), and most bases and posts include family-friendly support services. Each of the services requires single and dual-military parents to file a dependent care certificate that outlines plans for the care of children in case of parental deployment. However, more mothers than fathers are separated from the military for reasons related to dependents. For example, a review of all 1990 separations from the Navy showed that 0.91% of mothers and 0.05% of fathers were discharged for parenthood (P. J. Thomas & Thomas, 1992, cited in P. J. Thomas & Thomas, 1993).

Two outcomes often cited as related to work–family conflict for women in the gender literature, absenteeism and retention, have been explored in military settings. A 1994 study of 2,285 enlisted women and 3,104 enlisted men on active duty in the Navy concluded that the amount of time lost from the job did not differ for women and men, even when pregnancy and postpartum convalescence leave were included (P. J. Thomas & Thomas, 1994). Retention of women depends heavily on perceived familial compatibility and support. Pierce's (1998) study of active duty Air Force and guard or reserve women during the Persian Gulf War surprisingly found that personnel who stayed in the military 2 years later were more likely to have been activated, to have more dependent children, and to report greater familial disruption than those who left. Attrition rates were above the norm among deployed women who left their children in the care of ex-spouses (39%), among women who had a baby during or after the war (31%), and among women with an active duty partner (22%).

In conclusion, unlike civilian employers, the military has considered the marital, pregnancy, and parental status of women in recruitment and retention. Also, unlike much of the civilian sector, the military itself customarily provides some family-friendly facilities. Like the civilian workplace, issues of relocation, dual-career accommodations, and child care arrangements (although these needs can be more pressing in the military with the prospect of deployment) remain unresolved and do affect women's participation.

Gender Integration of Male-Dominated Occupations

Although the proportion of women in both the civilian and military workforces has grown remarkably during the past 30 years, it is the military that has led the way in the genuine integration of women across jobs. For example, Waite and Berryman's (1986) study using data from the National Longitudinal Survey of Youth Labor Market Behavior found that 34% of women in the military, but only 3% of women in the civilian sector, held jobs in which 90% or more of civilian workers were male. Conversely, 28% of civilian and only 3% of military workers worked in jobs that in the civilian sector were filled by 90% or more women. In other words, military women were more likely to do jobs stereotypically regarded as masculine and less likely to do feminine jobs than civilian women.

According to a GAO (1999b) report, enlisted women in 1998 served predominantly in functional support and administration (33%), including personnel, recruiting and counseling, law, supply administration, auditing and accounting, and general administration. The second largest concentration of enlisted women was in medical and dental specialties (17%). Over 40% of women officers worked in health care occupations, with administration (12%) taking second place. Comparing job patterns in 1990 to those in 1998, the GAO concluded that although a large percentage of women continued to cluster into more traditional jobs, women were making substantial inroads into more nontraditional fields such as aviation, surface warfare, air traffic control, and field artillery.

The GAO (1999b) cited two institutional barriers that inhibited the full integration of military women. The combat exclusion policy continues to bar women from some units, even though some jobs contained within them are open to women. For example, the Navy limits the number of women who can pursue medical corps training, because the Navy supplies these personnel to Marine Corps units that exclude women. Second, use of the Armed Services Vocational Aptitude Battery, recently revised (Department of Defense, 2004), to determine assignments for new recruits contains sections that measure exposure rather than aptitude and thus may disadvantage women with limited male-traditional experience.

Despite an abundance of opportunities to study jobs that have been newly opened to women, the process of gender integration in military settings has been woefully understudied, except at the military academies. Research by Stiehm (1981) and DeFleur (1985) at the U.S. Air Force Academy, Stevens and Gardner (1987) at the Coast Guard Academy, and Adams and colleagues (e.g., Yoder, Adams, & Prince, 1983) and the GAO (1994) at the U.S. Military Academy at West Point, New York (Army), found patterns that fit well with Kanter's (1977a, 1977b) seminal descriptions of proportional underrepresentation (*tokenism*) and its consequences. The military academies were first opened to women in 1976, and since then, each has matriculated 10% to 15% women in subsequent classes.

Across these studies, there was no clear evidence of lowered performance in women. Rather, clear and consistent patterns surfaced that painted markedly different pictures of the experiences of women and men who were coexisting in the same setting. Unlike men, accounts of women's experiences included intense and undiminished media attention, men's fears that women would benefit from preferential treatment, stresses and performance pressures, social isolation, and feelings of less peer acceptance (Yoder et al., 1983). Women have struggled with overprotection and marginalization, and they have grappled with the disjunction between gendered expectations and the masculinized demands of their work, such that women's leadership abilities have been evaluated less favorably by others. In a more recent study of West Point cadets (GAO, 1994), women were selected for top leadership positions at lower rates than men in seven of eight semesters. Across all studies, women's attrition rates were significantly higher than those for men. These findings of heightened visibility, marginalization, and role deviance are consistent with what has been documented across a wide range of male-dominated civilian occupations by tokenism researchers (for a review, see Yoder, 2002).

Workplace Discrimination

As in the civilian sector, gender discrimination is exceedingly difficult to document, appearing more readily in the aggregate than at an individual level (Rutte, Diekmann, Polzer, Crosby, & Messick, 1994). Three areas of potential gender discrimination toward military women have been explored: promotions, sexual harassment, and the blatant exclusion of lesbians.

Promotion Perceptions and Patterns

In a review of research on service members' perceptions of gender inequities, the GAO (1998c) concluded that some inequities were believed to exist, especially regarding local assignment policies and practices established by unit commanders. Aggregate data on promotion patterns for 1993 through 1997 did not substantiate these perceptions at the broader level of the military as a whole, but some exceptions were evident within branches (GAO, 1998a).

Across the services, promotion of officers was conducted by centralized boards, and enlisted personnel were advanced through examination or board recommendations (GAO, 1998c). Rates for women's and men's promotion were similar in 82% of the boards and examinations reviewed, with 15% of the remainder favoring women and 3% preferring men. Only the Army exhibited more significant differences advantaging men. Looking at professional military education selection, routinely conducted by boards, comparable rates occurred 46% of the time, with 29% of the remainder favoring women and 25% preferring men. Within services, the Army and Navy tended to favor men; the Air Force and Marines, women. For key military assignments, the Marine Corps and Navy relied on centralized boards and the Army and Air Force on decentralized boards. Across all services, 53% of selections showed similar rates for women and men, with 15% preferring women, and 32% favoring men. The Air Force and Navy had more significant differences that advantaged men, the Army slightly favored women, and no significant differences in assignment rates were found in the Marine Corps. In sum, the aggregated promotion data militarywide did not document clear patterns and practices of gender discrimination, at least within broad categories of promotion.

More consistent suggestions of gender inequities emerged from analyses of the promotion materials submitted on behalf of women and men. A series of studies conducted by the Department of Defense throughout the 1980s and 1990s questioned the gender neutrality of performance appraisals (GAO, 1995). For example, analyses conducted by the Navy Personnel Research and Development Center (NPRDC) in 1994 found that women's fitness reports contained significantly more references to personality traits and that women were devalued as leaders (cited in GAO, 1995). Two 1983 NPRDC experiments presented raters with masculinized and feminized narratives and found that promotion choices favored men and masculine evaluations of women. These findings are consistent with research on hiring biases for civilian workers (e.g., Glick, Zion, & Nelson, 1988). Most troubling was an exploration by the Army Research Institute (cited in GAO, 1995) of the reasons why promotable women left the Army, including (beyond career opportunities, family issues, and monetary issues) equitable treatment and equal opportunity issues, gender-based discrimination, and sexual harassment and policy.

Paralleling the civilian work force, women's perceptions of gender inequities can be contradicted by using broad clusters of job categories and by avoiding the top tiers of the hierarchy (Federal Glass Ceiling Commission, 1995; Thompson, 2000). More subtle indicators of biases in promotion, as found in informal performance appraisals and exit interviews, prove more consistent with women's perceptions, suggesting that gender discrimination remains but at a more elusive, covert level (Benokraitis, 1997).

Sexual Harassment

Department of Defense policies prohibiting discrimination first appeared in the 1940s, but sex discrimination did not make its way into these policies until 1970 and sexual harassment until 1979 (Bastian, Lancaster, & Reyst, 1996; Hay & Elig, 1999). The first militarywide sexual harassment survey was conducted in 1988 (Hay & Elig, 1999), modeling surveys of workers for the federal government. Targets of reported sexual harassment were found to be more widespread among military (64%) than civilian federally employed (42%) women.

Publicity surrounding the 1991 Tailhook incident (Lancaster, 1999) sparked a 1995 survey of 22,372 women and 5,924 men (Bastian et al., 1996). About 69% of respondents were enlisted personnel, 29% were commissioned officers, and 2% were warrant officers, proportionally representing the four service branches and the Coast Guard (Hay & Elig, 1999). The survey contained three parts: (a) a replication of the 1988 behavioral indexes of sexual harassment, (b) a modified version of the Sexual Experiences Questionnaire (SEQ; Fitzgerald, Magley, Drasgow, & Waldo, 1999), and (c) potential correlates and detailed questions probing an incident chosen by the respondent to represent the "situation that had the greatest effect" on him or her (Hay & Elig, 1999).

Direct comparisons of responses from 1988 to 1995 suggested declines in overall prevalence rates, from 64% to 55% of women and from 17% to 14% of men (Lancaster, 1999). However, the more comprehensive and behavior-based measures of sexual harassment using the SEQ revealed that 64% of men but only 24% of women reported no experiences with sexual harassment during the past 12 months of their military service (Fitzgerald, Magley, et al., 1999). Almost all incidents of sexual harassment reported by men involved gender harassment in which men felt treated unfavorably because of their sex (Fitzgerald, Magley, et al., 1999). Gender harassment also dominated women's responses but was combined with unwanted sexual attention for one of every four women. Sexual coercion, whereby sexual favors were demanded, affected more women than men and always in combination with the other types of harassment for both sexes. Independent research in 1991 with women cadets at the military academies estimated almost universal prevalence rates of 93% to 97% (GAO, 1994), as did another survey of combat support and combat service support units (L. N. Rosen & Martin, 1998).

Repeating a pattern often found with civilian respondents (Saunders, cited in Koss et al., 1994), sexually harassing behaviors were not always acknowledged by their targets as harassing. Of those military personnel who checked at least one item of the SEQ and thus indicated that behaviorally they had been the target of sexual harassment, 33% of the women and 76% of the men reported that they had not been sexually harassed (Bastian et al., 1996; see also L. N. Rosen & Martin, 1998).

The 1995 survey included questions hypothesized to correlate with experiences of sexual harassment. Harassment was related to negative job attitudes, jeopardized psychological well-being, and reduced health satisfaction, even after controlling for job effects (Fitzgerald, Drasgow, & Magley, 1999). Negative psychosocial reactions to harassment included productivity problems, unfavorable attitudes toward the military, emotional distress, and disrupted relations with family. Harassment was found to occur less frequently in gender-balanced work groups and in settings where personnel believed that the organization's upper echelons would enforce sexual harassment policies. Of the military's efforts to implement practices to respond effectively to harassment, provide resources like counseling for targets, and train its personnel about harassment, only perceived implementation was related to reduced incidence of harassment, especially for women. Additionally, a lack of perceived organizational implementation directly contributed to negative job-related outcomes for targets beyond the effects of experiencing the harassment itself (J. H. Williams, Fitzgerald, & Drasgow, 1999). Specifically, targets who felt that the military did little to respond to charges of harassment suffered reduced commitment to the military, work dissatisfaction, and dissatisfaction with supervisors.

Discrimination Against Lesbians

The military's formal policy surrounding gays in the military was rooted in the 1919 Articles of War, which banned sodomy (Herek, 1993, 1996). Policy later shifted from loosened screening during World War II and local discretion in the 1970s to a uniform policy in 1982 stating that "homosexuality is incompatible with military service" (Department of Defense Directive 1332.14, 1982, cited in Herek, 1996, p. 6).

During the 1980s, the military discharged 16,919 women and men under the separation category of homosexuality (Herek, 1996; see also V. L. Friedl, 1996). Although White women composed only 6% of military personnel, they accounted for 20% of those discharged for homosexuality. The Navy was the most active, accounting for only 27% of the active force but 51% of women and men discharged for homosexuality. Discharge rates for purported lesbians were out of proportion in the Marine Corps, where 28% of discharges for homosexuality involved women, who composed just 5% of the corps. In 1993, the current "don't ask, don't tell" policy (Herek, 1996; Ransom, 2001) differentiated between homosexual acts and an "abstract desire" or orientation toward homosexuality, removed questions about sexual orientation from enlistment, allowed association with same-sex gays and lesbians so long as the individual did not share their propensity, and reinstated discretion for commanding officers (P. J. Thomas & Thomas, 1996). Separations continued to disproportionately affect women; although women represented only 10% of military personnel in 1992, they accounted for 23% of discharges for homosexuality (P. J. Thomas & Thomas, 1996).

On March 24, 2000, the Department of Defense Inspector General's office released a survey of 71,000 service members that put these issues back in the public spotlight (Garamone, 2000). More than 80% of those surveyed reported that they had heard derogatory names, jokes, or remarks regarding homosexuals in the past year. Fully 85% believed that other service members as well as military leaders tolerated such offensive language. Furthermore, 37% said that they had personally experienced or witnessed homophobic harassment, most frequently in the form of offensive speech (88%) and less frequently as hostile gestures (35%), threats or intimidation (20%), graffiti (15%), vandalism of a service member's property (8%), physical assault (9%), limitation or denial of training or career opportunities (9%), and disciplinary actions or punishments (10%). Coworkers were identified as the most frequent sources of harassment (by 61% of those reporting exposure), followed by immediate supervisors (11%). Finally, 5% of respondents felt that the chain of command tolerated such overt harassment.

Women's Health Care

Studies of physical health care utilization with civilian samples consistently have found that women's usage exceeds men's, and the same pattern has been confirmed in the military (Woods, 1995). For example, in a sample of women on active duty or as active members of the guard or reserve forces during the Persian Gulf War, 76% used military health care services for the treatment of gender-specific health problems (Pierce, Antonakos, & Deroba, 1999). Similarly, an analysis of 20 ships' sick-call logs during June 1989 revealed that the age-adjusted visit ratio for women to men was 1.44:1. Excluding female-specific visits, the ratio remained unbalanced at 1.21:1 (Nice & Hilton, 1994), with gender differences for illness but not for injury or general health services. Looking at Naval women's data, significantly higher sick-call visits occurred for women in nontraditional as compared with traditional occupations (Nice & Hilton, 1994).

Uniform measures of both job-specific and general fitness are needed to counter gender inequities (GAO, 1998b). In contrast to male standards, which routinely were calibrated with actual performance data, women's criteria were often subjectively estimated, extrapolated from male standards, or based on command judgment. Body fat measures, which ignored racial differences in bone density, have been criticized for overstating the body fat of minority personnel. Furthermore, a National Academy of Sciences (1998) report suggested that setting a high body fat limit for women selects strong women who lack endurance; alternatively, establishing a low body fat standard values endurance over strength. A study of cadets at the military academies concluded that women's fitness performances from 1977 to 1987 showed greater improvement over time both for individual women compared with individual men and across later versus earlier cohorts (Baldi, 1991).

Turning to stress and substance use, a 1995 Department of Defense survey of health behaviors of more than 16,000 military personnel disclosed that military women (5.3%) reported drinking less heavily in the past month than military men (18.8%) but had similar rates of illicit drug use in the past year (5.3% for women; 6.7% for men) and cigarette smoking in the past month (26.3% for women; 32.7% for men; R. M. Bray, Fairbank, & Marsden, 1999). The amount of stress associated with being a woman in the military was predictive of both illicit drug and cigarette use. Unlike men, stress at work or in the family was not associated with substance use among women.

CONCLUSION

Research with women in the military necessarily reflects the uniqueness of the military context, yet many of the general patterns we have explored in this chapter are consistent with, and often extend, research conducted with employed women in civilian settings. Public attitudes about the work roles open to military women are congruent with general popular support for women in a myriad of male-dominated jobs and occupations. Work and family issues continue to challenge civilian and military women, men, and their employers, raising concerns about relocation, dual-career accommodations, and child care arrangements. The military, which has a long, accepted, and indeed expected history of providing benefits for the dependents of its employees, is poised to make a real difference for working mothers and fathers by expanding family-supportive facilities and policies.

Although the civilian workplace remains largely gender segregated, there are few remaining barriers to women's pioneering inclusion in jobs and occupations in the military. As combat restrictions continue to fall and jobs held exclusively by men open up to women, the possibilities for social scientific research are invaluable to military and gender psychologists alike. It is incumbent on these researchers to regard gender integration as more than simple access and to study the full, day-to-day processes of integration. The military also offers a lucrative setting for piloting interventions to facilitate women's genuine integration.

Questions have been raised in the military about the existence of gender discrimination involving promotional glass ceilings and concrete walls, sexual harassment, and the exclusion of gay men and lesbians. Nowhere is the productivity of cooperation between military and gender psychologists clearer than with their joint efforts to study sexual harassment. Finally, the involvement of mental and physical health psychologists could contribute more to the understanding of health utilization, physical training, and substance use.

Throughout this chapter, we described ways in which both psychology and the military have been responsive to changing societal images of women and women's roles in the workplace. At times, each also has been at the forefront of these societal changes—for example, by leading the way toward integrating women into formerly male-dominated occupations. The most successful integration of psychology with the military may be in extending the understanding of sexual harassment. A richer understanding of the lives, goals, needs, and successes of military women would be enhanced by further collaboration along these lines.

16

MILITARY FAMILY RESEARCH

MADY WECHSLER SEGAL

Research on military families has increased rapidly in the past 20 years as methodologies have become increasingly sophisticated and knowledge has accumulated. The research area is multidisciplinary, with contributions from psychologists, sociologists, social workers, psychiatrists, anthropologists, political scientists, economists, and historians. The armed forces themselves have supported much of the research, both via the activities of their uniformed and civilian personnel and through grants and contracts to civilian researchers. This research has led to a firmer understanding of the military family lifestyle, with its benefits and special challenges, and of the ways families adapt to military life.

The armed forces began to pay attention to family needs in the 1960s and established organizations, such as Army Community Service, that provide various services to help families adapt to the demands of military life, including information and referral services, lending closets, and family counseling. The recognition by policymakers of the difficulties military families face has fostered the development of programs and services for families.

The writing of this chapter was supported by the U.S. Army Research Institute for the Behavioral and Social Sciences under contracts DASW 0100K0016 and W74V8H-05-K-0007. The views expressed in this chapter are those of the author and not necessarily of the Army Research Institute, the Department of the Army, or the Department of Defense.

Research on military families has been spurred by concern for families, but a perhaps even greater impetus was the recognition and empirical demonstration of the relationship between family adaptation and both military readiness and retention of service members. Indeed, advocates for families sought to study these relationships precisely to provide proof to the military of the benefit that would accrue to the organization from helping service members meet their family needs. Such research was a major focus of the Army Family Research Program supported by the Army Research Institute for the Behavioral and Social Sciences in the 1980s and 1990s. The research has shown that service members' and spouses' perceptions of how much their leaders care about families are positively related to service members' satisfaction with, and commitment to, the military (e.g., Bourg & Segal, 1999; M. W. Segal & Harris, 1993). Much research shows that the more supportive a military spouse is of the service member's remaining in the service, the higher the retention intentions and actual retention behavior of the service member (e.g., Etheridge, 1989; Griffith, Rakoff, & Helms, 1992; Orthner, 1990). Research has even found that when unmarried soldiers who are in strong relationships perceive that their partners support their making a career in the Army, the soldiers' intention to remain in the service is influenced positively (M. W. Segal & Harris, 1993, p. 17).

Research also has shown that family issues affect the readiness of service members, both as individuals and military units (for a review of some this research, see M. W. Segal & Harris, 1993). For example, supportive relationships between soldiers and their spouses are a factor in soldiers' presence for duty and deployments and contribute to their desire to do their best work (Kirkland & Katz, 1989). Deployed soldiers who are worried about their families do not perform up to their capability and are serious threats to unit performance and safety. Both individual and unit readiness are positively affected by the perception that supervisors show support for families (M. W. Segal & Harris, 1993). Unit readiness is also positively affected when a family support group is available, activities for families are provided, and supervisors allow time off work to attend to family matters (M. W. Segal & Harris, 1993).

MILITARY FAMILY DEMOGRAPHY

One of the major reasons for greater attention to military families by both policymakers and researchers is the rise in the proportion of service members who are married and increases in the numbers of military family members. For most of U.S. history, the military forces consisted primarily of young single men. The few women in the military were not allowed to remain in service with children and, at times, were not even allowed to be married. Men needed their commanding officer's permission to get married.

Over the past 40 years, the proportion of military personnel who are married has risen (D. R. Segal et al., 1976). This increase has resulted in large part from the increased emphasis on retaining trained and experienced personnel and attempts to prevent great turnover of personnel. The movement away from conscription in 1973 intensified the need to attract and retain qualified personnel. From 1990 to 2001, there was a slight increase in the percentage of military personnel who were single without children, but there was also an increase in the percentage of single parents (for a detailed analysis of the demographics of U.S. military families, see D. R. Segal & Segal, 2004, from which all of the information in this section is adapted).

More than half of today's military members are married (about 51%). Of those who are married, 73% have children (Military Family Resource Center, 2001). As of 2001, there were 1,369,167 active duty service members and 1,882,081 family members (Military Family Resource Center, 2001). Family members include spouses, children, and adult dependents (such as siblings or parents). There are more than 77,000 dual military couples (i.e., who are both service members), and 43.8% of them have children. The number of single parents has risen to more than 64,000 single fathers and more than 23,000 single mothers (although not all of these single parents have their children living with them; Military Family Resource Center, 2001). Despite the fact that military women (11.3%) are more likely to be single parents than are men (5.5%), the much larger numbers of men than women in the armed forces means that there are more single fathers than single mothers.

Within enlisted and officer categories, higher ranking personnel are more likely to be married, because rank and age are closely related. Only lower ranking enlisted personnel are more likely to be single than married. Enlisted personnel are less likely to be married than officers, reflecting their younger ages. In each of the services, men are more likely to be married than women of the same rank. Large gender differences among senior enlisted personnel and officers reflect the difficulty for women of balancing work and family life in the military, especially for those with children. Among married military women, substantial proportions are married to military men.

Although higher ranking (and therefore older) personnel are more likely to be married, there is a special concern about the high rate of marriage among junior enlisted personnel compared with their civilian counterparts. Junior enlisted wages and benefits are not designed to support a family, and the news media and Congress have cited the eligibility of some enlisted families for civilian welfare benefits as a disgrace. Enlisted men in the four lowest pay grades are almost twice as likely to be married (25.3%) as 18- to 24-year-old civilian male high school graduates (13.1%). Although the ratio is smaller for women, junior enlisted women are also more likely to be married (30.7%) than their civilian peers (24.0%).

There are some differences among military personnel in marital status as a function of race and ethnicity. Similar to the situation in civilian families, Black military women are less likely (35.2%) than other women to be single and childless and more likely to be single parents (24.4%). Black women are also more likely to be in dual military marriages with children (13.6%) and less likely to be in dual military marriages without children (7.1%); this difference may at least partly reflect the higher ranks of Black women, who remain in service longer than White women. Like Black women, Black military men are more likely (36.4%) than other men to be single and childless. They are also more likely (9.1%) than Asian and Pacific Islander, White, and Hispanic men to be single parents; men in other racial and ethnic groups share this higher rate of sole parenthood.

MILITARY FAMILY LIFESTYLE

The military and the family have been described as "greedy institutions" because of the extent of the time, energy, and identity commitment they require of individuals (Coser, 1974; M. W. Segal, 1986b, 1989). Military family life involves the intersection of these two greedy institutions, with both positive and negative consequences for the way the military encompasses the service member and his or her family. Among the beneficial aspects of the lifestyle are the financial and social service benefits (e.g., housing, medical care, post exchanges, commissaries, recreational facilities) and the social benefits of sense of community. The armed forces also provide formal services to help families cope with the challenges of military family life.

Negative aspects of the military lifestyle include risk of injury or death, long duty hours and shift work, frequent moves, postings in foreign countries, separations and reunions, normative constraints on family members, the hierarchical nature of the organization, the isolated nature of the locations of many military bases, and a masculine culture that constrains valuing family (M. W. Segal, 1986a, 1999). Some of these negative aspects also have some positive effects, as will be described in the following sections covering some of the major characteristics of the military family lifestyle.

Family Separations and Reunions

Service members often have to leave their families when they go for training or sea duty or are deployed for missions. In addition, families sometimes choose to be separated rather than to move. Data on the frequency of family separations show that even before deployments to Afghanistan and Iraq, separations were common. A 1999 survey showed that 73% of active duty service members (81% in the Army) surveyed had been away from their permanent duty station at least one night because of military duties during the previous 12 months, 39% had been away at least five separate times, and 35% had been away for at least 3 months (Gaines, Deak, Helba, & Wright, 2000).

Separations have both positive and negative effects on families. On the positive side, relationships often become more flexible and stronger because of the changes separations necessitate (e.g., D. R. Segal et al., 1993). Both separations and the subsequent reunions require adaptations on the part of service members and their spouses and children (Rohall, Segal, & Segal, 1999; L. N. Rosen & Durand, 2000). Although most families are able to cope with separations, negative effects are common. The most frequently experienced problems are worry about each other, physical illness of the spouse at home, affective conditions (e.g., loneliness, depression, and anger), practical aspects of maintaining the home and car alone, the difficulties inherent in the role of single parent, financial strain, and adjustment when the service member returns (D. B. Bell, Stevens, & Segal, 1996; Schumm, Bell, Segal, & Rice, 1996; Schumm, Bell, & Tran, 1993). Certain family events (e.g., the birth of a child) pose more problems when a partner is away from home. Substantial proportions of families with children report at least moderate trouble with their children during separations (Coolbaugh & Rosenthal, 1992).

Being able to communicate with their families is a major morale issue while service members are away from home. Service members are separated not only from their spouses and children but also from their extended families, including their parents, siblings, and grandparents. In addition, financial hardships can result from separations because of the cost of telephone calls and extra child care (Applewhite & Segal, 1990; D. B. Bell & Teitelbaum, 1993; Ender, 1995). Other added expenses include the purchase of personal items (e.g., insect repellent, sun block, baby wipes) and relocation of the family to be closer to relatives during the deployment (D. B. Bell, Stevens, & Segal, 1996). Financial problems can also include loss of pay from extra jobs, loss of income from spouses' jobs, and delays in receiving military pay (Community and Family Support Center, 1991).

Reunion following separation is a stressful period for many families. Service members, spouses, and children often experience psychological and social changes during the separation necessitating renegotiation of relationships and roles when the absent member returns (Schumm et al., 1996; D. R. Segal et al., 1993).

In the early part of the 21st century, deployments have been frequent, including to war zones. Although families experience problems during peacetime separations, wartime deployments (and even peacekeeping operations) cause added stress for families, who worry about service members' safety. Repeated deployments take a further toll on families (Rohall et al., 1999; D. R. Segal et al., 2003). The current frequent and lengthy deployments of service members are likely to lead to future problems in recruitment and retention of service members who become dissatisfied with their inability to maintain a reasonable family life.

Geographic Mobility

Geographic mobility in the military also has both positive and negative effects on families. On the plus side is the opportunity to travel and live in different places, with their concomitant cultural differences, even within the United States. Such experiences can have a broadening effect on service members, their spouses, and especially their children (Ender, 2002).

The negative effects on families of frequent moves have also been identified and documented, including disruptions in social networks, schooling, and spouse employment (M. W. Segal & Harris, 1993). The large geographic size of the United States can make moves especially difficult if they involve great distances. Children's adjustment includes social integration into a new peer group as well as the challenges of inconsistencies in school curricula among jurisdictions. For civilian spouses of military personnel, the necessity of living in the local labor markets around military installations and the frequency of tied migration are detrimental to employment.

Spouse Employment Outcomes and Satisfaction

A special concern for families is spouse employment, which affects financial well-being and satisfaction with military life, which in turn affect retention. Spouse employment satisfaction has been shown to be a major determinant of family satisfaction with military life (e.g., Orthner, 1990). Military wives, like other American women, derive their personal identities less from their husbands than in the past, and they have increased their participation in the labor force. What affects spouse support for retention is not whether or not the spouse is employed but rather whether the spouse's employment outcomes (whether he or she is employed and wants to be or unemployed and doesn't want to be, type of work available, pay, etc.) meet his or her expectations (Scarville, 1990).

Spouse employment issues have become paramount, with increases in the proportion of spouses who desire employment and the difficulty many have finding appropriate jobs and developing careers. Military spouses experience higher rates of unemployment and lower earnings than their peers married to civilians (Hosek, Asch, Fair, Martin, & Mattock, 2002; Payne, Warner, & Little, 1992). Their difficulties are attributable in part to geographic mobility and in part to the areas where they live. Research has shown the negative effects on employment outcomes of the tied migration to which wives and husbands of military personnel are subject (Cooney, 2003).

Cooney (2003), using data from the 1992 Department of Defense Survey of Spouses, reported that 53.5% of military wives were employed compared with 72.5% of military husbands (this difference resulted primarily from the rates for White wives and husbands; the differences between employment of Black wives and husbands was not significant). Not surprisingly, more military wives (36.7%) were out of the labor force than military husbands (10.2%). Unemployment rates were 9.8% for the wives and 17.3% for the husbands.

Civilian men married to military women were found to be more dissatisfied with their employment opportunities than civilian women married to military men (Cooney, 2003), consistent with their higher rates of unemployment. Some research also shows that these men tend to be treated by others in the military community in a manner similar to that found by Kanter (1977b) to be common for women who represent a token minority in their occupation; such treatment includes disproportionate attention, exaggerating military culture in their presence, social isolation, and mistaken identity (i.e., being taken to represent the average for one's gender—in the case of civilian husbands of military women, being assumed to be in the military; Bourg, 1995).

Cooney (2003) also found that the tied migration of military spouses was directly responsible for detriments in spouse employment outcomes: "Net of several factors related to employment and earnings, increased levels of geographic mobility are associated with increased difficulty in finding employment, increased dissatisfaction with employment opportunities, decreased levels of employment, and lower annual earnings" (Cooney, 2003, Abstract), with results varying by gender, race, and officer versus enlisted status.

Recent studies also have demonstrated that the local labor markets containing military bases carry with them higher unemployment for women and lower returns on human capital investments such as education and job experience. Women living in local labor markets with high proportions of military personnel are more likely to be unemployed and to earn less money than women living in areas without military installations (Booth, Falk, Segal, & Segal, 2000). This finding holds when other relevant variables are controlled, and women whose husbands are in the armed forces suffer from even higher unemployment and lower wages than other women in the same areas (Booth, 2000, 2003).

The armed services have attempted to address spouse employment issues, but the focus has been on providing services to wives to make them more employable, such as workshops on how to write a resume and how to dress for and behave in interviews. Also provided at some military installations are listings of jobs available in the local area. Although job listings and employment skills training are useful, they do not help if appropriate jobs do not exist in the local labor market. Some recent initiatives have involved the military working with civilian employers to increase the jobs available to military spouses (Palmer, 2003). Future research will need to ascertain the effects of these efforts.

Family Violence

The prevalence of spouse abuse and child maltreatment (both abuse and neglect) in both civilian society and the military poses serious problems. The military may be particularly vulnerable to family violence because of the relative youth of most military personnel, their separation from the social support of extended family and long-term friends, the stresses of the military lifestyle, and the culture of an institution whose focus is the (albeit legitimate) use of violence. Indeed, the first question about family violence generally asked by the public (including reporters) is whether rates are higher in the military. Although some comparative research exists, the methodologies do not permit direct comparisons or firm conclusions; further research is needed that uses consistent definitions of instances of abuse. The answer is also different for child maltreatment and spouse abuse; the data that do exist show that child maltreatment rates and the severity of abuse are much lower in the military than in civilian society (for a review of the issues and the evidence, see Brewster, 2000). For spousal abuse, some research shows higher rates in the military, but the results change when differences in age and race between military and civilian samples are controlled: It appears that rates of mild to moderate violence are similar but that rates of severe physical aggression are higher in the military (Brannen & Hamlin, 2000).

Military programs and services to address family violence are given the label of *family advocacy*, which may be considered a misnomer and certainly hinders the location of research and resource materials by those not familiar with the terminology. The armed services have well-developed programs for the prevention, identification, and treatment of child maltreatment, with high standards for reporting and response (see Brewster, 2000, for a description of programs and policies), that began in the early 1980s under Department of Defense directives in response to congressional inquiries and mandate. Indeed, the highly developed reporting system in the military is one of the reasons that it is difficult to compare rates of violence between the military and civilian society. Both research and clinical experience have contributed to the design of programs and their evaluation. New initiatives have included programs to support new parents (Military Family Resource Center, 2000; M. Salas & Besetsny, 2000).

To help understand violence in military families, research is needed on the effects of specific aspects of the lifestyle, such as moving and separation. Also, family violence may vary as a function of the service member's military occupation. Family composition in civilian families appears to affect the risk of violence; similar research should be conducted with military families. Given some evidence that a high percentage of entering service members are victims of abuse, the possibility of intergenerational transmission of abuse needs to be examined.

Family Support Services

The military provides a wide variety of both formal and informal support services to service members and their families. Some of these programs date back centuries to the "camp followers" who provided armies with support such as cooking, supplies, and mending of clothing; indeed, senior enlisted wives were among those who provided such services (D. B. Bell & Iadeluca, 1987). As the military community grew in size following World War II, and as the numbers of enlisted personnel who were married and/or had children increased, the need for broader and more formal support led to the development of Army Community Service in 1965 and then to similar organizations in the other services. The 1980s witnessed grassroots activism by military spouses (Stanley, Segal, & Laughton, 1990) that spurred greater attention to military families by policy makers and unit leaders.

Family support provided by the military includes both formal programs and informal processes, occurs at multiple levels (e.g., servicewide, installation based, and unit based), and covers both general services (designed to be used by all service members—e.g., schools, libraries, recreation facilities) and targeted services (designed for those with more specific needs—e.g., counseling services, medical care for disabled family members; M. W. Segal & Harris, 1993). Such programs provide medical care, housing, information and referral, financial counseling, furniture loans, emergency food, emergency loans, and sponsors at the new location after moving.

Both formal programs and more informal support, such as perceived support by unit leaders for families, have been found to be important to the organizational commitment, readiness, and retention of service members (Bourg & Segal, 1999; M. W. Segal & Harris, 1993). The most successful unit-based family support groups are those that provide ongoing advocacy and assistance and that facilitate opportunities for family members to develop informal support among themselves (Bell, Schumm, Segal, & Rice, 1996; Schumm, Bell, Milan, & Segal, 2000). The approach that military organizations tend to use—hierarchical and formal—is often not the most effective way to support families (M. W. Segal, 1999). Research is still needed to measure the effects of particular policies, programs, and procedures on family adaptation to the military lifestyle and to analyze differential effects on different kinds of families.

CONCLUSION

Researchers know a great deal more about military families now than they did in past eras. Much research has been conducted about the lifestyle of these families and their experiences with the "greedy" nature of the military institution. Researchers know more about family adaptation to these lifestyle demands and about family effects on service members' readiness and retention. Some research has addressed the issues of differences by gender and rank. However, very little is known about how military families' experiences and reactions differ by race and ethnicity. Given the overrepresentation (compared with the civilian population) of Black personnel in the military, research is needed to address these differences.

Much more research is needed on families with members in the reserves and the National Guard. The total force concept has become a reality, and the U.S. military relies substantially on these service members (for an overview of some of the family issues of reserve component families, see Pryce, Ogilvy-Lee, & Pryce, 2000). Reserve and National Guard members have been subjected to unprecedented numbers of repeated deployments for wartime missions as well as peace enforcement and peacekeeping missions. Researchers know that separations strain these families, but they do not know the cumulative effects of these call-ups and the frequency and length of the deployments on these service members and their families. Indeed, researchers do not yet know enough about the effects of these repeated deployments on all active duty service members and their families.

The total force concept also includes civilian employees of the Department of Defense and the services and even contractors working with the military. Although the families of civilian employees are not directly included in the military, some aspects of the work demands may affect families. For example, many civilian workers have been deployed recently to hostile fire zones and have become victims of kidnappings and attacks. Research is needed that examines how families are affected by these separations, what kinds of support services would help, and what agencies are the most appropriate providers.

Continued research is needed on many aspects of the family lives of active duty service members, reservists, members of the National Guard, civilian employees, and military retirees. It remains to be seen if the armed forces will continue to support such research.

VI

CONCLUSIONS

Psychology and the military continue in the forefront of responses to the changing national security environment. Among the strengths of psychology are training in problem solving, methodological skills, critical thinking, conceptualization of problems, and the search for solutions. There is also a growing emphasis in military psychology on wellness, competence, resilience, human strengths, and positive psychology.

Numerous presidents of the American Psychological Association, starting with Robert M. Yerkes in 1917, have advocated that psychologists participate in addressing society's needs. Psychology has made important contributions to a key societal need, homeland defense, both within the military and through such programs as the American Psychological Association Resilience Initiative.

17

PSYCHOLOGY'S STRATEGIC POSITION FOR TODAY'S NATIONAL SECURITY CONCERNS

A. DAVID MANGELSDORFF

The nature of threats to U.S. national security has changed significantly. External security threats have changed from the forces of other nation-states (in the world wars, Korea, Vietnam, the Cold War, and Iraq) to cyberterrorism and urban warfare by terrorists who can strike anywhere. Whereas historically the military was the first line of defense, there is now a greater need for multiagency and multinational cooperation and for the creation of community disaster response plans. Public opinion and political will are more important to the success of the military and national security than in the past. As the military adjusts to its changing roles in the new national security environment, psychologists have much to offer. This chapter reviews recent advances in military operations and disaster response and psychology's roles in addressing them. In addition, it poses questions that will need to be examined in the near future.

The views of the author are his own and do not purport to represent those of the Department of Defense, the Department of the Army, or Baylor University.

PSYCHOLOGISTS IN UNIFORM

At the beginning of World War I, there were no provisions and no authorizations for psychologists in uniform. Some psychologists were commissioned in the Sanitary Corps under the direction of the Army Surgeon General. Other psychologists were recruited into the Army as enlisted men with the promise of commissions that did not occur. The Navy did not view psychologists or psychiatrists as contributing professionals (D. P. Gray, 1997; McGuire, 1990). There was no provision for training psychologists in the Reserves before World War II (Seidenfeld, 1966). During World War II, the Navy offered commissions to clinical psychologists in the Medical Department; the Army began a clinical psychology program in 1944.

With the reorganization of the national security organizations in 1947, uniformed psychology grew slowly. Common challenges were recruiting and training psychologists, developing meaningful career paths, implementing promotions (Mangelsdorff, 1984, 1989a, 1990), providing opportunities for command (Mangelsdorff, 1989b), and retaining a critical mass of psychologists with a career progression (Mangelsdorff, 1984, 1989a, 2000). Internship programs, postgraduate fellowships, nontraditional positions (Roberts & Barko, 1986), and new career opportunities have evolved for uniformed psychologists, especially in Special Operations Command. More psychologists are remaining in the services for their full careers. Career and executive development have evolved in the services, and mentoring is encouraged (W. B. Johnson & Harper, 2004).

Psychologists, paraprofessionals, enlisted counselors, and other mental health providers are now being deployed forward with service members (whether on ships, with combat units, or with mental health teams). Prevention and active outreach are the goals. Future missions will rely on smaller, more cohesive units performing unconventional operations (Mangelsdorff, 1999a) to conduct joint service and multinational operations worldwide. North Atlantic Treaty Organization forces working together during Operations Desert Shield and Desert Storm, in former Yugoslavia, and in Afghanistan have provided a good model for such operations in terms of standardization of equipment, operating procedures, training, command structures, communication protocols, and multinational units. Small units of Special Operations Command are being used more frequently in unconventional and multinational operations worldwide.

In August 2003, General Peter J. Schoomaker was named Army chief of staff; Schoomaker spent much of his career in joint and Special Operations Commands; his breadth of experience will be invaluable in transforming the Army to address the new security challenges. The military is repositioning its bases and personnel worldwide; troops will serve shorter tours and will rotate between training centers more often. The duration of combat tours will be shortened; units will rotate more frequently. Significant transformations are ongoing in the armed services and security organizations. Will they be the appropriate mix?

LESSONS LEARNED

Surprise attacks on U.S. interests (e.g., the attack on Pearl Harbor in 1941 and the attacks on the World Trade Center and Pentagon in 2001) created significant realignments of the national security organizations. The creation of the Department of Homeland Security in 2003 was the largest change in the security structure of the federal government since the 1947 reorganization. The report of the 9/11 Commission (National Commission on Terrorist Attacks Upon the United States, 2004) recommended even further changes in policies and organizations. However, following these recommendations would not fix the lack of coordination between the separate federal and security organizations; the same dynamics characteristic of the 1947 reorganization ended with political compromises. The 1986 Goldwater–Nichols Department of Defense Reorganization Act (P.L. 99-433), for example, went beyond organizational restructuring; it changed career incentives, requiring future uniformed service leaders to learn to work together. The *National Security Strategy* (2002) is the culmination of evolving local, state, and federal interventions, regulations, and programs from post–World War II through the 2001 terrorist attacks into a shared national strategy that is mutually supported by local and state governments and the private sector. How will the new organizations and structures align with the recommendations of the 9/11 Commission? What will be the final security objectives, and will sufficient political will exist for meaningful changes?

The challenges of poor organizational structures, ineffective multiagency coordination, incompatible communication systems, lack of accountability, and varying command structures that have plagued the response to homeland disasters (fires, hurricanes, storms, earthquakes, the 2001 terrorist attacks) also affect combat operations with joint service participation. The following critical issues are involved:

- What organization is in charge?
- What is the command structure?
- What are the communication lines?
- How is the operation to proceed?
- What are the operational objectives?
- Who is accountable for what?
- What are the alternatives?
- What will be the follow-on?

Training and realistic rehearsals for both military operations and critical events in the civilian world (e.g., disasters) will allow agencies to practice in advance using the communications, command and control procedures, logistics, and operations necessary for an optimal response. Further analyses of training exercises and operations are required with respect to the performance achieved and ways to improve it.

NATIONAL SECURITY CHALLENGES

National security challenges include ensuring sufficient force strength, adequate preparation and training for personnel, and supportive public opinion.

Ensuring Sufficient Force Strength

The size of the armed services in peacetime has been small relative to the U.S. population available; the forces have been downsized following each major conflict. When confronted by each successive external threat from nation-states, the United States has remobilized its population to meet the military personnel needs required. The volume of troops mobilized in 1917 created unique personnel and security needs. During World War I, more than 4.7 million persons served. During World War II, the armed services sought fit men and women of all races to fight and provide homeland support; 16.1 million persons served, and the entire nation supported the war efforts. Only 8.7 million persons served in Vietnam. With the all-volunteer force starting in 1973, the services needed to attract more recruits in different ways; they brought in more women and minorities. In the first Persian Gulf War, 2.2 million served. Thus, fewer citizens have served in successive conflicts in terms of total service members and percentage of national population (Directorate for Information Operations and Reports, 2005).

Demographic trends have had some implications for recruiting adequate numbers of personnel. Since the 1950s, the number of children born in the United States declined from 3.5 to 2.1 per woman (U.S. Census Bureau, 2004); fewer youths were thus available for the services to recruit. With the all-volunteer force, the military-eligible demographics changed; the forces had to use more women, more married enlisted personnel, and more educated personnel. For the Persian Gulf Wars, the forces were older, more married, and better educated, and they contained more women and more reservists; these trends will continue in spite of the continued downsizing of the armed forces that started in 1990. The total force personnel policy, which integrates active duty, Reserve, and National Guard forces, was possible because adequate Reserve forces were available (Mangelsdorff, 1999b; Mangelsdorff & Moses, 1993). With the global war on terrorism, more reservists are being mobilized and deployed for longer tours. In the future, the forces may not be able to recruit adequate numbers of active duty and Reserve component personnel; in 2005, the Reserves fell short of recruiting adequate numbers of personnel. Successful recruiting depends on public support for war efforts, unemployment, and the lack of well-compensated civilian job alternatives. Will sufficient numbers of citizens (and new immigrants) join the armed services to meet recruiting goals?

Globally, the demographic challenge for security and military recruits is particularly acute for many developed nations (e.g., Italy, Germany, Japan) with low birth rates (fewer than 1.8 children per woman). The demographics affect the labor force available, the tax base for social support programs, worker productivity, retirement ages, and the personnel available to join the armed services. The increased use of National Guard and Reserve forces, immigrants, frequent tours, and repeated deployments are creating significant personnel challenges.

Preparation and Training

Selecting and training military personnel are costly ventures. High casualty rates without reinforcements strain operational readiness; therefore, retaining well-prepared personnel is essential, particularly when human resources are scarce. Military training has evolved from apprentice models to group lectures and small group discussions to computer simulations and distance learning (Fletcher & Rockway, 1986; Lavisky, 1976; Olsen & Bass, 1982; Paparone, 2001; Quick, Joplin, Nelson, Mangelsdorff, & Fiedler, 1996; R. A. Reiser, 2001a, 2001b; Roberts & Barko, 1986; Seglie & Selby-Cole, 2000; Wisher & Olson, 2002; Wisher, Sabol, Moses, & Ramsberger, 2002). Computer simulations and distance learning using the Internet are changing how training is conducted in the services, in business, and in society.[1] Each of the armed services is developing digital networks to support education, training, command and control, security, information management, and operational activities. Digitization of the battlefield affects how units train and fight. It provides information about the battlefield (enemy and friendly troop locations; resources to recognize and protect assets) to all command levels in real time. The process was demonstrated with the tactics and operations used in Iraq in April 2003. Concerns with respect to the digitization efforts include the following:

- Will there be adequate bandwidth to support the volume of communications?
- Will frontline commanders in battle have access to timely reports and updates?

---

[1] The ARPANET was developed by the Advanced Research Projects Agency in 1972 to support security, communications, assessment, and training; it became the Internet. Information management and technology are changing how the armed services train and operate.

- Will commanders invest in the training and upkeep required to keep the information systems operational?
- Will backups be available when the systems are down? How can technology help in urban warfare?

Another technology application benefiting the military is *telehealth* (the use of electronic information and telecommunications technologies to support long-distance health care; see the Web site for the Office for the Advancement of Telehealth: http://telehealth.hrsa.gov/). Telehealth and clinical practice are converging (Jerome et al., 2000). Telehealth is a logical choice for rural, isolated populations and has proved effective in providing biofeedback to deployed troops (Earles, Folen, & James, 2001), and the military can be expected to pioneer new applications of telehealth to support deployed service members.

The services promote realistic training with a great deal of supervision, feedback, and assessment. Military training is risky and stressful, so service members are trained in psychological support and "buddy aid." Training to physical and mental fitness standards and promoting realistic expectations improve performance. Realistic military training comes at the price of about 2,000 deaths per year (1980–2004; Directorate of Information and Operational Reports, 2005) from training accidents and injuries; this statistic has been decreasing since 1980, in part because of increased attention to health and safety.

The Army is learning while doing, promoting lessons learned while engaged in ongoing operations (Bonk & Wisher, 2000; Seglie & Selby-Cole, 2000). The Army is tasked with being a learning organization (Senge, 1990); the Center for Army Lessons Learned at Fort Leavenworth, Kansas, which collects, catalogues, analyzes, and disseminates lessons learned from Army actions, is the focal point for collection of after action reports (Morrison & Meliza, 1999) and lessons learned (Army Regulation 11-33). The Army Knowledge Online portal is among the largest worldwide; computer-based training programs are being proliferated. An Army digital Warrior Knowledge Network is being developed to support leader development and education. Online practice communities provide models for knowledge sharing and learning (Kilner, 2002), but do service members have time to access, analyze, and understand the information available? New technology, changing missions, increased training requirements, and increased urban warfare conflicts combine with the need for information management, access, and overload and secure communication to challenge the decision-making and leadership skills of service members at all levels. As the military shrinks in size while undertaking increased missions, doing more with less is the norm.

The Internet is changing global societies and security concerns by creating new security risks. More and more commerce is conducted using the Internet. The national power grids depend on reliable, secure communications. Information warfare, commerce, power grids, infrastructures, and security systems depend heavily on information technology and space-based capabilities. What backup systems are available if these communication systems are not? The cable and telephone companies' increased use of fiber optics technology creates communications vulnerabilities; the repeated hurricanes in the summers of 2004 and 2005 knocked out power and communications for thousands of families in Florida, Mississippi, Louisiana, Texas, and other southeastern states. Existing power and communications backups are not adequate. The number of hurricanes striking the U.S. coasts has risen in the 21st century as part of a multidecade cycle of more hurricanes with increased frequency and intensity. With increasing urbanization on the coasts, these storms will continue to cause serious disruptions to cities and infrastructures.

Protecting the Mental Health of the Force

Behavioral and social scientists have studied adjustment to military life extensively. Personnel databases have been used to develop prediction models for classification, health, disease, performance, education, attitudes, and retention factors. The armed services encourage and promote healthy lifestyles of service members and their families; physical training and mental fitness are advocated in accord with the goals of *Healthy People: The Surgeon General's Report on Health Promotion and Disease Prevention* (1979). Realistic training results in a number of accidental deaths, though the trends have been declining. Baseline physiological and psychological measures are collected before and during overseas deployments and on return to assess changes in service members' health.

Because of the great number of peacemaking, peacekeeping, and humanitarian assistance deployments since 1990, service members have been separated from their families more frequently and for longer periods. How much disruption of family support and stable relationships can service members tolerate? The ongoing Army transformation program plans to stabilize units for longer periods to keep personnel together and to increase family stability and predictability (Gayton, 2004; Schoomaker & Vassalo, 2004). Support services are developed at active duty installations, but generally Reserves and National Guard family members are not provided the same access to health care facilities and support. Reserves and National Guard members are nevertheless required to maintain fitness and health standards while working in their civilian careers. With the Reserve mobilizations to Afghanistan and Iraq in 2003, significant numbers of Reserve Component personnel failed to meet deployment standards. Planning is needed to meet the interests of the Reserve Components and family members in the future.

The Military Health System is the only health maintenance organization with a readiness wartime mission; it cares for about 8.9 million potential users at about 150 military treatment facilities. The military health system assesses the performance of its treatment facilities and the attitudes of its beneficiaries (Mangelsdorff, 1979, 1994; Mangelsdorff & Finstuen, 2003; Mangelsdorff, Finstuen, Larsen, & Weinberg, 2005; Mangelsdorff, Rogers, Finstuen, & Pryor, 2004). It protects those who serve (Institute of Medicine, 1999, 2000, 2001) and evaluates national healthcare trends (Davies & Felder, 1990; Kohn, Corrigan, & Donaldson, 1999). Like a volunteer fire department, the Military Health System trains realistically to perform both its war and peacetime missions. The active duty strength has declined since 1990, but the number of eligible beneficiaries has not declined; the military health workload has thus increased as the number of health providers has decreased. The Veterans Administration estimated that there were about 25.5 million eligible veterans as of 2004, with over 7 million enrolled (Principi, 2004). Partnership between Department of Defense and Veterans Administration treatment facilities is increasing as part of the transformation. The Department of Defense has led in developing management techniques, theory, and practice. Innovations are under way in project management, applications of operations research, program evaluation and review techniques, management information systems, systems analysis, use of computers, systems management, line and staff organization, executive development, and the Delphi forecast (D. S. Brown, 1989; O'Connor & Brown, 1980; Smalter & Ruggles, 1966).

Innovations in health care practice in the military have been numerous. The Department of Defense Psychopharmacology Demonstration Program showed that uniformed psychologists can add psychotropic medications to their treatment options; this model is being expanded in the civilian sector (Newman, Phelps, Sammons, Dunivin, & Cullen, 2000). Psychologists are also more involved in primary care. A significant portion of demand for health care has an emotional basis (Cummings & VandenBos, 1981; Kroenke, Arington, & Mangelsdorff, 1990; Kroenke & Mangelsdorff, 1989; Kroenke, Wood, Mangelsdorff, Meier, & Powell, 1988; Matarazzo, 1980; O'Donohue, Ferguson, & Cummings, 2002; VandenBos & DeLeon, 1988). Appropriate assessment and treatment of psychological distress is changing the practice of health care. Telecommunications will play a greater part in health care delivery; electronic medical records will also permit multiple practitioners to examine the health history and course of disease over numerous deployments.

The health records of military personnel, although in general better than those of many civilians, have much room for improvement. Sometimes medical encounters during deployments are not recorded. There is no standard format for capturing the medical histories of service members; frequently the Department of Veterans Affairs must reenter the data, if it is available. Exposure to hazardous substances during overseas deployments, for example, may put service members' health at risk, and a systematic approach is required to check for background exposures and for long-term follow-up. In the Veterans Benefits Improvement Act of 1994 (GI Bill of Rights; P.L. 103-

446), the Department of Veterans Affairs asked the Institute of Medicine to review its Uniform Case Assessment Protocol for Persian Gulf War veterans (Institute of Medicine, 1998). The program is to provide a systematic, comprehensive medical protocol for the diagnosis of health problems of Persian Gulf theater veterans.

The Institute of Medicine (1999, 2000) has delineated strategies and practice guidelines to protect the health of deployed U.S. forces. The National Center for Post Traumatic Stress Disorder (2004a) is preparing to offer more mental health and rehabilitative services for service members returning from Iraq and Afghanistan. The Department of Veterans Affairs is realigning its facilities, closing underutilized hospitals, and opening outpatient clinics in areas with large veteran populations (Principi, 2004) and has developed guidelines for the management of acute and chronic stress (Department of Veterans Affairs, 2004).

Predeployment screening guidelines for mental health may have allowed some soldiers with mental health problems to be inappropriately deployed to Iraq in 2003 (Berkowitz, 2004). Surveys of soldiers' mental health in the combat zone showed some low morale and high stress along with access barriers to behavioral health services. These troops perceived that seeking mental health care was a sign of weakness. The soldiers at risk for low morale and high stress were most likely from the Reserves, National Guard, and support (transportation) units. More aggressive psychological support outreach programs have been recommended (Holloway, 2004a, 2004b; Mangelsdorff & Bartone, 1997). By placing mental health practitioners on ships, at training centers, and forward with deployed units, psychology can be brought to service members. The physical and mental health of deployed and returning service members needs continued monitoring (Adler, Wright, Huffman, Thomas, & Castro, 2002; Wright et al., 2005).

The armed services have a long history of providing community response to disasters (Archer, 2003; Caplan, 1964; Ritchie & Hoge, 2002; Tyler & Gifford, 1990; Wright, Ursano, Bartone, & Ingraham, 1990). Interventions to support the mental health of survivors of the attack on the Pentagon on September 11, 2001, have provided valuable models for treatment (Cozza, Huleatt, & James, 2002; Hoge et al., 2002; Ritchie & Hoge, 2002). In Operation Solace (one intervention offered at the Pentagon), for example, the mental health response team provided high-risk individuals and groups with an active outreach program of mental health surveillance and monitoring. Similar interventions and dynamics occurred during and after the attacks on the World Trade Center in New York City. Support and care were also critical for first responders and volunteer helpers. The Defense Science Board (2004) reported that Department of Defense roles and missions in homeland security extend beyond homeland defense to include support to civil authorities for emergency preparedness and incident response. State-by-state vulnerability assessments and joint training of Department of Defense units, the

National Guard, local emergency organizations, and first responders are needed.

Public Opinion

Congress determines the size and composition of the armed services, and the military executes the policies of the nation's politicians. The military is thus under civilian control. Because the American public elects the officials who control the military, support by the public for service members and first responders is critical. The military strategist von Clausewitz (1832/1968) noted the importance of national will in sustaining the military. Kipling (1899) described British soldiers like Tommy Atkins who served the nation with duty, honor, and courage; unfortunately, Kipling observed, a soldier was only appreciated as "'saviour of 'is country' when the guns begin to shoot" (Kipling, 1899, p. 101). Public opinion polls continuously track attitudes toward the military and patriotism both in society and in the armed forces.

The relationship between the size of the armed forces and the level of support by the U.S. public for the military were related for most of the 20th century. For example, the *Time* magazine Man of the Year (*Time Magazine Man of the Year*, 1951) for 1950 was the "American fighting-man"; there was obvious support for the victorious allies. Attitudes toward service personnel declined during the 1960s and 1970s with the politics of the Vietnam War. However, in response to the 2001 terrorist attacks and the wars in Afghanistan and Iraq, *Time* magazine again named the American soldier as its Person of the Year for 2003 (*Time Magazine Person of the Year*, 2003).

Support for the duty and sacrifices of the members of the armed forces and first responders is vital to sustaining their dedication and commitment. When national will and support are lacking, service members are at heightened risk for becoming stress casualties. A report on the National Vietnam Veterans Readjustment Study (Kulka, 1990) tracked the readjustment of veterans during the Vietnam War era; the psychological impact of the war and its aftermath still remains controversial. The National Center for Post Traumatic Stress Disorder, established in 1990, assesses the functioning of veterans and women in the military (see http://www.ncptsd.va.gov/index.html). Research by these and other organizations has shown that serving in the armed services creates demands on the health and wellness of the service members. The health of peacekeepers, Persian Gulf War veterans, and other deployed groups needs to be systematically monitored (Pflanz & Sonnek, 2002). Life course development and military service have been explored (Elder, Gimbel, & Ivie, 1991; Gade, 1991). Life course development research has demonstrated that military service during World War II was a valuable component in the growth and development of many American men. With peacekeepers and reservist groups being deployed more frequently and for longer tours, soldiers' mental and physical health may become at risk. More longitudinal research is needed for assessing the effects.

AMERICAN PSYCHOLOGICAL ASSOCIATION RESPONSES TO NATIONAL SECURITY CHALLENGES

Starting in 1917, Robert M. Yerkes, president of the American Psychological Association (APA), helped to organize psychology's resources in support of the war efforts. During and after World War II, psychologists actively promoted psychology's contributions to the well-being of service members. The Committee of the National Research Council prepared works like *Psychology for the Fighting Man* (National Research Council, 1943) and *Psychology for the Returning Serviceman* (Child & Van de Water, 1945). The majority of the psychological literature published during and following World War II dealt with security issues (Driskell & Olmstead, 1989). Military and federal institutions actively supported psychological research and development programs that have benefited veterans, military progress, development, and society.

Several APA presidents have advocated psychologists' responsibility to address society's needs. G. A. Miller (1969) urged psychologists to use their expertise to help society and share their knowledge. DeLeon (1986, 2002) emphasized the importance of integrating behavioral sciences and communications technology throughout the nation's science, health policy, and health care systems. DeLeon (1988), in accordance with the APA bylaws, encouraged psychologists to participate in public service; serve the public good; and "give psychology away." Zimbardo (2004) asked whether psychology was making a significant difference; he cited examples from the psychological testing and assessment research on military personnel selection, psychological stress, humanizing work, human factors, posttraumatic stress disorder, trauma, and health. Zimbardo (2002) advocated working to improve the quality of life of service members and urged psychologists to contribute to an ongoing APA database (see http://www.psychologymatters.org). Seligman's (2002) "positive psychology" focuses on the resilience of human beings; the armed services advocate resilient behaviors.

Ronald F. Levant and Laura Barbanel spearheaded the APA Subcommittee on Psychology's Response to Terrorism (Levant, Barbanel, & DeLeon, 2004). The APA Task Force on Promoting Resilience in Response to Terrorism (APA Help Center, 2004; APA Practice Directorate, 2004; Newman, 2005) has produced fact sheets to assist psychologists in fostering resilience in clients following a terrorist attack; the APA Practice Directorate and a public education campaign have been distributing the materials. Terrorism is extremely complex (Moghaddam & Marsella, 2003); developing an understanding of this phenomenon requires concerted efforts by individuals and agencies. It will take the efforts of many social, behavioral, and political scientists to support national will and defeat terrorism.

A National Center on Disaster Psychology and Terrorism is being established (APA Monitor, 2002) to prepare to help victims of future tragedies. In 2004, the Department of Homeland Security released a broad area announcement for a university-based Center of Excellence in Behavioral and Social Aspects of Terrorism and Counter-Terrorism (Department of Homeland Security, Broad Area Announcement, 2004). The Department of Homeland Security has already established selected centers of excellence as required by the Homeland Security Act of 2002 (P.L. 107-296), harnessing the nation's scientific knowledge and technological expertise to protect America.

CONCERNS AND FUTURE CONSIDERATIONS

Much science-based knowledge is available to assist in the development of public policy responses to disasters and stress (Ahearn & Cohen, 1984; DeLeon, 1988; Hartsough & Myers, 1985; Levant, 2003; Levant et al., 2004; Lorion, Iscoe, DeLeon, & VandenBos, 1996; Martinez, Ryan, & DeLeon, 1995). Under a statement of understanding (American Red Cross, 1991), APA works with the American Red Cross to provide volunteers (as agents of the Red Cross) for disaster mental health services. In 1992, the Disaster Response Network of the APA was launched to respond to local and national disasters. Network members have responded to numerous events, including the Oklahoma City bombing; Hurricanes Hugo, Ivan, Jeanne, Katrina, Rita, and Wilma; the Northridge, California, earthquake; and the attacks of September 11, 2001, on New York City and the Pentagon. The Disaster Response Network has contributed more than 12,000 days helping the American Red Cross (APA Practice Directorate, 2004). The terrorist attacks on New York and Washington required new considerations for handling casualties and responders (Castellano, 2003; Hammond & Brooks, 2001; Kendra & Wachtendorf, 2003; Simon & Teperman, 2001).

A public health approach to the psychological consequences of terrorism should be incorporated into disaster response planning (Butler, Panzer, & Goldfrank, 2003; Smelser, & Mitchell, 2002). Mental health disaster response plans have been proposed on numerous occasions (Aguilera & Planchon, 1995; G. A. Jacobs, 1995; Kronenfeld & Whicker, 1984; Lystad, 1988; Mangelsdorff, 1985; Sowder & Lystad, 1986; G. M. Summers & Cowan, 1991; Tierney, 2000). Federal involvement in mental health aspects of disaster relief work occurs under the auspices of the National Institute of Mental Health and includes funding for mental health crisis counseling and training, establishment of the Center for Mental Health Studies of Emergencies, and research funding and publication. Local health care facilities need to conduct hazard vulnerability analyses (Mangelsdorff, Savini, & Doering, 2001) and to incorporate mental health as part of their disaster planning and training. Disaster mental health plans need to be updated to reflect the current security concerns.

From an occupational health psychology perspective, what is needed is continual surveillance of how service members, first responders, and their family members adjust to the stressful demands of their professions. A. J. W. Taylor (1982, 1983, 1987) described the groups at risk after catastrophic events, particularly rescue and recovery workers and those involved with body handling and victim identification, as "hidden victims." With the intense media and Internet coverage of recent disasters, the possible exposed community increases with the additional media viewers. Cyberterrorism adds to the anguish of citizens. By disseminating the APA resiliency information and providing public awareness programs, psychologists can show what psychology has to offer (Newman, 2005).

Health care practitioners need to be familiar with the best practice guidelines for dealing with mass violence and early intervention (Department of Veterans Affairs, 2004; National Institute of Mental Health, 2002; Ritchie et al., 2004). The war-related risks for military deployments have been well described (Friedman, 2004; Hoge et al., 2004; Institute of Medicine, 1999, 2000). The perception that active duty personnel seeking mental health assistance fear stigma is troubling but not unique to the armed services. The consequences of traumatic stress on public health must be addressed from a societal and community perspective rather than just the individual clinical level (Butler et al., 2003; Smelser & Mitchell, 2002). After massive casualty events, the magnitude of need generally exceeds the resources available; community interventions are the best way to meet this level of need. New educational support and outreach strategies are required. In addition, mental health practitioners are encouraged to study the practice guidelines available (Friedman, 2004; Hoge et al., 2004; National Center for PTSD, 2004; National Institute of Mental Health, 2002; Operation Iraqi Freedom Mental Health Advisory Team, 2004; Ritchie et al., 2004).

PSYCHOLOGISTS BEYOND THE MILITARY

The APA resilience initiative (Newman, 2005) is critical to educating and supporting the public. In the face of the global war on terrorism and the natural disasters affecting the United States in the 21st century, behavioral and social scientists, health practitioners, academics, policymakers, security personnel, and community leaders need to work together to develop and implement new solutions for the security challenges facing the nation. Psychologists with military experience can assist with community disaster response in terms of providing planning, training, evaluation, public awareness, and outreach programs. The nation needs the active involvement and commitment of psychologists, whether as service members, first responders, Red Cross or Disaster Research Network volunteers, consultants, practitioners, educators, or concerned citizens. Psychologists, working together, can help build community resilience to address the current security challenges.

ADDITIONAL RESOURCES

Alluisi, E. A. (1994). APA Division 21: Roots and rooters. In H. L. Taylor (Ed.), *Who made distinguished contributions to engineering psychology?* (pp. 4–22). Washington, DC: Division 21, American Psychological Association.

Anastasi, A. (1961). *Psychological testing.* New York: Macmillan.

Beach, E. L. (1986). *The United States Navy: 200 Years.* New York: Henry Holt.

Benjamin, L. T., Jr. (Ed.). (1997). *A history of psychology.* New York: McGraw-Hill.

Benjamin, L. T., Jr. (2005). A history of clinical psychology as a profession in America (and a glimpse at its future). *Annual Reviews Clinical Psychology, 1,* 1–30.

Bingham, W. V. (1947). Military psychology in war and peace. *Science, 106,* 155–160.

Bray, C. W. (1948). *Psychology and military proficiency: A history of the Applied Psychology Panel of the National Defense Research Committee.* Princeton, NJ: Princeton University.

Britt, S. H., & Morgan, J. D. (1946). Military psychologists in World War II. *American Psychologist, 1,* 423–437.

Capshew, J. H. (1992). Psychologists on site: A reconnaissance of the historiography of the laboratory. *American Psychologist, 47,* 132–142.

Cronin, C. (Ed.). (1998). *Military psychology: An introduction.* Needham Heights, MA: Simon & Schuster.

Dewsbury, D. A. (Ed.). (1997–2000). *Unification through division: Histories of the divisions of the American Psychological Association.* Washington, DC: American Psychological Association.

Eitelberg, M. J. (1988). *Manpower for military occupations.* Washington, DC: Department of Defense.

Flanagan, J. C. (1949). Problems of personnel psychology in the armed forces. *Bulletin of U.S. Army Medical Department, 9,* 173–180.

Flanagan, J. C. (Ed.). (1952). *Psychology in the world emergency.* Pittsburgh, PA: University of Pittsburgh Press.

Hausman, W., & Rioch, D. M. (1967). Military psychiatry: A prototype of social and preventive psychiatry in the United States. *Archives of General Psychiatry, 16,* 727–739.

Hoffman, L. E. (1992). American psychologists and wartime research on Germany, 1941–1945. *American Psychologist, 47,* 264–273.

Hunt, W. A. (1975). Clinical psychology in 1944–45. *Journal Clinical Psychology, 31,* 173–178.

Hunter, E. J. (1982). *Families under the flag: A review of military family literature.* New York: Praeger.

Hunter, W. S. (1946). Psychology in the war. *American Psychologist, 1,* 479–492.

Jenkins, J. G. (1947). New opportunities and new responsibilities for the psychologist. In G. A. Kelly (Ed.), *New methods in applied psychology (Proceedings of the Maryland*

Conference on Military Contributions to Methodology in Applied Psychology, November 27–28, 1945) (pp. 1–13). College Park, MD: University of Maryland.

Ingraham, L. H., & Manning, F. J. (1981). Cohesion: Who needs it, what is it and how do we get it to them? *Military Review, 61*(6), 2–12.

Koch, S., & Leary, D. E. (Eds.). (1992). *A century of psychology as science.* Washington, DC: American Psychological Association.

Lambert, M. J., & Lambert, J. M. (1999). Use of psychological tests for assessing treatment outcome. In M. E. Maruish (Ed.), *The use of psychological testing for treatment planning and outcome assessment* (2nd ed., pp. 115–151). Mahwah, NJ: Erlbaum.

Mangelsdorff, A. D. (1992, August). *Contributions of military to the development of psychology: A centennial perspective.* Symposium chaired at the 100th Annual Convention of the American Psychological Association, Washington, DC.

Mangelsdorff, A. D. (1997, August). *Evolution of military psychology and Division 19— Past, present, and future.* Symposium chaired at the 105th Annual Convention of the American Psychological Association, Chicago.

Napoli, D. S. (Ed.). (1981). *Architects of adjustment: History of the psychological profession in the United States.* Port Washington, NY, and London: Kennikat Press.

Pratt, C. C. (Ed.). (1941). Military psychology. *Psychological Bulletin, 38,* 309–507.

Reister, F. A. (Ed.). (1975). *Medical statistics in World War II.* Washington, DC: U.S. Government Printing Office.

Routh, D. K. (1994). *Clinical psychology since 1917: Science, practice, and organization.* New York: Plenum Press.

Seidenfeld, M. A. (1944). Clinical psychology in Army hospitals. *Psychological Bulletin, 41,* 510–514.

Sperling, P. I. (1968). A new direction for military psychology: Political psychology. *American Psychologist, 23,* 97–103.

Stevens, S. S. (Ed.). (1951). *Handbook of experimental psychology.* New York: Wiley.

Street, W. R. (1994). *A chronology of noteworthy events in American psychology.* Washington, DC: American Psychological Association.

Swank, R. L., & Marchard, W. E. (1946). Combat neuroses, development of combat exhaustion. *Archives of Neurology and Psychiatry, 55,* 236–247.

Taylor, H. L. (Ed.). (1994). *Who made distinguished contributions to engineering psychology.* Washington, DC: American Psychological Association.

Wilkins, W. L. (1972). Psychiatric and psychological research in the Navy before World War II. *Military Medicine, 137,* 228–231.

Yerkes, R. M. (1919). Report of the Psychology Committee of the National Research Council. *Psychological Review, 26,* 83–149.

Yerkes, R. M. (1945). Post-war psychological services in the armed forces. *Psychological Bulletin, 42,* 396–398.

Yerkes, R. M., & Bridges, J. W. (1921). The point scale. *Boston Medical and Surgical Journal, 23,* 857–866.

REFERENCES

Adkins, J. A. (1999). Promoting organizational health: The evolving practice of occupational health psychology. *Professional Psychology: Research and Practice, 30*, 129–137.

Adler, A. B., Wright, K. M., Huffman, A. H., Thomas, J. L., & Castro, C. A. (2002). Deployment cycle effects on the psychological screening of soldiers. *U.S. Army Medical Department Journal, 4/5/6*, 31–37.

Aguilera, D. M., & Planchon, L. A. (1995). The American Psychological Association–California Psychological Association Disaster Response Project: Lessons from the past, guidelines for the future. *Professional Psychology: Research and Practice, 26*, 550–557.

Ahearn, F. L., Jr., & Cohen, R. E. (Eds.). (1984). *Disasters and mental health: An annotated bibliography*. Rockville, MD: National Institute of Mental Health.

Ahrenfeldt, R. H. (1958). *Psychiatry in the British Army in the Second World War*. New York: Columbia University Press.

Alcott, E. B., & Williford, R. C. (1986). *Aerospace Medical Division: Twenty-five years of excellence*. Brooks Air Force Base, TX: History Office, Aerospace Medical Division.

Allender, L., Lockett, J., Headley, D., Promisel, D., Kelley, T., Salvi, L., et al. (1994). *HARDMAN II and IMPRINT verification, validation & accreditation report, integration*. Aberdeen Proving Ground, MD: Methods Branch, U.S. Army Research Laboratory, Human Research & Engineering Directorate.

Alluisi, E. A. (1981, June 10–12). *Image Generation/Display Conference II: Closing comments* (AFHRL-TP-81-28). Brooks Air Force Base, TX: Air Force Human Resources Laboratory.

Altman, I., & Haythorn, W. W. (1967). The ecology of isolated groups. *Behavioral Science, 12*, 169–182.

Ambler, R. K., & Guedry, F. E. (1966). *Technical report: Validity of a Brief Vestibular Disorientation Test in screening pilot trainees*. Pensacola, FL: Naval Aerospace Medical Research Laboratory.

Ambler, R. K., & Guedry, F. E. (1974). *Technical report: The Brief Vestibular Disorientation Test as an assessment tool for non-pilot aviation personnel*. Pensacola, FL: Naval Aerospace Medical Research Laboratory.

American Institutes for Research. (1996). *Fifty years of behavioral and social science research*. (Available from the American Institutes for Research, 1000 Thomas Jefferson Street, NW, Washington, DC 20007-3835)

American Institutes for Research. (2003). *John C. Flanagan: Pioneer and AIR founder*. Retrieved August 19, 2003, from http://www.air.org/overview/flanagan.htm

American Psychiatric Association. (1944). *One hundred years of American psychiatry*. New York: Columbia University Press.

American Psychological Association Disaster Response Network. (2004). *Fact sheet.* Retrieved September 1, 2004, from http://www.apa.org/practice/drnindex.html

American Psychological Association Help Center. (2004). *About us.* Retrieved September 1, 2004, from http://helping.apa.org/about

American Psychological Association Monitor. (2002). *Psychology to play key role in national center on terrorism.* Retrieved September 1, 2004, from http://www.apa.org/monitor/dec02/terrorism.html

American Psychological Association Practice Directorate. (2004). *APA practice.* Retrieved September 1, 2004, from http://www.apa.org/practice/

American Psychological Association Public Policy Office. (2002). *Combating terrorism: Responses from the behavioral sciences.* Retrieved May 19, 2004, from http://www.apa.org/ppo/issues/svignetteterror2.html

American Red Cross. (1991). *Statement of understanding between the American Psychological Association and the American National Red Cross.* Washington, DC: Author.

American Red Cross. (1998). *Disaster mental health services* (Red Cross Publication No. 3043). Washington, DC: Author.

Ancoli-Israel, S., Cole, R., Alessi, C., Chambers, M., Moorcroft, W., & Pollak, C. P. (2003). The role of actigraphy in the study of sleep and circadian rhythms. *Sleep, 26,* 342–392.

Anderson, R. S. (Ed.). (1966). *Neuropsychiatry in World War II: Volume I. Zone of the interior.* Washington, DC: Department of the Army. (Available from the U.S. Superintendent of Documents, Government Printing Office, Washington, DC 20402)

Annis, J. F., McDaniel, J. W., & Krauskopf, P. (1991). Male and female strength for performing common industrial tasks in different postures. In W. Karwowski & J. W. Yates (Eds.), *Advances in industrial ergonomics and safety* (Vol. 3, pp. 193–200). London: Taylor & Francis.

Appel, J. W. (1966). Preventive psychiatry. In A. J. Glass & R. J. Bernucci (Eds.), *Neuropsychiatry in World War II* (pp. 373–415). Washington, DC: U.S. Government Printing Office.

Appel, J. W., & Beebe, G. W. (1946). Preventive psychiatry: An epidemiologic approach. *Journal of the American Medical Association, 103,* 196–199.

Applewhite, L. W., & Segal, D. R. (1990). Telephone use by peacekeeping troops in the Sinai. *Armed Forces & Society, 17,* 117–126.

Archer, S. E. (2003). Civilian and military cooperation in complex humanitarian operations. *Military Review, 83*(2), 32–41.

Armstrong, H. G. (1936). The loss of tactical efficiency of flying personnel in open cockpit aircraft due to cold temperatures. *Military Surgeon, 79,* 133–140.

Army–Navy Medical Service Corps Act of 1947, ch. 459, 61 Stat. 734 (1947) (enacted).

Arthur, R. J., Gunderson, E. K., & Richardson, J. W. (1966). The Cornell Medical Index as a mental health survey instrument in the naval population. *Military Medicine, 131,* 605–610.

Artiss, K. (1963). Human behavior under stress—from combat to social psychiatry. *Military Medicine, 128,* 1011–1015.

Ashburn, P. M. (1929). A history of the medical department of the United States Army. Washington, DC: U.S. Government Printing Office.

ASVAB Working Group. (1980). *History of the Armed Services Vocational Aptitude Battery (ASVAB) 1974–1980.* Unpublished report to the Principal Deputy Assistant Secretary of Defense (Manpower, Reserve Affairs, and Logistics), Washington, DC.

Bachrach, A. J., & Egstrom, G. H. (1987). *Stress and performance in diving.* Flagstaff, AZ: Best.

Bacon, B. L., & Staudenmeier, J. J. (2003). A historical overview of combat stress control units of the U.S. Army. *Military Medicine, 168,* 689–693.

Bailey, P. (1918). War neuroses, shell shock, and nervousness in soldiers. *Journal of the American Medical Association, 71,* 2148–2153.

Bailey, P., Williams, F. E., Komora, P. O., Salmon, T. W., & Fenton, N. (Eds.). (1929). *The Medical Department of the United States Army in the World War: Volume 10. Neuropsychiatry.* Washington, DC: U.S. Government Printing Office.

Bair, J. T. (1952). *Memorandum report: The characteristics of the wanted and unwanted pilot in training and in combat.* Pensacola, FL: Naval Aerospace Medical Research Laboratory.

Bair, J. T., & Hollander, E. P. (1953). Studies in motivation of student aviators at the Naval School of Aviation Medicine. *Journal of Aviation Medicine, 24,* 514–517, 522.

Baldi, K. A. (1991). An overview of physical fitness of female cadets at the military academies. *Military Medicine, 156,* 537–539.

Balkin, T. J., Braun, A. R., Wesensten, N. J., Jeffries, K., Varga, M., Baldwin, P., et al. (2002). The process of awakening: A PET study of regional brain activity patterns mediating the reestablishment of alertness and consciousness. *Brain, 125,* 2308–2319.

Balkin, T. J., Thorne, D., Sing, H., Thomas, M., Redmond, D., Wesensten, N., et al. (2000). *Effects of sleep schedules on commercial motor vehicle driver performance* (FMCSA Tech. Rep. No. DOT-MC-00-133). Washington, DC: Department of Transportation, Federal Motor Carrier Safety Administration.

Banderet, L. E., & Burse, R. L. (1991). Effects of high terrestrial altitude on military performance. In R. Gal & A. D. Mangelsdorff (Eds.), *Handbook of military psychology* (pp. 233– 254). Chichester, England: Wiley & Sons.

Bank, A. (1986). *From OSS to Green Berets.* Novato, CA: Presidio Press.

Banks, L. M. (1995). *The Office of Strategic Services psychological selection program.* Ft. Leavenworth, KS: Command & General Staff College.

Barlow, D. H. (2000). Evidenced-based practice: A world view. *Clinical Psychology: Science and Practice, 7,* 241–242.

Barnes, J. K., Woodward, J. J., Smart, C., Otis, G. A., & Huntington, D. L. (Eds.). (1870–1888). *The medical and surgical history of the war of the rebellion (1861–65)*. Washington, DC: U.S. Government Printing Office.

Barret, J. E., Barret, J. A., Oxman, T. E., & Gerber, P. D. (1988). The prevalence of psychiatric disorders in primary care practice. *Archives of General Psychiatry, 45,* 1100–1106.

Bartone, P. (1998). Stress in the military setting. In C. Cronin (Ed.), *Military psychology: An introduction* (pp. 112–146). Needham Heights, MA: Simon & Schuster.

Baskir, L. M., & Strauss, W. A. (1978). *Chance and circumstance: The draft, the war and the Vietnam generation.* New York: Vintage Books.

Bastian, L. D., Lancaster, A. R., & Reyst, H. E. (1996, December). *Department of Defense 1995 sexual harassment survey* (Report No. 96-014). Arlington, VA: Defense Manpower Data Center.

Beare, A. N., Bondi, K. R., Biersner, R. J., & Naitoh, P. (1981). Work and rest on nuclear submarines. *Ergonomics, 24,* 593–610.

Beckett, M. B., & Hodgdon, J. A. (1984). *Techniques for measuring body circumferences and skinfold thicknesses* (NHRC Tech. Rep. No. 84-39). San Diego, CA: Naval Health Research Center.

Bell, D. B., & Iadeluca, R. B. (1987). *The origins of volunteer support for Army family programs.* Alexandria, VA: U.S. Army Research Institute for the Behavioral and Social Sciences.

Bell, D. B., Schumm, W. R., Segal, M. W., & Rice, R. E. (1996). The family support system for the MFO. In R. H. Phelps & B. J. Farr (Eds.), *Reserve component soldiers as peacekeepers* (pp. 355–394). Alexandria, VA: U.S. Army Research Institute for the Behavioral and Social Sciences.

Bell, D. B., Stevens, M. L., & Segal, M. W. (1996). *How to support families during overseas deployments: A sourcebook for service providers.* Alexandria, VA: U.S. Army Research Institute for the Behavioral and Social Sciences.

Bell, D. B., & Teitelbaum, J. M. (1993, October). *Operation Restore Hope: Preliminary results of a survey of Army spouses at Ft. Drum, New York.* Paper presented at the Inter-University Seminar on Armed Forces and Society Biennial Conference, Baltimore.

Bell, F. E., III. (1974). *Advanced simulation in undergraduate pilot training (ASUPT)* (Technical Fact Sheet AFHRL/FT-TN 73-01). Brooks Air Force Base, TX: Air Force Human Resources Laboratory.

Bellamy, R. F., & Llewellyn, C. H. (1990). Preventable casualties: Rommel's flaw, Slim's edge. *Army, 40*(5), 52–56.

Benjamin, L. T., Jr., & Baker, D. B. (2004). *From séance to science: A history of the profession of psychology in America.* Belmont, CA: Wadsworth.

Bennett, W., Jr., Ruck, H. W., & Page, R. C. (Eds.). (1996). Military occupational analysis [Special issue]. *Military Psychology, 8*(3).

Benokraitis, N. V. (1997). Sex discrimination in the 21st century. In N. V. Benokraitis (Ed.), *Subtle sexism: Current practice and prospects for change* (pp. 5–33). Thousand Oaks, CA: Sage.

Bensel, C. K. (1997). Soldier performance and functionality: Impact of chemical protective clothing. *Military Psychology, 9,* 287–300.

Berkowitz, B. (2004). *The military's mounting mental health problems.* Retrieved September 1, 2004, from http://www.alternet.org/story/18556

Berkshire, J. R., & Nelson, P. D. (1958). *Technical report: Leadership peer ratings related to subsequent proficiency in the fleet.* Pensacola, FL: Naval Aerospace Medical Research Laboratory.

Berkshire, J. R., Wherry, R. J., Jr., & Shoenberger, R. W. (1965). Secondary selection in naval aviation training. *Educational & Psychological Measurement, 25,* 191–198.

Berkum, M. M., Bialek, H. M., Kern, R. P., & Yagi, K. (1962). Experimental studies of psychological stress. *Psychological Monographs, 76*(15, Whole No. 534).

Berman, W. C. (1970). *The politics of civil rights in the Truman administration.* Columbus: Ohio State University Press.

Bernstein, B. J. (1970). The ambiguous legacy: The Truman administration and civil rights. In B. J. Bernstein (Ed.), *Politics and policies of the Truman administration* (pp. 271–296). Chicago: University of Chicago Press.

Bevan, W. (1980). On getting in bed with a lion. *American Psychologist, 35,* 779–789.

Bevan, W. (1982). A sermon of sorts in three parts. *American Psychologist, 37,* 1303–1322.

Bibb, S. C. (2002). *Healthy People 2000* and population health improvement in the Department of Defense Military Health System. *Military Medicine, 167,* 552–555.

Biersner, R. J., Gunderson, E. K., & Rahe, R. H. (1972). Relationships of sports interests and smoking to physical fitness. *Journal of Sports Medicine and Physical Fitness, 12,* 107–110.

Bing, M. N., & Eisenberg, K. L. (2003, May). *The prediction and prevention of attrition in U.S. Navy submarines.* Paper presented at the Department of Defense, Human Factors Engineering, Technical Advisory Group, 49th Semiannual Conference, Augusta, GA.

Bing, M. N., & Eisenberg, K. L. (2004, August). *Development and validation of the SubMarine Attrition Risk Test (SMART).* Paper presented at the 112th Annual Conference of the American Psychological Association, Honolulu, HI.

Bittner, A. C., Carter, R. C., Kennedy, R. S., & Harbeson, M. M. (1986). Performance evaluation tests for environmental research: Evaluation of 114 measures. *Perceptual & Motor Skills, 63,* 683–708.

Blaiwes, A. S., Puig, J. A., & Regan, J. J. (1973). Transfer of training and the measurement of training effectiveness. *Human Factors, 15,* 523–533.

Blaiwes, A. S., & Regan, J. J. (1970). *Technical report: An integrated approach to the study of learning, retention, and transfer.* Orlando, FL: Naval Training Devices Center.

Bonk, C. J., & Wisher, R. A. (2000). *Applying collaborative and e-learning tools to military distance learning: A research framework* (ARI Interim Report). Alexandria, VA: Army Research Institute.

Booth, B. H. (2000). *The impact of military presence in local labor markets on unemployment rates, individual earning and returns to education.* Unpublished doctoral dissertation, University of Maryland, College Park.

Booth, B. H. (2003). Contextual effects of military presence on women's earnings. *Armed Forces & Society, 30,* 25–52.

Booth, B. H., Falk, W. W., Segal, D. R., & Segal, M. W. (2000). The impact of military presence in local labor markets on the employment of women. *Gender & Society, 14,* 318–332.

Booth-Kewley, S., Andrews, A. M., Shaffer, R. A., Gilman, P. A., Minagawa, R. Y., & Brodine, S. K. (2001). One-year follow-up evaluation of the Sexually Transmitted Diseases/Human Immunodeficiency Virus Intervention Program in a Marine Corps sample. *Military Medicine, 166,* 987–995.

Booth-Kewley, S., Larson, G. E., & Ryan, M. A. K. (2002). Predictors of Navy attrition I: Analysis of 1-year attrition. *Military Medicine, 167,* 760–769.

Boring, E. G. (1950). *A history of experimental psychology* (2nd ed.). New York: Appleton-Century-Crofts.

Borman, W. C. (1982). Validity of a behavioral assessment for predicting military recruiter performance. *Journal of Applied Psychology, 67,* 3–9.

Borman, W. C., & Fischl, M. A. (1980). *Evaluation of an Army assessment program* (Tech. Rep. 57). Minneapolis, MN: Personnel Decisions Research Institute.

Borman, W. C., & Motowidlo, S. J. (1993). Expanding the criterion domain to include elements of contextual performance. In N. Schmitt & W. C. Borman (Eds.), *Personnel selection in organizations* (pp. 71–98). San Francisco: Jossey-Bass.

Borman, W. C., Rosse, R. L., & Rose, S. R. (1982). *Predicting performance in recruiter training: Validity of assessment in the recruiter development center* (PDRI Tech. Rep. No. 73). Minneapolis, MN: Personnel Decisions Research Institute.

Bourg, C. (1995, August). *Male tokens in a masculine environment: Men with military mates.* Paper presented at the 1995 Annual Meeting of the American Sociological Association, Washington, DC.

Bourg, M. C., & Segal, M. W. (1999). The impact of family supportive policies and practices on organizational commitment to the army. *Armed Forces & Society, 25,* 633–652.

Bowden, M. (1999). *Black Hawk down: A story of modern war.* New York: Atlantic Monthly Press.

Boyer, C. B., Shafer, M. A., Shaffer, R. A., Brodine, S. K., Ito, S. I., Yninguez, D. L., et al. (2001). Prevention of sexually transmitted diseases and HIV in young military men: Evaluation of a cognitively based skills-building intervention. *Sexually Transmitted Diseases, 28,* 349–355.

Brannen, S. J., & Hamlin, E. R., II. (2000). Understanding spouse abuse in military families. In J. A. Martin, L. N. Rosen, & L. R. Sparacino (Eds.), *The military family: A practice guide for human service providers* (pp. 169–183). Westport, CT: Praeger.

Bray, R. M., Fairbank, J. A., & Marsden, M. E. (1999). Stress and substance abuse among military women and men. *American Journal of Drug & Alcohol Abuse, 25,* 239–256.

Bray, R. M., Hourani, L. L., Rae, K. L., Dever, J. A., Brown, J. M., Vincus, A. A., et al. (2003). *2002 Department of Defense survey of health related behaviors among military personnel* (Report No. RTI/7841/006 FR). Research Triangle Park, NC: Research Triangle Institute.

Bray, R. M., & Marsden, M. E. (2000). Trends in substance use among U.S. military personnel: The impact of changing demographic composition. *Substance Use and Misuse, 35,* 949–969.

Breaugh, J. A. (1983). Realistic job previews: A critical appraisal and future research directions. *Academy of Management Review, 8,* 612–619.

Brewster, A. L. (2000). Responding to child maltreatment involving military families. In J. A. Martin, L. N. Rosen, & L. R. Sparacino (Eds.), *The military family: A practice guide for human service providers* (pp. 185–196). Westport, CT: Praeger Publishers.

Britt, T. W., Adler, A. B., & Bartone, P. T. (2001). Deriving benefits from stressful events: The role of engagement in meaningful work and hardiness. *Journal of Occupational Health Psychology, 6,* 53–63.

Brodie, G., & Brodie, F. (1962). *From crossbow to H-bomb.* New York: Dell.

Brooks, J. E. (1992). *A research needs analysis for U.S. Army Special Operations Forces: Final report* (ARI Research Report No. 1631). Alexandria, VA: U.S. Army Research Institute for the Behavioral and Social Sciences.

Brower, J. M. (2003, November 19). Distributed mission training. *Military Training Technology: Online Edition, 8*(4). Available at http://www/military-training-technology.com/article.cfm?DocID=>272>

Brown, D. S. (1989). *Management concepts and practices.* Washington, DC: National Defense University.

Brown, J. S., & Burton, R. R. (1975). Multiple representations of knowledge for tutorial reasoning. In D. G. Bobrow & A. Collins (Eds.), *Representation and understanding* (pp. 311–749). New York: Academic Press.

Brown, J. S., & Burton, R. R. (1978). *An investigation of computer coaching for informal learning activities* (Tech. Rep.). Lowry Air Force Base, CO: Air Force Human Resources Laboratory.

Bryan, G. L. (1972). Evaluation of basic research in the context of mission orientation. *American Psychologist, 27,* 947–950.

Burlingame, G. M., Lambert, M. J., Reisinger, C. W., Neff, W. M., & Mosier, J. (1995). Pragmatics of tracking mental health outcomes in a managed care setting. *Journal of Mental Health Administration, 22,* 226–236.

Butler, A. S., Panzer, A. M., & Goldfrank, L. R. (Ed.). (2003). *Preparing for the psychological consequences of terrorism: A public health strategy.* Washington, DC: Institute of Medicine, National Academy Press.

Caldwell, J. A., & Caldwell, J. L. (2003). *Fatigue in aviation: A guide to staying awake at the stick*. Burlington, VT: Ashgate Publishing.

Caldwell, J. A., & Caldwell, J. L. (2005). Fatigue in military aviation: An overview of U.S. military approved pharmacological countermeasures. *Aviation, Space, and Environmental Medicine, 76*(7, Section 2, Suppl.), C-39–C-51.

Caldwell, J. A., Caldwell, J. L., Smith, J. K., & Brown, D. L. (2004). Modafinil's effects on simulator performance and mood in pilots during 37 h without sleep. *Aviation, Space, and Environmental Medicine, 75*, 771–776.

Caldwell, J. L., Caldwell, J. A., & Salter, C. A. (1997). Effects of chemical protective clothing and heat stress on Army helicopter pilot performance. *Military Psychology, 9*, 315–328.

Camfield, T. M. (1992). The American Psychological Association and World War I: 1914 to 1919. In R. B. Evans, V. S. Sexton, & T. C. Cadwallader (Eds.), *The American Psychological Association: A historical perspective* (pp. 91–118). Washington, DC: American Psychological Association.

Campbell, J. P. (1990). An overview of the Army Selection and Classification Project (Project A). *Personnel Psychology, 43*, 231–239.

Campbell, J. P., & Knapp, D. J. (Eds.). (2001). *Exploring the limits in personnel selection and classification*. Mahwah, NJ: Erlbaum.

Campbell, R. C., Ford, J. P., & Campbell, C. H. (1978). *Development of a workshop on construction and validation of skill qualification tests* (Publication No. FR-WD KY-78-2). Alexandria, VA: Human Resources Research Organization.

Caplan, G. C. (1964). *Principles of preventive psychiatry*. New York: Basic Books.

Capshew, J. H., & Hilgard, E. R. (1992). The power of service: World War II and professional reform in the American Psychological Association. In R. B. Evans, V. S. Sexton, & T. C. Cadwallader (Eds.), *The American Psychological Association: A historical perspective* (pp. 149–175). Washington, DC: American Psychological Association.

Capshew, J. H., & Laszlo, A. C. (1986). "We would not take no for an answer": Women psychologists and gender politics during World War II. *Journal of Social Issues, 42*(1), 157–180.

Carbone, E. G., Cigrang, J. A., Todd, S. L., & Fiedler, E. R. (1999). Predicting outcome of military basic training for individuals referred for psychological evaluation. *Journal of Personality Assessment, 72*, 256–265.

Carr, W., Phillips, B., & Drummond, S. P. A. (2003, August). *Operational sleep and fatigue: Effects, measurement, and models*. Paper presented at the annual meeting of the American Psychological Association, Toronto, Ontario, Canada.

Carr, W., Phillips, B. D., & Drummond, S. P. A. (2004). SAFTE–FAST model assessment: Individual differences. *Sleep, 27*(Suppl. 1), A156.

Carroll, J. D., Julesz, B., Mathews, M. V., Rothkopf, E. Z., Sternberg, S., & Wish, M. (1984). Behavioral science. In S. Millman (Ed.), *A history of engineering and science in the Bell System: Communication sciences (1925–1980)* (pp. 431–471). Short Hills, NJ: AT&T Bell Laboratories.

Carter, R. C. (1978). Knobology underwater. *Human Factors, 21,* 293–302.

Castellano, C. (2003). Large group crisis intervention for law enforcement in response to the September 11 World Trade Center mass disaster. *International Journal of Emergency Mental Health, 5,* 211–215.

Center of Military History. (1990). *Merrill's marauders.* Washington, DC: Center of Military History, Department of the Army.

Centers for Disease Control and Prevention. (2004). *Timeline.* Retrieved September 1, 2004, from http://www.cdc.gov/od/oc/media/timeline.htm

Chaffee, R. B. (1983). *Completed suicide in the Navy and Marine Corps* (NHRC Tech. Rep. No. 82-17). San Diego, CA: Naval Health Research Center.

Chapanis, A., Garner, W. R., & Morgan, C. T. (1949). *Applied experimental psychology: Human factors in engineering design.* New York: Wiley.

Child, I. L., & Van de Water, M. (Eds.). (1945). *Psychology for the returning serviceman.* Washington, DC: Infantry Journal.

Christal, R. E. (1988). Theory-based abilities measurement: The learning abilities measurement program (AMD 25th Anniversary Lectures). *Aviation, Space and Environmental Medicine, 59,* A52–A58.

Cigrang, J. D., Carbone, E. G., Todd, S., & Fiedler, E. (1998). Mental health attrition from Air Force basic military training. *Military Medicine, 163,* 834–838.

Cigrang, J. D., Todd, S., & Carbone, E. G. (2000). Stress management training for military trainees returned to duty after a mental health evaluation: Effect on graduation rates. *Journal of Occupational Health Psychology, 5,* 48–55.

Clausewitz, K. von (1968). *On war.* New York: Penguin Classics. (Original work published 1832)

Coates, J. B., Jr. (Ed.). (1954–1976). The Medical Department of the United States Army in World War II. Washington, DC: U.S. Government Printing Office.

Coffman, E. M. (2003). *The Regulars: The American Army 1898–1941.* Cambridge, MA: Belknap Press of Harvard University.

Committee on Military Nutrition Research, Food and Nutrition Board. (2001). *Caffeine for the sustainment of mental task performance: Formulations for military operations.* Washington, DC: Institute of Medicine, National Academy Press.

Committee on Training in Clinical Psychology. (1947). Recommended graduate training program in clinical psychology. *American Psychologist, 2,* 539–558.

Committee on Training in Clinical Psychology. (1948). Clinical training facilities: 1948. *American Psychologist, 3,* 317–318.

Community and Family Support Center. (1991). *Family factors in Operation Desert Shield: Phase I report* (Unpublished manuscript). Alexandria, VA: U.S. Army, Community and Family Support Center.

Conway, T. L. (1989). Behavioral, psychological, and demographic predictors of physical fitness. *Psychological Reports, 65,* 1123–1135.

Conway, T. L., & Cronan, T. A. (1988). Smoking and physical fitness among Navy shipboard men. *Military Medicine, 153,* 589–594.

Conway, T. L., & Cronan, T. A. (1992). Smoking, exercise, and physical fitness. *Preventive Medicine, 21,* 723–734.

Conway, T. L., Cronan, T. A., & Peterson, K. A. (1989). Circumference-estimated percent body fat vs. weight indices: Relationships to physical fitness. *Aviation, Space, and Environmental Medicine, 60,* 433–437.

Conway, T. L., Hervig, L. K., & Vickers, R. R., Jr. (1989). Nutrition knowledge among Navy recruits. *Journal of the American Dietetic Association, 89,* 1624–1628.

Conway, T. L., Trent, L. K., & Conway, S. W. (1989). *Physical readiness and lifestyle habits among U.S. Navy personnel during 1986, 1987, and 1988* (Publication No. A219355 NHRC-89-24). San Diego, CA: Naval Health Research Center.

Conway, T. L. (1998). Tobacco use and the United States Military: A long-standing problem. *Tobacco Control, 7,* 219–221.

Coolbaugh, K., & Rosenthal, A. (1992). *Family separations in the Army.* Alexandria, VA: U.S. Army Research Institute for the Behavioral and Social Sciences.

Cooney, R. (2003). *Moving with the military: Race, class, and gender differences in the employment consequences of tied migration.* Unpublished doctoral dissertation, University of Maryland, College Park.

Cornish, D. (1956). *The Sable Arm: Black troops in the Union Army.* Lawrence: University of Kansas.

Coser, L. A. (1974). *Greedy institutions: Patterns of undivided commitment.* New York: Free Press.

Cote, D. O., Krueger, G. P., & Simmons, R. R. (1985). Helicopter copilot workload during nap-of-the-earth flight. *Aviation Space and Environmental Medicine, 56,* 153–157.

Cozza, S. J., Huleatt, W. J., & James, L. C. (2002). Walter Reed Army Medical Center's mental health response to the Pentagon attack. *Military Medicine, 167*(Suppl. 4), 12–16.

Crawford, M. P. (1962). Concepts of training. In R. M. Gagne (Ed.), *Psychological principles in systems development* (pp. 301–341). New York: Holt, Rinehart & Winston.

Crawford, M. P. (1970). Military psychology and general psychology. *American Psychologist, 25,* 328–336.

Crawford, M. P. (1984). Highlights in the development of the Human Resources Research Organization (HumRRO). *American Psychologist, 39,* 1267–1271.

Cronan, T. A., Conway, T. L., & Hervig, L. K. (1989). Evaluation of smoking interventions in recruit training. *Military Medicine, 154,* 371–375.

Cronan, T. A., Conway, T. L., & Kaszas, S. L. (1991). Starting to smoke in the Navy: When, where, and why? *Social Science and Medicine, 33,* 1349–1353.

Crowley, J. S. (1991). *Human factors of night vision devices: Anecdotes from the field concerning visual illusions and other effects* (USAARL Technical Report No. 91-15). Fort Rucker, AL: U.S. Army Aeromedical Research Laboratory.

Cummings, N. A., & VandenBos, G. R. (1981). The twenty year Kaiser Permanente experience with psychotherapy and medical utilization: Implications for national health policy and national health insurance. *Health Policy Quarterly, 1,* 159–175.

DaCosta, J. M. (1871). On irritable heart; a clinical study of a form of functional cardiac disorder and its consequences. *American Journal of the Medical Sciences, 61,* 17–52.

Dallenbach, K. M. (1946). The Emergency Committee of Psychology, National Research Council. *American Journal of Psychology, 59,* 496–582.

Darley, J. G. (1957). Psychology and the Office of Naval Research: A decade of development. *American Psychologist, 12,* 305–323.

Davies, N. E., & Felder, L. H. (1990). Applying brakes to the runaway American health care system. *Journal of the American Medical Association, 263,* 73–76.

Deeter, D. P., & Gaydos, J. C. (Eds.). (1993). *Textbook of military medicine: Part 3, Volume 2. Occupational health: The soldier and the industrial base.* Washington, DC: U.S. Government Printing Office.

Defense Manpower Commission. (1976). *Defense manpower: The keystone of national security. Report to the President and Congress.* Washington, DC: U.S. Government Printing Office.

Defense Manpower Data Center. (2002). *Information and technology for better decision making.* Washington, DC: Author.

Defense Science Board. (2004). *DoD roles and missions in homeland security: Volume 2. Supporting reports.* Retrieved October 1, 2004, from http://www.acq.osd.mil/dsb/reports/2004-05-vol_ii.final_part_a.pdf

Defense Science Board. (n.d.). *Defense Science Board.* Retrieved December 9, 2005, from http://www.acq.osd.mil/dsb/history.htm

DeFleur, L. B. (1985). Organizational and ideological barriers to sex integration in military groups. *Work and Occupations, 12,* 206–228.

DeLeon, P. H. (1986). Increasing the societal contribution of organized psychology. *American Psychologist, 41,* 466–474.

DeLeon, P. H. (1988). Public policy and public service: Our professional duty. *American Psychologist, 43,* 309–315.

DeLeon, P. H. (2002). Presidential reflections: Past and future. *American Psychologist, 57,* 425–430.

Dempsey, C. A. (1985). *50 years of research on man in flight.* Wright–Patterson Air Force Base, OH: U.S. Air Force.

Department of the Army. (1961). *Manual for administering and scoring the Special Forces selection battery* (Department of the Army Pamphlet 611-140). Washington, DC: Headquarters, Department of the Army.

Department of the Army. (1989). *Field Manual 21-78: Resistance and escape.* Washington, DC: Headquarters, Department of the Army.

Department of the Army. (1992). *Field Manual 21-76: Survival and evasion.* Washington, DC: Headquarters, Department of the Army.

Department of the Army. (2004). *Army Regulation 11-33: Army Lessons Learned Program: System development and application*. Washington, DC: Author.

Department of Defense. (2000, April). *Personal communiqué from the Office of the Assistant Secretary of Defense*. Washington, DC: Author.

Department of Defense. (2004, February 6). *Defense Department updates ASVAB norms*. Retrieved December 9, 2005, from http://www.defenselink.mil/releases/20040206-0332.html

Department of Homeland Security. (2004). *Homeland Security Centers (HS Centers) Program Broad Agency Announcement*. Retrieved September 1, 2004, from http://www.dhs.gov/interweb/assetlibrary/S_T_BAA06July2004.pdf

Department of the Navy. (1998). *Base realignment & closure IV: FY 1999 budget estimates and justification data submitted to Congress*. Washington, DC: Author.

Department of Veterans Affairs. (2004). *Management of post traumatic stress disorder in primary care clinical practice guideline*. Retrieved September 25, 2004, from http://www.oqp.med.va.gov/cpg/PTSD/PTSD_Base.htm

Deutsch, A. (1944). *One hundred years of American psychiatry*. New York: Columbia University Press.

Dienstfrey, S. J. (1988). Women veterans' exposure to combat. *Armed Forces and Society, 14*, 549–558.

Dinges, D. F., Maislin, G., Krueger, G. P., Brewster, R. M., & Carroll, R. R. (2005, April). *Pilot test of fatigue management technologies* (FMCSA Tech. Rep. No. RT-05-002). Washington, DC: U.S. Department of Transportation, Federal Motor Carrier Safety Administration.

Directorate for Information Operations and Reports. (2005). *U.S. active duty military deaths per 100,000 serving—1980 through 2004*. Retrieved December 9, 2005, from http://web1.whs.osd.mil/mmid/casualty/death_rates1.pdf

Disaster Relief Act, P.L. 81-875, 81st Cong. (1950) (enacted).

Dobak, W., & Phillips, T. (2001). *The Black regulars, 1866–1898*. Norman: University of Oklahoma Press.

Doll, D. T., & Hanna, T. E. (1989). Enhanced detection with bimodal sonar displays. *Human Factors, 31*, 539–550.

Doll, D. T., Hanna, T. E., & Russotti, J. S. (1992). Masking in three-dimensional auditory displays. *Human Factors, 34*, 255–265.

Driskell, J. E., & Olmstead, B. (1989). Psychology and the military: Research applications and trends. *American Psychologist, 44*, 43–54.

Druckman, D., & Bjork, R. A. (Eds.). (1991). *In the mind's eye: Enhancing human performance*. Washington, DC: National Academy Press.

Druckman, D., & Bjork, R. A. (Eds.). (1994). *Learning, remembering, believing: Enhancing human performance*. Washington, DC: National Academy Press.

Druckman, D., Singer, J. E., & van Cott, H. (Eds.). (1997). *Enhancing organizational performance*. Washington, DC: National Academy Press.

Druckman, D., & Swets, J. A. (Eds.). (1988). *Enhancing human performance: Issues, theories, and techniques*. Washington, DC: National Academy Press.

Drummond, S. P. A., Salamat, J. S., Brown, G. G., Dinges, D. F., & Gillin, J. C. (2003). *Brain regions underlying differential PVT performance. Sleep, 26*(Suppl. 1), A183.

Dunivin, D. (2003). Experiences of a Department of Defense prescribing psychologist: A personal account. In M. T. Sammons, R. L. Levant, & R. U. Paige (Eds.), *Prescriptive authority for psychologists: A history and guide* (pp. 103–116). Washington, DC: American Psychological Association.

Dyregrov, A. (1997). The process in psychological debriefings. *Journal of Traumatic Stress, 10,* 589–605.

Dyregrov, A. (2002). *CISD, psychological debriefing, & group help after critical incidents: Reference list compiled and updated by Dr. Philos. Atle Dyregrov.* Retrieved September 1, 2004, from http://www.trauma-pages.com/cisd-2001.htm

Earles, J., Folen, R. A., & James, L. C. (2001). Biofeedback using telemedicine: Clinical applications and case illustrations. *Behavioral Medicine, 27,* 77–82.

Echevarria, A. J., II. (2004). *Toward an American way of war.* Carlisle Barracks, PA: Strategic Studies Institute, Army War College.

Egbert, R. L., Meeland, T., Cline, V. B., Forgy, E. W., Spickler, M. W., & Brown, C. (1958). *FIGHTER I: A study of effective and ineffective combat performers* (Special Report 13). Alexandria, VA: Human Resources Research Organization.

Ehrhardt, L. E., Cavallero, F. R., & Kennedy, R. S. (1975). Effect of predictor display on carrier landing performance: Laboratory mechanization. *Catalog of Selected Documents in Psychology, 5,* 220–221.

Eitelberg, M. J., Laurence, J. H., & Waters, B. K. (1984). *Screening for service: Aptitude and education criteria for military entry.* Washington, DC: Office of the Assistant Secretary of Defense (Manpower, Installations and Logistics).

Elder, G. H., Jr., Gimbel, C., & Ivie, R. (1991). Turning points in life: The case of military service and war. *Military Psychology, 3,* 215–232.

Ellis, J. A. (Ed.). (1986). *Military contributions to instructional technology.* New York: Praeger.

Elsmore, T. F., Hegge, F. W., Naitoh, P., & Kelly, T. (1995). *Technical report: A comparison of the effects of sleep deprivation on synthetic work performance and a conventional performance assessment battery.* San Diego, CA: Naval Health Research Center.

Ely, H. E. (1925). *Memo to Chief of Staff [Army]: The uses of Negro manpower in war.* Carlisle Barracks, PA: Army War College.

Ender, M. G. (1995). G.I. phone home: The use of telecommunications by the soldiers of Operation Just Cause. *Armed Forces & Society, 21,* 435–454.

Ender, M. G. (Ed.). (2002). *Military brats and other global nomads: Growing up in organization families.* Westport, CT: Praeger.

Endsley, M. R. (1995). Towards a theory of situation awareness. *Human Factors, 37,* 32–64.

Engelman, R. C., & Joy, R. J. T. (1975). *Two hundred years of military medicine.* Fort Detrick, MD: Historical Unit, U.S. Army.

Englert, D. R., Hunter, D. L., & Sweeney, B. J. (2003). Mental health evaluations of U.S. Air Force basic military training and technical training students. *Military Medicine, 168*, 904–910.

Englund, C. E., Reeves, D. L., Shingledecker, C. A., Thorne, D. R., Wilson, K. P., & Hegge, F. W. (1987). *Unified Tri-Service Cognitive Performance Assessment Battery (UTC-PAB)* (AD No. A181480; NHRC/JWGD3 MILPERF Report No. 87-10). San Diego, CA: Naval Health Research Center.

Ericsson, K. A., & Kintsch, W. (1995). Long-term working memory. *Psychological Review, 102*, 211–245.

Establishing the President's Committee on Equality of Treatment and Opportunity in the Armed Services, Exec. Order No. 9981, Fed. Reg. 13:4313 (1948).

Etheridge, R. M. (1989). *Family factors affecting retention: A review of the literature.* Alexandria, VA: U.S. Army Research Institute for the Behavioral and Social Sciences.

Evans, R. B., Sexton, V. S., & Cadwallader, T. C. (Eds.). (1992). *The American Psychological Association: A historical perspective.* Washington, DC: American Psychological Association.

Farnsworth, D. (1951). *Technical report: Proposed armed forces color vision test for screening.* Groton, CT: Naval Submarine Medical Research Laboratory.

Farnsworth, D. (1952). *Technical report: Developments in submarine and small vessel lighting.* Groton, CT: Naval Submarine Medical Research Laboratory.

Federal Emergency Management Agency. (2004). *About FEMA.* Retrieved September 1, 2004, from http://www.fema.gov/about

Federal Glass Ceiling Commission. (1995, March). *Good for business: Making full use of the nation's human capital.* Washington, DC: U.S. Department of Labor.

Federico, P. A., Bickel, R. R., Ullrich, T. E., Bridges, B., & Van de Wetering, B. (1989). *BATMAN (Battle-Management Assessment System) & ROBIN (Raid Originator Bogie Ingress) rationale, software design, and database descriptions* (Publication No. NPRDC-TN89-18). San Diego, CA: Navy Personnel Research and Development Center.

Finch, G., & Cameron, F. (Eds.). (1956). *Symposium on Air Force human engineering, personnel, and training research.* Washington, DC: National Academy of Sciences and National Research Council.

Fiske, D. W. (1946). Naval aviation psychology: III. The special services group. *American Psychologist, 1*, 544–548.

Fiske, D. W. (1947). Naval aviation psychology: IV. The central research group. *American Psychologist, 2*, 67–72.

Fitts, P. M. (1946). German applied psychology during World War II. *American Psychologist, 1*, 151–161.

Fitts, P. M. (1947). Psychological research on equipment design in the AAF. *American Psychologist, 2*, 93–98.

Fitzgerald, L. F., Drasgow, F., & Magley, V. J. (1999). Sexual harassment in the Armed Forces: A test of an integrated model. *Military Psychology, 11*, 329–343.

Fitzgerald, L. F., Magley, V. J., Drasgow, F., & Waldo, C. R. (1999). Measuring sexual harassment in the military: The Sexual Experiences Questionnaire (SEQ–DoD). *Military Psychology, 11*, 243–263.

Flanagan, J. C. (Ed.). (1948). *The Aviation Psychology Program in the Army Air Forces* (Report No. 1). Washington, DC: U.S. Government Printing Office.

Fleming, J. (1996). *Intelligent computer-aided training testbed (ICATT)*. Brooks Air Force Base, TX: Armstrong Laboratory, U.S. Air Force Scientific Advisory Board Human Centered Technology Panel Science and Technology Review.

Fletcher, J. D., & Rockway, M. R. (1986). Computer-based training in the military. In J. A. Ellis (Ed.), *Military contributions to instructional technology* (pp. 171–222). New York: Praeger.

Foa, E. B., Keane, T. M., & Friedman, M. J. (2000). *Effective treatments for PTSD: Practice guidelines from the International Society for Traumatic Stress Studies.* New York: Guilford Press.

Foley, T. (1984a, September). History of Air Force clinical psychology to 1978. *The Forum,* p. 5.

Foley, T. (1984b, December). History of Air Force clinical psychology to 1978. *The Forum,* p. 9.

Foley, T. (1985). History of Air Force clinical psychology to 1978. *The Forum,* p. 5.

Ford, L. A., Campbell, R. C., Knapp, D. J., & Walker, C. B. (1999). *21st century soldiers and noncommissioned officers: Critical predictors of performance* (ARI Tech. Rep. 1102). Arlington, VA: U.S. Army Institute for the Behavioral and Social Sciences.

Frank, L. H., Luz, J. T., & Crooks, B. M. (1999). *Command history for Naval Health Research Center, San Diego, CA.* San Diego, CA: Naval Health Research Center.

Franklin, J. H. (1997). *From slavery to freedom: A history of African Americans* (7th ed.). New York: McGraw-Hill.

Franz, S. I. (1912). *Handbook of mental examination methods.* New York: Journal of Nervous and Mental Disease Publishing.

Freidl, K. E. (2005). Biomedical research on health and performance of military women: Accomplishments of the Defense Women's Health Research Program (DWHRP). *Journal of Women's Health, 14*, 764–802.

Friedl, V. L. (1996). *Women in the United States military 1901–1995: A research guide and annotated bibliography.* Westport, CT: Greenwood Press.

Friedman, M. J. (2004). Acknowledging the psychiatric cost of war. *New England Journal of Medicine, 351*, 75–77.

Gade, P. A. (1991). Military service and the life-course perspective: A turning point for military personnel research. *Military Psychology, 3*, 187–200.

Gade, P. A., & Drucker, A. J. (2000). A history of Division 19 (Military Psychology). In D. A. Dewsbury (Ed.), *Unification through division: Histories of the divisions of the American Psychological Association* (Vol. 5, pp. 9–32). Washington, DC: American Psychological Association.

Gagné, R. M. (1962). Military training and principles of learning. *American Psychologist, 17*, 83–91.

Gaines, C., Deak, M. A., Helba, C., & Wright, L. C. (2000). *Tabulations of responses from the 1999 survey of active duty personnel* (Vols. 1–2). Arlington, VA, Defense Manpower Data Center.

Gal, R. (1986). *A portrait of the Israeli soldier*. Westport, CT: Greenwood Press.

Gal, R., & Mangelsdorff, A. D. (Eds.). (1991). *Handbook of military psychology*. Chichester, England: Wiley.

Garamone, J. (2000, March 24). Survey details harassment, Cohen calls for action plan. *American Forces Press Service*. Retrieved March 25, 2000, from http://www.defenselink.mil/news/mar2000/n03242000_20003243.html

Garland, F. C., Mayers, D. L., Hickey, T. M., Miller, M. R., Shaw, E. K., Gorham, E. D., et al. (1989). Incidence of human immunodeficiency virus seroconversion in U.S. Navy and Marine Corps personnel 1986 through 1988. *Journal of the American Medical Association, 262*, 3161–3165.

Gayton, S. J. (2004). Have we finally found the manning holy grail? *Military Review, 84*(2), 17–20.

Geldard, F. A. (1953). Military psychology: Science or technology? *American Journal of Psychology, 66*, 335–348.

General Accounting Office. (1994). *DoD service academies: More actions needed to eliminate sexual harassment* (Chapter Report 1/31/94, GAO/NSIAD-94-6). Washington, DC: Author.

General Accounting Office. (1995). *Equal opportunity: DoD studies on discrimination in the military* (Letter Report 4/07/95, GAO/NSIAD-95-103). Washington, DC: Author.

General Accounting Office. (1998a). *Gender issues: Analysis of promotion and career opportunities data* (Letter Report 5/26/98, GAO/NSIAD-98-157). Washington, DC: Author.

General Accounting Office. (1998b). *Gender issues: Improved guidance and oversight are needed to ensure validity and equity of fitness standards* (Chapter Report 11/17/98, GAO/NSIAD-99-9). Washington, DC: Author.

General Accounting Office. (1998c). *Gender issues: Information to assess service members' perceptions of gender inequities* (Letter Report 11/18/98, GAO/NSIAD-99-27). Washington, DC: Author.

General Accounting Office. (1999a). *Gender issues: Perceptions of readiness of selected units* (Letter Report 5/13/99, GAO/NSIAD-99-120). Washington, DC: Author.

General Accounting Office. (1999b). *Gender issues: Trends in the occupational distribution of military women* (Letter Report 9/14/99, GAO/NSIAD-99-212). Washington, DC: Author.

Ginn, R. V. N. (1978). Of purple suits and other things: An Army officer looks at unification of the Department of Defense medical services. *Military Medicine, 143*, 15–24.

Ginn, R. V. N. (1997). *The history of the U.S. Army Medical Service Corps*. Washington, DC: Office of the Surgeon General and Center of Military History.

Glass, A. J. (1953). Preventive psychiatry in the combat zone. *Armed Forces Medical Journal, 4,* 683–692.

Glass, A. J. (1971). Military psychiatry and changing systems of mental health care. *Journal of Psychiatric Research, 8,* 499–512.

Glass, A. J., & Bernucci, R. J. (Eds.). (1966). *Neuropsychiatry in World War II: Volume 1. Zone of the interior.* Washington, DC: U.S. Government Printing Office.

Glass, A. J., & Bernucci, R. J. (Eds.). (1973). *Neuropsychiatry in World War II: Volume 2. Overseas theaters.* Washington, DC: U.S. Government Printing Office.

Glenn, J. F., Burr, R. E., Hubbard, R. W., Mays, M. Z., Moore, R. J., Jones, B. H., & Krueger, G. P. (Eds.). (1991). *Sustaining health and performance in the desert: Environmental medicine guidance for operations in Southwest Asia* (USARIEM Tech. Note No. 91-1 and 91-2, pocket version). Natick, MA: U.S. Army Research Institute of Environmental Medicine. (DTIC Document Reproduction Service Nos. ADA229643 and ADA229846)

Glick, P., Zion, C., & Nelson, C. (1988). What mediates sex discrimination in hiring decisions? *Journal of Personality and Social Psychology, 55,* 178–186.

Godbee, D. C., & Odom, J. W. (1997). Utilization of Special Forces medical assets during disaster relief: The Hurricane Andrew experience. *Military Medicine, 162,* 92–95.

Goldstein, J. (2001). *War and gender: How gender shapes the war system and vice versa.* New York: Cambridge University Press.

Goldwater–Nichols Department of Defense Reorganization Act, P.L. 99-433, 99th Cong. (1986) (enacted).

Gorman, P. F. (1992). *The secret of future victories* (Paper P-2653). Alexandria, VA: Institute for Defense Analyses.

Graham, W. F., Hourani, L. L., Sorenson, D., & Yuan, H. (2000). A366-590 Demographic differences in body composition of Navy and Marine Corps personnel: Findings from the perception of Wellness and Readiness Assessment. *Military Medicine, 165,* 60–69.

Grau, L. W., & Jorgensen, W. A. (1997). Beaten by the bugs: The Soviet–Afghan war experience. *Military Review, 77*(6), 30–37.

Gray, D. P. (1997). *Many specialties, one corps: A pictorial history of the U.S. Navy Medical Service Corps.* Washington, DC: Department of the Navy.

Gray, T. H., Chun, E. K., Warner, H. D., & Edwards, J. L. (1981). *Advanced flight simulation utilization in A-10 conversion and air-to-surface attack training* (Publication No. AFHRL-TR-80-20). Williams Air Force Base, AZ: Air Force Human Resources Laboratory.

Graybiel, A., Clark, B., MacCorquodale, K., & Hupp, D. I. (1946). Role of vestibular nystagmus in visual perception of moving targets in the dark. *American Journal of Psychology, 59,* 259–266.

Green, B. F., & Mavor, A. S. (Eds.). (1994). *Modeling cost and performance for military enlistment.* Washington, DC: National Academy Press.

Green, R. J., Self, H. C., & Ellifritt, T. S. (1995). *50 years of human engineering.* Wright–Patterson Air Force Base, OH: Crew Systems Directorate, Armstrong Laboratory, Directorate Materiel Command.

Greene, R. L. (1991). *The MMPI-2/MMPI: An interpretive manual.* Needham Heights, MA: Allyn & Bacon.

Grefer, J. E., & Harris, D. M. (2003). *Evaluation of the USMC Operational Stress Control and Readiness "OSCAR" Pilot Program.* Alexandria, VA: Centers for Naval Analyses.

Griffith, J. D., Rakoff, S. H., & Helms, R. F. (1992). *Family and other impacts on retention.* Alexandria, VA: U.S. Army Research Institute for the Behavioral and Social Sciences.

Grinker, R. F., & Spiegel, J. P. (1944). *Men under stress.* New York: McGraw-Hill.

Gropman, A. (1998). *The Air Force integrates 1945–1964* (2nd ed.). Washington, DC: Smithsonian Institution Press.

Guedry, F. E., Benson, A. J., & Moore, H. J. (1982). Influence of visual display and frequency of whole-body angular oscillation on incidence of motion sickness. *Aviation Space & Environmental Medicine, 53,* 564–569.

Gum, D. R., Albery, W. B., & Basinger, J. D. (1975). *Advanced simulation in undergraduate pilot training: An overview* (Publication no. AFHRL-TR-75-59(1)). Wright–Patterson Air Force Base, OH: Air Force Human Resources Laboratory.

Gunderson, E. K., & Arthur, R. J. (1969). A brief mental health index. *Journal of Abnormal Psychology, 74,* 100–104.

Gunderson, E. K. E., & Crooks, B. M. (Eds.). (1999). *Naval Health Research Center: Thirty year review.* San Diego, CA: Naval Health Research Center.

Gunderson, E. K. E., & Hourani, L. L. (2003). The epidemiology of personality disorders in the U.S. Navy. *Military Medicine, 168,* 575–582.

Gunderson, E. K. E., Looney, J. G., & Goffman, J. M. (1975). *A comparative study of prognosis in major mental disorders* (NHRC Tech. Rep. No. 75-80). San Diego, CA: Naval Health Research Center.

Guttmacher, M. (1966). The mental hygiene consultation services. In R. S. Anderson (Ed.), *Neuropsychiatry in World War II: Volume 1. Zone of the interior* (pp. 349–372). Washington, DC: Office of the Surgeon General, Department of the Army.

Haddock, C. K., Weg, M. V., DeBon, M., Klesges, R. C., Talcott, G. W., Lando, H., & Peterson, A. (2001). Evidence that smokeless tobacco is a gateway for smoking initiation in young adult males. *Preventive Medicine, 32,* 262–267.

Hammond, J., & Brooks, J. (2001). The World Trade Center attack: Helping the helpers: The role of critical incident stress management. *Critical Care, 5,* 315–317.

Harrell, T. W. (1992). Some history of the Army General Classification Test. *Journal of Applied Psychology, 77,* 875–878.

Harris, B. C., Simutis, Z. M., & Gantz, M. M. (2002). *Women in the U.S. Army: An annotated bibliography* (Special Report No. 48). Alexandria, VA: U.S. Army Research Institute for the Behavioral and Social Sciences.

Harris, J. D. (1958). Hearing. *Annual Review of Psychology, 9,* 47–70.

Harris, J. D. (1964). A factor analytic study of three signal detection abilities. *Journal of Speech & Hearing Research, 7,* 71–78.

Harris, J. D. (1972). Audition. *Annual Review of Psychology, 24,* 313–346.

Harris, J. D., Pickler, A. G., Hoffman, H. S., & Ehmer, R. H. (1958). The interaction of pitch and loudness discriminations. *Journal of Experimental Psychology, 56,* 232–238.

Harris, J. D., & Sergeant, R. L. (1971). Monaural/binaural audible angles for a moving sound source. *Journal of Speech & Hearing Research, 14,* 618–629.

Hartsough, D. M., & Myers, D. G. (1985). *Disaster work and mental health: Prevention and control of stress among workers.* Rockville, MD: National Institute of Mental Health.

Hay, M. S., & Elig, T. W. (1999). The 1995 Department of Defense sexual harassment survey: Overview and methodology. *Military Psychology, 11,* 233–242.

Headley, D. B. (1982). Effects of atropine sulfate and pralidoxime chloride on visual, physiological, performance, subjective, and cognitive variables in man: A review. *Military Medicine, 147,* 122–132.

Health Promotion Program. (1986). *Directive 1010.10 NOTAL.* Washington, DC: Department of Defense.

Healthy people: The Surgeon General's report on health promotion and disease prevention. (1979). Washington, DC: U.S. Government Printing Office.

Helmkamp, J. C., & Kennedy, R. D. (1996). Causes of death among U.S. military personnel: A 14-year summary, 1980–1993. *Military Medicine, 161,* 311–317.

Hendrix, W. H. (2003). Psychological fly-by: A brief history of industrial psychology in the U.S. Air Force. *APS Observer, 16,* 11.

Herbold, J. R. (1986). AIDS policy development with the Department of Defense. *Military Medicine, 151,* 623–627.

Herek, G. M. (1993). Sexual orientation and military service: A social science perspective. *American Psychologist, 48,* 538–549.

Herek, G. M. (1996). Social science, sexual orientation, and military personnel policy. In G. M. Herek, J. B. Jobe, & R. M. Carney (Eds.), *Out in force: Sexual orientation in the military* (pp. 3–14). Chicago: University of Chicago Press.

Hilgard, E. R. (1987). *Psychology in America: A historical survey.* San Diego, CA: Harcourt Brace Jovanovich.

Hilton, T. T., & Dolgin, D. L. (1991). Pilot selection in the military of the free world. In R. Gal & A. D. Mangelsdorff (Eds.), *Handbook of military psychology* (pp. 81–101). Chichester, England: Wiley.

Hodgdon, J. A., & Beckett, M. B. (1984). *Prediction of percent body fat for U.S. Navy men from body circumferences and height* (NHRC Tech. Rep. No. 84–11). San Diego, CA: Naval Health Research Center.

Hogan, D. W. (1992). *U.S. Special Operations in World War II.* Washington, DC: Center of Military History, Department of the Army.

Hoge, C. W., Castro, C. A., Messer, S. C., McGurk, D., Cotting, D. I., & Koffman, R. L. (2004). Combat duty in Iraq and Afghanistan, mental health problems, and barriers to care. *New England Journal of Medicine, 351*, 13–22.

Hoge, C. W., Orman, D. T., Robichaux, R. J., Crandell, E. O., Patterson, V. J., Engel, C. C., et al. (2002). Operation Solace: Overview of the mental health intervention following the September 11, 2001 Pentagon attack. *Military Medicine, 167*(Suppl. 9), 44–47.

Hoiberg, A. (1991). Military psychology and women's role in the military. In R. Gal & A. D. Mangelsdorff (Eds.), *Handbook of military psychology* (pp. 725–739). Chichester, England: Wiley.

Hoiberg, A., Berard, R. H., Watten, R. H., & Caine, C. (1984). Correlates of weight loss in treatment and at follow-up. *International Journal of Obesity, 8*, 457–465.

Hoiberg, A., & McNally, M. S. (1991). Profiling overweight patients in the U.S. Navy: Health conditions and costs. *Military Medicine, 156*, 76–82.

Hoiberg, A., & Pugh, W. M. (1978). Predicting Navy effectiveness: Expectations, motivation, personality, aptitude, and background variables. *Personnel Psychology, 31*, 841–852.

Holland, V. M., Kaplan, J. D., & Sabol, M. A. (1999). Preliminary tests of language learning in a speech–interactive graphics microworld. *CALICO Journal, 16*, 339–359.

Hollander, E. P. (1954). Studies of leadership among naval aviation cadets. *Journal of Aviation Medicine, 25*, 164–170, 200.

Holloway, J. D. (2004a). Army uncovers mental health-service gap. *Monitor on Psychology, 35*(7), 36–37.

Holloway, J. D. (2004b). Helping with post-conflict adjustment. *Monitor on Psychology, 35*(2), 32–33.

Holm, J. (1992). *Women in the military: An unfinished revolution.* Novato, CA: Presidio Press.

Holsenbeck, L. S. (1994). Joint Task Force Andrew: The 44th Medical Brigade mental health staff officer's after action review. *Military Medicine, 159*, 186–191.

Homeland Security Act, P.L. 107-296, 107th Cong., H.R. 5005 (2002) (enacted).

Horley, G. L., Dousa, W. J., Phillabaum, R. A., Lince, D. L., & Brainerd, S. T. (1978). *Human Engineering Laboratory Battalion Artillery Test No. 6 (HELBAT 6): Digital links in fire direction* (HEL Tech. Rep. No. TM-22-78). Aberdeen Proving Ground, MD: U.S. Army Human Engineering Laboratory. (DTIC Document Reproduction Service No. ADC018100)

Hosek, J., Asch, B. C., Fair, C., Martin, C., & Mattock, M. (2002). *Married to the military: The employment and earnings of military wives compared with those of civilian wives.* Santa Monica, CA: RAND.

Houghton Mifflin. (2000). *The American heritage dictionary of the English language* (4th ed.) [Electronic version]. Retrieved June 17, 2005, from http://dictionary.reference.com/search?q=new%20age

Hourani, L. L., & Hilton, S. M. (1999). *Department of the Navy Suicide Incident Report (DONSIR): Preliminary findings January–June 1999* (NHRC Tech. Rep. No. 99-5E). San Diego, CA: Naval Health Research Center.

Hourani, L. L., Jones, D., Kennedy, K., & Hirsch, K. (1999). *Review article: Update on suicide assessment instruments and methodologies* (NHRC Tech. Rep. No. 99-31). San Diego, CA: Naval Health Research Center.

Hourani, L. L., Warrack, A. G., & Coben, P. A. (1997). *Suicide in the U.S. Marine Corps, 1990–1996* (NHRC Tech. Rep. No. 97-32). San Diego, CA: Naval Health Research Center.

Hourani, L. L., Yuan, H., & Bray, R. M. (2003). Psychosocial and health correlates of types of traumatic event exposures among U.S. military personnel. *Military Medicine, 168,* 736–743.

Hovland, C. I., Lumsdaine, A. A., & Sheffield, F. D. (1949). *Studies in social psychology in World War II: Vol. 3. Experiments in mass communication.* Princeton, NJ: Princeton University Press.

Hughes, R., Brooks, R., Graham, P., Sheen, R., & Pickens, T. (1982). Tactical ground attack: On the transfer of training from flight simulator to operational red flag range exercise. In *Proceedings of the 26th Annual Meeting of the Human Factors Society* (pp. 596–600). Santa Monica, CA: Human Factors Society.

Human Resources Research Organization. (2004). *HumRRO.* Retrieved September 1, 2004, from http://www.humrro.org/corpsite/html/about/about_us.html

Hurrell, R. M., & Lukens, J. H. (1994). Attitudes toward women in the military during the Persian Gulf war. *Perceptual and Motor Skills, 78,* 99–104.

Hurtado, S. L., & Conway, T. L. (1996). Changes in smoking prevalence following a strict no-smoking policy in U.S. Navy recruit training. *Military Medicine, 161,* 571–576.

Hurtado, S. L., Shaffer, R. A., Schuckit, M. A., Simon-Arndt, C. M., Castillo, E. M., Minagawa, R. Y., et al. (2003). *Evaluation of an alcohol misuse prevention program in a military population* (NHRC Tech. Rep. No. 03-26). San Diego, CA: Naval Health Research Center.

Hurtado, S. L., Simon-Arndt, C. M., Patriarca-Troyk, L. A., & Highfill-McRoy, R. M. (2004). *Effectiveness of an alcohol abuse secondary prevention program among Marine Corps aviation personnel* (NHRC Tech. Rep. No. 04-21). San Diego, CA: Naval Health Research Center.

Hutchins, C. W., Jr., & Kennedy, R. S. (1965). *Technical report: The relationship between past history of motion sickness and attrition from flight training.* Pensacola, FL: Naval Aerospace Medical Research Laboratory.

Hutt, M. L. (1947). An analysis of duties performed by clinical psychologists in the Army. *American Psychologist, 2,* 52–56.

Incentives for the Use of Health Information Technology and Establishing the Position of the National Health Information Technology Coordinator, Exec. Order No. 13, 335, 69 C.F.R. 24059 (2004).

Ingraham, L. H., & Manning, F. J. (1984). *Boys in the barracks: Observations on American military life.* Philadelphia: Institute for the Study of Human Issues.

Institute of Medicine. (1998). *Adequacy of the VA Persian Gulf Registry and Uniform Case Assessment Protocols.* Washington, DC: Author.

Institute of Medicine. (1999). *National Center for Military Deployment Health Research.* Washington, DC: National Academy Press. Retrieved September 1, 2004, from http://www.nap.edu/books/0309066301/html/9.html

Institute of Medicine. (2000). *Protecting those who serve: Strategies to protect the health of deployed U.S. forces.* Washington, DC: National Academy Press.

Institute of Medicine. (2001). *Crossing the quality chasm: A new health system for the 21st century.* Washington, DC: National Academy Press.

Ireland, M. W. (Ed.). (1921–1929). *The Medical Department of the United States Army in the World War.* Washington, DC: U.S. Government Printing Office.

Isler, W. C., Oordt, M., Hunter, C. L., & Rowan, A. (2005, April). *Perceptions and practice management of insomnia in primary care.* Poster session presented at the Annual Meeting of the Society of Behavioral Medicine, Boston.

Jacobs, G. A. (1995). The development of a national plan for disaster mental health. *Professional Psychology: Research and Practice, 26,* 543–549.

Jacobs, T. O., & Jaques, E. (1987). Leadership in complex systems. In J. Zeidner (Ed.), *Human productivity enhancement* (pp. 7–65). New York: Praeger.

Jacobson, J. Z. (1951). *Scott of Northwestern: The life story of a pioneer in psychology and education.* Chicago: Louis Mariano.

James, J. J., Frelin, A. J., & Jeffrey, R. J. (1982). Disease and nonbattle injury rates and military medicine. *Medical Bulletin of the U.S. Army, Europe, 39*(8), 17–27.

Janowitz, M. (1960). *The professional soldier.* New York: Free Press.

Jenkins, J. G. (1942). Utilization of psychologists in the United States Navy. *Psychological Bulletin, 39,* 371–375.

Jenkins, J. G. (1945). Naval aviation psychology: I. The field service organization. *Psychological Bulletin, 42,* 631–637.

Jenkins, J. G. (1946). Naval aviation psychology: II. The procurement and selection organization. *American Psychologist, 1,* 45–49.

Jenkins, J. G. (1948). The nominating technique as a method of evaluating air group morale. *Journal of Aviation Medicine, 19,* 12–19.

Jerome, L. W., DeLeon, P. H., James, L. C., Folen, R., Earles, J., & Gedney, J. J. (2000). The coming of age of telecommunications in psychological research and practice. *American Psychologist, 55,* 407–421.

Jobes, D. A., & Drozd, J. F. (2004). The CAMS approach to working with suicidal patients. *Journal of Contemporary Psychotherapy, 34,* 73–85.

Johnson, L. C. (1967). Sleep and sleep loss: Their effect on performance. *Naval Research Reviews, 20,* 16–22.

Johnson, L. C. (1982). Sleep deprivation and performance. In W. B. Webb (Ed.), *Biological rhythms, sleep, and performance* (pp. 111–141). New York: Wiley.

Johnson, L. C., & Naitoh, P. (1974). The operational consequences of sleep deprivation and sleep deficit. *AGARDograph, 193*, 1–43.

Johnson, R. F., & Kobrick, J. L. (1997). Effects of wearing chemical protective clothing on rifle marksmanship and on sensory and psychomotor tasks. *Military Psychology, 9*, 301–314.

Johnson, W. B., & Harper, G. P. (2004). *Becoming a leader the Annapolis way.* New York: McGraw-Hill.

Jones, D. E., & Lee, J. J. (2002, October). Take the tough cases to sea. *United States Naval Institute Proceedings, 128*(10), 61–64.

Jones, F. D., Sparacino, L. R., Wilcox, V. L., & Rothberg, J. M. (Eds.). (1994). *Military psychiatry: Preparing in peace for war* (Textbook of Military Medicine, Part I). Washington, DC: U.S. Government Printing Office.

Jones, F. D., Sparacino, L. R., Wilcox, V. L., Rothberg, J. M., & Stokes, J. W. (Eds.). (1995). *War psychiatry* (Textbook of Military Medicine, Part I). Washington, DC: U.S. Government Printing Office.

Joy, R. J. T. (1994). Medicine in the Armed Forces of the United States of America. In *The Oxford companion to medicine* (2nd ed., pp. 73–80). Oxford, England: Oxford University.

Jung, T. P., Makeig, S., Stensmo, M., & Sejnowski, T. (1997). Estimating alertness from the EEG power spectrum. *IEEE Transactions on Biomedical Engineering, 44*, 60–69.

Kamimori, G. H., Karyekar, C. S., Otterstettere, R., Cox, D. S., Balkin, T. J., Belenky, G. L., & Eddington, N. D. (2002). The rate of absorption and relative bioavailability of caffeine administered in chewing gum versus capsules to normal healthy volunteers. *International Journal of Pharmaceutics, 234*, 159–167.

Kanter, R. M. (1977a). *Men and women of the corporation.* New York: Basic Books.

Kanter, R. M. (1977b). Some effects of proportions on group life: Skewed sex ratios and responses to token women. *American Journal of Sociology, 82*, 965–990.

Kaplan, J. (1985). Successful manning of new defense systems. *Program Manager— The Journal of the Defense Systems Management College, 14*(3), 11–15.

Kaplan, J., & Hartel, C. (1988). MANPRINT methods: Development of Hardman III. *Proceedings of the 27th Annual U.S. Army Operations Research Symposium, 1*, 3-427–3-436.

Kaplan, J. D., & Holland, V. M. (1995). Application of learning principles to the design of a 2nd language tutor. In V. M. Holland, J. D. Kaplan, & M. Sams (Eds.), *Intelligent language tutors: Theory shaping technology* (pp. 273–287). Mahwah, NJ: Erlbaum.

Kaplan, Z., Iancu, J., & Bodner, E. (2001). Review of psychological debriefing after extreme stress. *Psychiatric Services, 52*, 824–827.

Kasl, S. V., & Cobb, A. (1966). Health behavior, illness behavior, and sick role behavior: I. Health and illness behavior. *Archives of Environmental Health, 12*, 246–266.

Katz, L. C., & Grubb, G. N. (2003). *Enhancing U.S. Army aircrew coordination training* (ARI Special Report No. 56). Alexandria, VA: U.S. Army Research Institute for the Behavioral and Social Sciences.

Keenan, P. A., Felber, H. R., & Dugan, B. A. (1997). *Virginia state police selection* (HumRRO Final Report FR-EADD-97-35). Alexandria, VA: Human Resources Research Organization.

Keene, J. D. (1994). Intelligence and morale in the Army of a democracy: The genesis of military psychology during the First World War. *Military Psychology, 6,* 235–253.

Keller, F. S., Christo, I. J., & Schoenfeld, W. N. (1946). Studies in International Morse Code: V. The effect of the "phonetic" equivalent. *Journal of Applied Psychology, 30,* 265–270.

Kellogg, R. S., Prather, D. C., & Castore, C. H. (1980). Simulated A-10 combat environment. In *Proceedings of the 24th Annual Meeting of the Human Factors Society* (pp. 573–577). Santa Monica, CA: Human Factors Society.

Kelly, H. O., & Robinson, S. (2003). Science advocacy weekend workshop and congressional briefing focus on military psychology. *Psychological Science Agenda, 17*(1), 1–2.

Kenardy, J. (2000). The current status of psychological debriefing. *British Medical Journal, 321,* 1032–1033.

Kendra, J. M., & Wachtendorf, T. (2003). Elements of resilience after the World Trade Center disaster: Reconstituting New York City's Emergency Operations Centre. *Disasters, 27*(1), 37–53.

Kennedy, R. S. (1971). A comparison of performance on visual and auditory monitoring tasks. *Human Factors, 13,* 93–97.

Kennedy, R. S., Turnage, J. J., & Lane, N. E. (1997). Development of surrogate methodologies for operational performance measurement: Empirical studies. *Human Performance, 10,* 251–282.

Kilbourne, B., Goodman, J., & Hilton, S. M. (1988). *Predicting functional versus organic psychotic diagnoses of hospitalized Navy personnel* (NHRC Tech. Rep. No. 88-46). San Diego, CA: Naval Health Research Center.

Kilner, P. (2002). Transforming Army learning through communities of practice. *Military Review, 82*(3), 21–27.

Kinney, J. A. S. (1963). Night vision sensitivity during prolonged restriction from sunlight. *Journal of Applied Psychology, 47,* 65–67.

Kinney, J. A., Luria, S. M., & Weitzman, D. O. (1969). Effect of turbidity on judgments of distance underwater. *Perceptual and Motor Skills, 28,* 331–333.

Kinney, J. A., & Miller, J. W. (1974). *Technical report: Judgments of the visibility of colors made from an underwater habitat.* Groton, CT: Naval Submarine Medical Research Laboratory.

Kipling, R. (1899). *Ballads and barrack-room ballads.* New York: F. R. Fenno.

Kirkland, F. R., & Katz, P. (1989). Combat readiness and the Army family. *Military Review, 69,* 64–74.

Kite, M. E. (2001). Changing times, changing gender roles: Who do we want women and men to be? In R. K. Unger (Ed.), *Handbook of the psychology of women and gender* (pp. 215–227). New York: Wiley.

Klesges, R. C., Haddock, C. K., Lando, H., & Talcott, G. W. (1999). Efficacy of forced smoking cessation and an adjunctive behavioral treatment on long-term smoking rates. *Journal of Consulting and Clinical Psychology, 67,* 952–958.

Knapp, D. J., & Campbell, R. C. (2004). *Army enlisted personnel competency assessment program Phase I: Needs analysis* (Tech. Rep. 1151). Arlington, VA: U.S. Army Research Institute for the Behavioral and Social Sciences.

Knox, K. L., Litts, D. A., Talcott, G. W., Feig, J. C., & Caine, E. D. (2003). Risk of suicide and related adverse outcomes after exposure to a suicide prevention programme in the U.S. Air Force: Cohort study. *British Medical Journal, 327,* 1–5.

Kobrick, J. L., & Johnson, R. F. (1991). Effects of hot and cold environments on military performance. In R. Gal & A. D. Mangelsdorff (Eds.), *Handbook of military psychology* (pp. 215–232). Chichester, England: Wiley.

Kocian, D. F. (1976). *A visually-coupled airborne system simulator (VCASS): An approach to visual simulation* (AMRL Tech. Rep. 77-1). Wright–Patterson Air Force Base, OH: Aerospace Medical Research Laboratory.

Kocian, D. F. (1996). *Helmet mounted sensory technology.* Wright–Patterson Air Force Base, OH: Armstrong Laboratory, U.S. Air Force Scientific Advisory Board Human Centered Technology Panel Science and Technology Review.

Kohn, L. T., Corrigan, J. M., & Donaldson, M. S. (Eds.). (1999). *To err is human: Building a safer health system.* Washington, DC: National Academy Press.

Kohout, J., & Wicherski, M. (2003, January). *1999 doctorate employment survey.* Retrieved May 22, 2004, from http://research.apa.org/des99report.html

Kolb, D., Baker, G. D., & Gunderson, E. K. E. (1983). Effects of alcohol rehabilitation treatment on health and performance of Navy enlisted men. *Drug and Alcohol Dependence, 11,* 309–319.

Kolb, D., & Gunderson, E. K. E. (1985). Research on alcohol abuse and rehabilitation in the U.S. Navy. In M. A. Shucket (Ed.), *Series in psychosocial epidemiology: Vol. 5. Alcohol patterns and problems.* New Brunswick, NJ: Rutgers University Press.

Kootte, A. F. (2002). Psychosocial response to disaster: The attacks on the Stark and the Cole. *Medicine, Conflict and Survival, 18*(1), 44–58.

Koss, M. P., Goodman, L. A., Browne, A., Fitzgerald, L. F., Keita, G. P., & Russo, N. F. (1994). *No safe haven: Male violence against women at home, at work, and in the community.* Washington, DC: American Psychological Association.

Kroenke, K., Arington, M. E., & Mangelsdorff, A. D. (1990). The prevalence of symptoms in medical outpatients and adequacy of therapy. *Archives of Internal Medicine, 150,* 1685–1689.

Kroenke, K., & Mangelsdorff, A. D. (1989). Common symptoms in ambulatory care: Incidence, evaluation, therapy and outcome. *American Journal of Medicine, 86,* 262–266.

Kroenke, K., Wood, D. R., Mangelsdorff, A. D., Meier, N. J., & Powell, J. B. (1988). Chronic fatigue in primary care: Prevalence, patient characteristics, and outcome. *Journal of the American Medical Association, 260*, 929–934.

Kronenfeld, J. J., & Whicker, M. L. (1984). *U.S. national health policy: An analysis of the federal role*. New York: Praeger.

Krueger, G. P. (1983). The role of the behavioral scientist in assessing the health hazards of developmental weapon systems. In A. W. Schopper & U. V. Nowak (Eds.), *Proceedings of the Army Medical Department Behavioral Sciences R&D Conference* (pp. 56–77). Fort Rucker, AL: U.S. Army Aeromedical Research Laboratory.

Krueger, G. P. (1986). *Publications of the Department of Human Behavioral Biology— 1958–1986, Walter Reed Army Institute of Research* (WRAIR BB Tech. Rep. No. 86-1). Washington, DC: Walter Reed Army Institute of Research.

Krueger, G. P. (1989). Sustained work, fatigue, sleep loss and performance: A review of the issues. *Work & Stress, 3*, 129–141. (DTIC Document Reproduction Service No. ADA215234)

Krueger, G. P. (1991a). Environmental factors and military performance. In R. Gal & A. D. Mangelsdorff (Eds.), *Handbook of military psychology* (pp. 207–385). Chichester, England: Wiley.

Krueger, G. P. (1991b). Sustained military performance in continuous operations: Combatant fatigue, rest and sleep needs. In R. Gal & A. D. Mangelsdorff (Eds.), *Handbook of military psychology* (pp. 255–277). Chichester, England: Wiley.

Krueger, G. P. (1993). Environmental medicine research to sustain health and performance during military deployment: Desert, arctic, high altitude stressors. *Journal of Thermal Biology, 18*, 687–690.

Krueger, G. P. (1998a). Military performance under adverse conditions. In C. Cronin (Ed.), *Military psychology: An introduction* (pp. 89–111). Needham Heights, MA: Simon & Schuster Custom Publishing.

Krueger, G. P. (1998b). Psychological research in the military setting. In C. Cronin (Ed.), *Military psychology: An introduction* (pp. 15–30). Needham Heights, MA: Simon & Schuster Custom Publishing.

Krueger, G. P., & Banderet, L. E. (1997). The effects of chemical protective clothing on military performance: A review of the issues. *Military Psychology, 9*, 255–286. (DTIC Document Reproduction Service No. ADA341415)

Krueger, G. P., Cardinal, D. T., & Stephens, M. E. (1992, September). *Publications and technical reports of the United States Army Research Institute of Environmental Medicine, 1961–1992* (USARIEM Tech. Note 92-3). Natick, MA: U.S. Army Research Institute of Environmental Medicine. (DTIC Document Reproduction Service No. ADA259790)

Kulka, R. A. (Ed.). (1990). *Trauma and the Vietnam War generation: Report of findings from the National Vietnam Veterans Readjustment Study*. New York: Brunner/Mazel.

Kyllonen, P. C. (1995). Cognitive abilities testing: An agenda for the 1990s. In M. G. Rumsey, C. B. Walker, & J. H. Harrison (Eds.), *Personnel selection and classification* (pp. 103–125). Hillsdale, NJ: Erlbaum.

Kyllonen, P. C. (1996, December 3–6). *Abilities testing.* Brooks Air Force Base, TX: Armstrong Laboratory, U.S. Air Force Scientific Advisory Board Human Centered Technology Panel Science and Technology Review.

Lancaster, A. R. (1999). Department of Defense sexual harassment research: Historical perspectives and new initiatives. *Military Psychology, 11,* 219–231.

LaRocco, J. M., Pugh, W. M., Jones, A. P., & Gunderson, E. K. E. (1977). *Situational determinants of retention decisions* (NHRC Tech. Rep. No. 77-3). San Diego, CA: Naval Health Research Center.

Larson, G. E., Booth-Kewley, S., Merrill, L. L., & Stander, V. A. (2001). Physical symptoms as indicators of depression and anxiety. *Military Medicine, 166,* 796–799.

Larson, G. E., Booth-Kewley, S., & Ryan, M. A. K. (2002). Predictors of Navy attrition: II. A demonstration of potential usefulness for screening. *Military Medicine, 167,* 770–776.

Larson, G. E., Booth-Kewley, S., Saccuzzo, D. P., Johnson, N. E., Farmer, W. L., & Alderton, D. L. (2004). *Positivity and the five-factor model of personality.* Manuscript under review.

Laskow, G. B., & Grill, D. J. (2003). The Department of Defense experiment: The Psychopharmacology Demonstration Project. In M. T. Sammons, R. L. Levant, & R. U. Paige (Eds.), *Prescriptive authority for psychologists: A history and guide* (pp. 77–102). Washington, DC: American Psychological Association.

Laurence, J. H., & Ramsberger, P. F. (1991). *Low aptitude men in the military: Who profits, who pays?* New York: Praeger.

Lavisky, S. (1976). Army research on training: How it all began. In S. S. Busnell, P. W. Caro, S. Lavisky, H. H. McFann, & J. E. Taylor (Eds.), *Instructional technology in the military* (Professional paper No. PP-2-76, pp. 9–14). Alexandria, VA: Human Resources Research Organization.

Laxar, K., Beare, A. N., Lindner, R., & Moeller, G. (1983). Judgments of relative motion in tactical displays. *Journal of Applied Psychology, 68,* 262–272.

Laxar, K., & Olson, G. M. (1978). Human information processing in navigational displays. *Journal of Applied Psychology, 63,* 734–740.

Leckie, W. (1967). *The buffalo soldiers: A narrative of the Negro cavalry in the West.* Norman: University of Oklahoma Press.

LeDuc, P. A., Greig, J. L., & Dumond, S. L. (2005). Involuntary eye responses as measures of fatigue in U.S. Army Apache aviators. *Aviation, Space, and Environmental Medicine, 76*(7, Section II, Suppl.), C-86–C-91.

Lee, U. (1966). *The employment of Negro troops.* Washington, DC: Center for Military History.

Legree, P. J., Fischl, M. A., Gade, P. A., & Wilson, M. (1998). Testing word knowledge by telephone to estimate general cognitive aptitude using an adaptive test. *Intelligence, 26,* 91–98.

Lentz, J. M., & Guedry, F. E. (1978). Motion sickness susceptibility: A retrospective comparison of laboratory tests. *Aviation Space & Environmental Medicine, 49,* 1281–1288.

Lentz, J. M., & Guedry, F. E. (1982). Apparent instrument horizon deflection during and immediately following rolling maneuvers. *Aviation Space & Environmental Medicine, 53,* 549–553.

Lesgold, A., Lajoie, S. P., Bunzo, M., & Eggan, G. (1992). SHERLOCK: A coached practice environment for an electronics troubleshooting job. In J. H. Larkin & R. W. Chabay (Eds.), *Computer-assisted instruction and intelligent tutoring systems: Establishing communication and collaboration* (pp. 201–238). Hillsdale, NJ: Erlbaum.

Levant, R. F. (2003). Promoting resilience in response to war and terrorism. *The Independent Practitioner.* Retrieved September 1, 2004, from http://www.division42.org/membersarea/ipfiles/summer_03/advocacy/update.html

Levant, R. F., Barbanel, L., & DeLeon, P. H. (2004). Psychology's response to terrorism. In F. M. Moghaddam & A. J. Marsella (Eds.), *Understanding terrorism* (pp. 265–282). Washington, DC: American Psychological Association.

Levy-Leboyer, C. (1988). Success and failure in applying psychology. *American Psychologist, 43,* 779–785.

Lewin, K. (1951). *Field theory in social science.* Chicago: University of Chicago Press.

Lewin, K., Lippitt, R., & White, R. (1939). Patterns of aggressive behavior in experimentally created "social climates." *Journal of Social Psychology, 10,* 271–299.

Lewin, M. A. (1998). Kurt Lewin: His psychology and a daughter's recollections. In G. A. Kimble & M. Wertheimer (Eds.), *Portraits of pioneers in psychology* (Vol. 3, pp. 105–118). Washington, DC: American Psychological Association.

Lewis, N. E., & Engle, B. (Eds.). (1954). *Wartime psychiatry: A compendium of the international literature.* New York: Oxford University Press.

Lieberman, H. R. (1990). Nutritional strategies to sustain aircrew performance during long duration flights. *Aviation, Space, and Environmental Medicine, 61,* 374–378.

Link, M. M. (1965). *Space medicine in Project Mercury.* Washington, DC: National Aeronautics and Space Administration.

Link, M. M., & Coleman, H. A. (1955). *Medical support of the Army Air Forces in World War II.* Washington, DC: Office of the Surgeon General, United States Air Force.

Lintz, L. M., Pennell, R., & Yasutake, J. Y. (1979). *Integrated test of the advanced instructional system (AIS)* (Tech. Rep. No. TR-79-40). Lowry Air Force Base, CO: Technical Training Division, Air Force Human Resources Laboratory.

Lippmann, W. (1922). The mental age of Americans. *New Republic, 32,* 213–215.

Litts, D. A., Moe, K. O., Roadman, C. H., Janke, R., & Miller, J. (1999). Suicide prevention among active duty Air Force personnel—United States, 1990–1999. *Morbidity and Mortality Weekly Report, 48,* 1053–1057.

Lock, J. D. (1998). *To fight with intrepidity: The complete history of the U.S. Army Rangers 1622 to present.* New York: Simon & Schuster.

Lockett, J. (2000). *Integrations methods branch* (Briefing package). Aberdeen Proving Ground, MD: Integration Methods Branch, U.S. Army Research Laboratory, Human Research & Engineering Directorate.

Long, G. M., Ambler, R. K., & Guedry, F. E. (1975). Relationship between perceptual style and reactivity to motion. *Journal of Applied Psychology, 60,* 599–605.

Lorion, R. P., Iscoe, I., DeLeon, P. H., & VandenBos, G. R. (Eds.). (1996). *Psychology and public policy: Balancing public service and professional needs.* Washington, DC: American Psychological Association.

Luria, S. M. (1990). More about psychology in the military. *American Psychologist, 45,* 96–97.

Luria, S. M., & Kinney, J. A. (1970, March 13). Underwater vision. *Science, 167,* 1454–1451.

Luria, S. M., Kinney, J. A., & Weissman, S. (1967). Estimates of size and distance underwater. *American Journal of Psychology, 80,* 282–286.

Lystad, M. (1988). (Ed.). *Mental health response to mass emergencies: Theory and practice.* New York: Brunner/Mazel.

MacGregor, M. C., Jr. (1981). *Integration of the armed forces, 1940–1945.* Washington, DC: Center for Military History

MacKinnon, D. W. (1980). *How assessment centers were started in the United States: The OSS Assessment Program.* Pittsburgh, PA: Development Dimensions International.

Macmillan, J. W. (1951). Comment: Basic research under ONR. *American Psychologist, 6,* 94.

Maier, M. H. (1993). *Military aptitude testing: The past fifty years.* Monterey, CA: Defense Manpower Data Center.

Makeig, S., & Inlow, M. (1993). Lapses in alertness: Coherence of fluctuations in performance and the EEG spectrum. *Electroencephalography and Clinical Neurophysiology, 86,* 23–35.

Mangelsdorff, A. D. (1979). Patient satisfaction questionnaire. *Medical Care, 17,* 86–90.

Mangelsdorff, A. D. (1984). Issues affecting Army psychologists' decisions to remain in the service: A follow-up study. *Professional Psychology: Research and Practice, 15,* 544–552.

Mangelsdorff, A. D. (1985). Lessons learned and forgotten: The need for prevention and mental health interventions in disaster preparedness. *Journal of Community Psychology, 13,* 239–257.

Mangelsdorff, A. D. (1989a). A cross-validation study of factors affecting military psychologists' decisions to remain in service: The 1984 active duty psychologists survey. *Military Psychology, 1,* 241–251.

Mangelsdorff, A. D. (1989b). *Establishment of a separate psychology service at Walter Reed Army Medical Center* (Report No. HR 89-007). Fort Sam Houston, TX: U.S. Army Health Care Studies and Clinical Investigation Activity. (DTIC Document Reproduction Service No. ADA211310)

Mangelsdorff, A. D. (1990). *A concordance of U.S. Army psychology conference proceedings from 1958 through 1988* (NTIS Report No. ADA231388). San Antonio, TX: Fort Sam Houston, Health Care Studies and Clinical Investigation Activity.

Mangelsdorff, A. D. (1994). Patient attitudes and utilization patterns in Army medical treatment facilities. *Military Medicine, 159,* 686–690.

Mangelsdorff, A. D. (Ed.). (1999a). Military cohesion [Special issue]. *Journal of Military Psychology, 11*(1).

Mangelsdorff, A. D. (1999b). Reserve components' perceptions and changing roles. *Military Medicine, 164,* 715–719.

Mangelsdorff, A. D. (2000). Military psychology: History of the field. In A. E. Kazdin (Ed.), *Encyclopedia of psychology* (Vol. 5, pp. 259–263). Washington, DC: American Psychological Association.

Mangelsdorff, A. D., & Bartone, P. (1997). (Eds.). *Psychological readiness for multinational operations: Directions for the 21st century* (Tech. Rep., Proceedings of the NATO Partnership for Peace Workshop). Heidelberg, Germany: North Atlantic Treaty Organization Council.

Mangelsdorff, A. D., & Finstuen, K. (2003). Patient satisfaction in military medicine: Status and an empirical test of a model. *Military Medicine, 168,* 744–749.

Mangelsdorff, A. D., Finstuen, K., Larsen, S., & Weinberg, E. (2005). Patient satisfaction in military medicine: Model refinement and assessment of Department of Defense effects. *Military Medicine, 170,* 309–314.

Mangelsdorff, A. D., & Moses, G. R. (1993). A survey of Army Medical Department Reserve personnel mobilized in support of Operation Desert Storm. *Military Medicine, 158,* 254–259.

Mangelsdorff, A. D., Rogers, J., Finstuen, K., & Pryor, R. (2004, Winter). U.S. Army–Baylor University health care administration program evidenced-based outcomes in the military health system. *Journal of Health Education and Administration, 21,* 81–89.

Mangelsdorff, A. D., Savini, G., & Doering, D. (2001). *Chemical and bioterrorism preparedness checklist.* Chicago: American Hospital Association.

Manning, L., & Wright, V. R. (2000). *Women in the military: Where they stand* (3rd ed.). Washington, DC: Women's Research and Education Institute.

Marriott, B. M. (Ed.). (1993). *Review of the results of nutritional intervention, U.S. Army Ranger training Class 11/92 (Ranger II).* Washington, DC: Committee on Military Nutrition, Institute of Medicine, National Academy Press.

Marshall, S. L. A. (1944). *Island victory.* New York: Penguin Books.

Marshall, S. L. A. (1947). *Men against fire: The problem of battle command in future war.* New York: William Morrow.

Marshall, S. L. A. (2000). *Men against fire: The problem of battle command.* Norman: University of Oklahoma Press.

Martin, E. L. (1981). *Training effectiveness of platform motion: Review of motion research involving the advanced simulator for pilot training and the simulator for air-to-air combat* (Publication No. AFHRL-TR-79-51). Williams Air Force Base, AZ: Air Force Human Resources Laboratory.

Martinez, R., Ryan, S. D., & DeLeon, P. H. (1995). Responding to trauma—Extraordinarily meaningful. *Professional Psychology: Research and Practice, 26,* 541–542.

Maruish, M. (1999). Introduction. In M. E. Maruish (Ed.), *The use of psychological testing for treatment planning and outcome assessment* (2nd ed., pp. 1–39). Mahwah, NJ: Erlbaum.

Matarazzo, J. D. (1980). Behavioral health and behavioral medicine: Frontiers for a new psychology. *American Psychologist, 35,* 807–817.

McCallum, M., Sandquist, T., Mitler, M., & Krueger, G. P. (2003, July). *Commercial transportation operator fatigue management reference* (U.S. Department of Transportation Tech. Rep.). Washington, DC: U.S. Department of Transportation Research and Special Programs Administration.

McCaughey, B. G. (1986). The psychological symptomatology of a U.S. Naval disaster. *Military Medicine, 151,* 162–165.

McCaughey, B. G. (1987). U.S. Navy Special Psychiatric Rapid Intervention Team (SPRINT). *Military Medicine, 152,* 133–135.

McClusky, M. R., Trepagnier, J. C., Cleary, F. K., & Tripp, J. M. (1975). *Development of performance objectives and evaluation of prototype performance objectives and evaluation of prototype performance tests for eight combat arms MOS* (Vols. 1–2; Publication No. FR-CD (C)-75-9). Alexandria, VA: Human Resources Research Organization.

McCormick, E. J. (1957). *Human engineering.* New York: McGraw-Hill.

McCrae, R. R. (1992). Editor's introduction to Tupes and Christal. *Journal of Personality, 60,* 217–219.

McCrae, R. R., & John, O. P. (1992). An introduction to the five-factor model and its applications. *Journal of Personality, 60,* 175–216.

McCullough, D. G. (1992). *Truman.* New York: Simon & Schuster.

McDaniel, J. W. (1988). COMBIMAN and CREW CHIEF. In K. H. E. Kroemer, S. H. Snook, K. Meadows, & S. Deutsch (Eds.), *Ergonomic models of anthropometry, human biomechanics and operator equipment interfaces* (pp. 55–60). Washington, DC: National Academy Press.

McDaniel, J. W. (1990). Models for ergonomic analysis and design COMBIMAN and CREW CHIEF. In W. Karwowski, A. M. Genardy, & S. S. Astour (Eds.), *Computer-aided ergonomics: A researcher's guide* (pp. 138–156). London: Taylor & Francis.

McDaniel, J. W. (1991). The development of computer models for ergonomic accommodation. In A. Mital & W. Karwowski (Eds.), *Workspace equipment and tool design* (pp. 29–66). Amsterdam: Elsevier.

McDonough, J. H. (2002). Performance impacts of nerve agents and their pharmacological countermeasures. *Military Psychology, 14,* 93–119.

McEvoy, G. M., & Cascio, W. F. (1985). Strategies for reducing employee turnover: A meta-analysis. *Journal of Applied Psychology, 70,* 342–353.

McFann, H. H., Hammes, J. A., & Taylor, J. E. (1955). *TRAINFIRE I: A new course in basic rifle marksmanship* (Tech. Rep. 22). Alexandria, VA: Human Resources Research Organization.

McGrath, J. E. (1970). *Social and psychological factors in stress.* New York: Holt, Rinehart & Winston.

McGuire, F. L. (1990). *Psychology aweigh! A history of clinical psychology in the United States Navy, 1900–1988.* Washington, DC: American Psychological Association.

McLellan, T. M., Bell, D. G., Lieberman, H. R., & Kamimori, G. H. (2003–2004, Winter). The impact of caffeine on cognitive and physical performance and marksmanship during sustained operations. *Canadian Military Journal: Military Medicine, 4,* 47–54.

McLellan, T. M., Kamimori, G. H., Voss, D. M., Bell, D. G., Cole, K. G., & Johnson, D. (2005). Caffeine maintains vigilance and improves run times during night operations for special forces. *Aviation, Space, and Environmental Medicine, 76,* 647–654.

McPherson, J. (1988). *Battle cry of freedom: The Civil War era.* New York: Oxford University Press.

McPherson, J. (1991). *The Negro's Civil War: How American Blacks felt and acted during the War for the Union.* New York: Oxford University Press.

Meglino, B. M., Ravlin, E. C., & DeNisi, A. S. (2001). A meta-analytic examination of realistic job preview effectiveness: A test of three counterintuitive propositions. *Human Resource Management Review, 10,* 407–434.

Meliza, L. L., & Tan, S. C. (1996). *SIMNET Unit Performance Assessment System (UPAS) Version 2.5 user's guide* (Research Rep. 96-05). Alexandria, VA: U.S. Army Research Institute for the Behavioral and Social Sciences. (DTIC Document Reproduction Service No. ADA318046)

Melton, A. W. (Ed.). (1947). *Apparatus tests* (Army Air Forces Aviation Psychology Program Report No. 4). Washington, DC: U.S. Government Printing Office.

Melton, A. W. (1957). Military psychology in the United States of America. *American Psychologist, 12,* 97–103.

Menninger, W. C. (1948). *Psychiatry in a troubled world.* Oxford, England: Macmillan.

Military Family Resource Center. (2000, March). *New parent support programs in the military services.* Retrieved September 1, 2004, from http://www.militaryhomefront.dod.mil/portal/page?_pageid=73,46033&_dad=itc&_schema=PORTAL&section_id=20.80.500.270.0.0.0.0.0&current_id=20.80.500.270.500.90.90.0.0.0&content_id=167748

Military Family Resource Center. (2001). *Profile of the military community: 2001 demographics.* Arlington, VA: Author.

Miller, G. A. (1969). Psychology as a means of promoting human welfare. *American Psychologist, 24,* 1063–1075.

Miller, J. G. (1946). Clinical psychology in the Veterans Administration. *American Psychologist, 1,* 181–189.

Miller, M. (1973). *Plain speaking: An oral biography of Harry S. Truman.* New York.

Milliken, C. S., Leavitt, W. T., Murdock, P., Orman, D. T., Ritchie, E. C., & Hoge, C. W. (2002). Principles guiding implementation of the Operation Solace plan:

"Pieces of PIES," therapy by walking around. *Military Medicine, 167*(Suppl. 9), 48–57.

Misra, R., & Panigrahi, B. (1995). Change in attitudes toward working women: A cohort analysis. *International Journal of Sociology and Social Policy, 15*, 1–20.

Mitchell, B. (1998). *Women in the military: Flirting with disaster.* Washington, DC: Regnery Publishing.

Mitchell, J. L., & Driskell, W. E. (1996). Military job analysis: A historical perspective [Special issue]. *Military Psychology, 8*(3).

Mitchell, J. T. (1983). When disaster strikes. *Journal of Emergency Medical Services, 8*, 36–39.

Moe, K. O., Lombard, T. N., Lombard, D. M., & Wilson, P. G. (1997, August). *Mental health and primary care prototype project.* Paper presented at the meeting of the American Psychological Association, Chicago.

Moeller, G., Chattin, C., Rogers, W., Laxar, K., & Ryack, B. (1981). Performance effects with repeated exposures to the diving environment. *Journal of Applied Psychology, 66*, 502–510.

Moes, G. S., Lall, R., & Johnson, W. B. (1996). Personality characteristics of successful Navy submarine personnel. *Military Medicine, 161*, 239–242.

Moghaddam, F. M., & Marsella, A. J. (2003). *Understanding terrorism.* Washington, DC: American Psychological Association.

Morgan, W. J. (1957). *The O.S.S. and I.* New York: Norton.

Morrison, J. E., & Meliza, L. L. (1999). *Foundations of the after action review process* (Special Rep. 42). Alexandria, VA: U.S. Army Research Institute for the Behavioral and Social Sciences.

Morse, V. A., Libby, M. A., & Harris, J. D. (1973). *Technical report: Pitch discrimination in background noises up to 95 db SPL.* Groton, CT: Naval Submarine Medical Research Laboratory.

Moskos, C. (1977). From institution to occupation: Trends in military organization. *Armed Forces & Society, 4*, 41–50.

Moskos, C. (1986). Institutional/occupational trends in armed forces: An update. *Armed Forces & Society, 12*, 377–382.

Moskos, C., & Butler, J. (1996). *All that we can be: Black leadership and racial integration the Army way.* New York: Basic Books.

Murray, J. D. (1978, February). *Report to the Surgeon General: Air Force clinical psychology survey* (Publication No. USAF SCN 77-151). Washington, DC: Office of the Surgeon General, Medical Plans and Health Programs Division, Personnel and Education Plans Branch.

Myers, C. S. (1915). A contribution to the study of shell shock. *Lancet, 188*, 316–320.

Myers, C. S. (1940). *Shell shock in France: 1914–1918.* New York: Macmillan.

Myers, R. B. (2004). *National military strategy of the United States of America.* Washington, DC: Joint Chiefs of Staff.

Naitoh, P. (1989). *Technical report: Minimal sleep to maintain performance*. San Diego, CA: Naval Health Research Center.

Naitoh, P., & Townsend, R. E. (1970). The role of sleep deprivation research in human factors. *Human Factors, 12,* 575–585.

Nalty, B. (1986). *Strength for the fight: A history of Black Americans in the military.* New York: Free Press.

National Academy of Sciences. (1998). *Assessing readiness in military women: The relationship of body composition, nutrition, and health.* Washington, DC: National Academy Press. Retrieved March 28, 2002, from http://www.nap.edu/catalog/6104.html

National Center for Post Traumatic Stress Disorder. (2004a). *Iraq war clinician's guide.* Retrieved September 25, 2004, from http://www.ncptsd.org//war/iraq_clinician_guide_v2/iraq_clinician_guide_v2.pdf

National Center for Post Traumatic Stress Disorder. (2004b). *National Center for PTSD.* Retrieved September 1, 2004, from http://www.ncptsd.va.gov

National Coalition on Health Care. (2004). *Building a better health care system: Specifications for reform.* Retrieved December 9, 2005, from http://www.nchc.org/materials/studies/reform.pdf

National Commission on Terrorist Attacks Upon the United States. (2004). *The 9/11 Commission report.* Retrieved September 30, 2004, from http://www.9-11commission.gov/report/911report.pdf

National Committee on Vital and Health Statistics. (2001). *50th anniversary symposium report.* Retrieved June 21, 2005, from http://www.cdc.gov/nchs/data/ncvhs/nchvs50th.pdf

National Defense Act of 1920, 41 Stat. 759.

National Institute for Occupational Safety and Health. (1996). *About NIOSH.* Retrieved September 1, 2004, from http://www.cdc.gov/niosh/about.html

National Institute of Mental Health. (2002). *Mental health and mass violence—Evidence based early psychological intervention for victims/survivors of mass violence: A workshop to reach consensus on best practices.* Washington, DC: U.S. Government Printing Office.

National Institute of Mental Health. (2004). *History.* Retrieved September 1, 2004, from http://www.nimh.nih.gov/About/history.cfm

National Institutes of Health. (1997, February 11–13). *Interventions to prevent HIV risk behaviors: NIH consensus statement online.* Retrieved September 1, 2004, from http://consensus.niv.gov/

National Institutes of Health. (2004). *About NIH.* Retrieved September 1, 2004, from http://www.nih.gov/about

National Oceanic and Atmospheric Administration. (2004). *NOAA history.* Retrieved September 1, 2004, from http://www.history.noaa.gov

National Research Council. (1943). *Psychology for the fighting man.* Washington, DC: Infantry Journal.

National security strategy. (2002). Washington, DC: Office of the President.

Naval Research Laboratory. (2004). *History*. Retrieved September 1, 2004, from http://www.nrl.navy.mil/content.php?p=history

Navy Fact File. (2004). *The expeditionary strike group*. Retrieved September 13, 2004, from http://www.chinfo.navy.mil/navpalib/news/.www/esg.html

Neel, S. (Ed.). (1973). *Vietnam studies: Medical support of the U.S. Army in Vietnam 1965–1970*. Washington, DC: U.S. Government Printing Office.

Nelson, P. D. (1971). Personnel performance prediction. In R. W. Little (Ed.), *Handbook of military institutions* (pp. 91–122). Beverly Hills, CA: Sage.

Newhouse, P. A., Penetar, D. M., Fertig, J. B., Thorne, D. R., Sing, H. C., Thomas, M. L., et al. (1992). Stimulant drug effects on performance and behavior after prolonged sleep deprivation: A comparison of amphetamine, nicotine, and deprenyl. *Military Psychology, 4*, 207–233.

Newman, R. (2005). APA's resilience initiative. *Professional Psychology, 36*, 227–229.

Newman, R., Phelps, R., Sammons, M. T., Dunivin, D. L., & Cullen, E. A. (2000). Evaluation of the Psychopharmacology Demonstration Project: A retrospective analysis. *Professional Psychology, 31*, 598–603.

Nice, D. S., & Hilton, S. (1994). Sex differences and occupational influences on health care utilization aboard U.S. Navy ships. *Military Psychology, 6*, 109–123.

Occupational Safety and Health Act, P.L. 91-596, 91st Cong., S. 2193 (1970) (enacted).

O'Connor, M. G., & Brown, D. S. (1980). Military contributions to management. *Defense Management Journal, 16*, 50–57.

O'Donahue, W. T., Ferguson, K. E., & Cummings, N. A. (2002). Introduction: Reflections on the medical cost offset effect. In N. A. Cummings, W. T. O'Donahue, & K. E. Ferguson (Eds.), *The impact of medical cost offset on practice and research: Making it work for you* (pp. 11–25). Reno, NV: Context Press.

O'Donnell, R. D., Moise, S., & Schmidt, R. M. (2005). Generating performance test batteries relevant to specific operational tasks. *Aviation, Space, and Environmental Medicine, 76*(7, Section II, Suppl.), C-24–C-30.

O'Donnell, V. M., Balkin, T. J., Andrade, J. R., Simon, L. M., Kamimori, G. H., Redmond, D. P., & Belenky, G. L. (1988). Effects of triazolam on performance and sleep in a model of transient insomnia. *Human Performance, 1*, 145–160.

Office of Naval Research. (2004). *History and mission of ONR*. Retrieved September 1, 2004, from http://www.onr.navy.mil/about/history/

Office of Strategic Services Assessment Staff. (1948). *Assessment of men: Selection of personnel for the office of strategic services*. New York: Rinehart & Company.

Older, H. J. (1947). Comment: In defense of military psychology. *American Psychologist, 2*, 105–106.

Olsen, J. R., & Bass, V. F. (1982). The application of performance based technology in the military: 1960–1980. *Performance and Instruction, 21*, 32–36.

Operation Iraqi Freedom Mental Health Advisory Team. (2004). *Report of 16 December 2003*. Retrieved September 25, 2004, from http://www.armymedicine.army.mil/news/mhat/mhat_report.pdf

Ormel, J., Von Korff, M., Ustun, T. B., Pini, S., Korten, A., & Oldehinkel, T. (1994). Common mental disorders and disability across cultures: Results from the WHO Collaborative Study on Psychological Problems in General Health Care. *Journal of the American Medical Association, 272,* 1741–1748.

Orthner, D. K. (1990). *Family impacts on the retention of military personnel.* Alexandria, VA: U.S. Army Research Institute for the Behavioral and Social Sciences.

Osborn, W. C., Campbell, R. C., & Ford, J. P. (1977). *Handbook for development of Skill Qualification Tests* (HumRRO Final Report FR-CD(L)-77-1). Alexandria, VA: Human Resources Research Organization.

Palmer, V. (2003, May 20). Corporate partnership aims to increase spouse jobs. *Army News.* Available at http://www.usarj.army.mil/archives/archives/2003/may/30/armyNews/story04.htm

Paparone, C. R. (2001). Piercing the corporate veil: OE and Army transformation. *Military Review, 81*(2), 78–82.

Parker, S. B., Griswold, L. M., & Roberts, D. E. (2003, November). *Effects of ginkgo biloba extract (EGB-761) administration on performance battery at moderate altitude in summer operations.* Paper presented at the Indo-U.S. Workshop: Enhancing Human Performance in Military Environments, New Delhi, India.

Parker, S. B., Walsh, B. J., & Roberts, D. E. (2003). Effects of acetazolamide administration on performance battery at moderate altitude. *Medicine and Science in Sports and Exercise, 35*(Suppl.), S163.

Parsons, H. M. (1972). *Man–machine system experiments.* Baltimore: The Johns Hopkins Press.

Pastel, R. (Ed.). (2001). International conference on the operational impact of psychological casualties from weapons of mass destruction [Special issue]. *Military Medicine, 166*(Suppl. 2).

Payne, D. M., Warner, J. T., & Little, R. D. (1992). Tied migration and returns to human capital: The case of military wives. *Social Science Quarterly, 73,* 324–339.

Penney, L. M., Horgen, K. E., & Borman, W. C. (1999). *An annotated bibliography of recruiting research conducted by the U.S. Army Research Institute for the Behavioral and Social Sciences* (PDRI Tech. Rep. No. 340). Tampa, FL: Personnel Decisions Research Institutes.

Pew, R. W. (1994). Paul Morris Fitts, 1912–1965. In H. L. Taylor (Ed.), *Division 21 members who made distinguished contributions to engineering psychology* (pp. 23–44). Washington, DC: American Psychological Association, Division 21, Division of Applied Experimental and Engineering Psychologists.

Phillips, J. M. (1998). Effects of realistic job previews on multiple organizational outcomes: A meta-analysis. *Academy of Management Journal, 41,* 673–690.

Pickren, W. E., & Schneider, S. F. (2005). *Psychology and the National Institute of Mental Health.* Washington, DC: American Psychological Association.

Pierce, P. F. (1998). Retention of Air Force women serving during Desert Shield and Desert Storm. *Military Psychology, 10,* 195–213.

Pierce, P. F., Antonakos, C., & Deroba, B. A. (1999). Health care utilization and satisfaction concerning gender-specific health problems among military women. *Military Medicine, 164*, 98–102.

Pflanz, S., & Sonnek, S. (2002). Work stress in the military: Prevalence, causes, and relationship to emotional health. *Military Medicine, 167*, 877–882.

Plag, J. A., & Goffman, J. M. (1968). Fleet effectiveness prediction studies at a recruit training command. *Naval Research Reviews, 21*(6), 18–25.

Povenmire, H. K., & Roscoe, S. N. (1973). Incremental transfer effectiveness of a ground-based general aviation trainer. *Human Factors, 15*, 534–542.

Premack, S. L., & Wanous, J. P. (1985). A meta-analysis of realistic job previews experiments. *Journal of Applied Psychology, 70*, 706–719.

President's Commission on Mental Health, Exec. Order No. 11,973, 42 C.F.R. 10677 (1977).

Principi, A. J. (2004). *The Department of Veterans Affairs CARES Business Plan Studies*. Retrieved September 1, 2004, from http://www.va.gov/cares

Pryce, J. G., Ogilvy-Lee, D., & Pryce, D. H. (2000). The "citizen–soldier" and Reserve Component families. In J. A. Martin, L. N. Rosen, & L. R. Sparacino (Eds.), *The military family: A practice guide for human service providers* (pp. 25–42). Westport, CT: Praeger.

Psychiatry Working Group. (1985). *Final report: A report prepared for the Deputy Surgeon General of the U.S. Air Force*. Washington, DC: U.S. Air Force.

Pugh, H. L. (Ed.). (1957). *The history of the Medical Department of the United States Navy, 1945–1955*. Washington, DC: U.S. Government Printing Office.

Pulakos, E. D., Tsacoumis, S., & Reynolds, D. H. (1992). *Development of a selection system and a career development program for DEA special agent supervisors and managers* (Publication No. HII FR-92-17). Alexandria, VA: HumRRO International.

Quarles, B. (1953). *The Negro in the Civil War*. Boston: Little Brown.

Quarles, B. (1961). *The Negro in the American Revolution*. Chapel Hill: University of North Carolina Press.

Quick, J. C. (1999a). Occupational health psychology: The convergence of health and clinical psychology with public health and preventive medicine in an organizational context. *Professional Psychology: Research and Practice, 30*, 123–128.

Quick, J. C. (1999b). Occupational health psychology: Historical roots and future directions. *Health Psychology, 18*, 82–88.

Quick, J. C., Barab, J., Fielding, J., Hurrell, J. J., Jr., Ivancevich, J. M., Mangelsdorff, A. D., et al. (1992). Occupational mental health promotion: A prevention agenda based on education and treatment. *American Journal of Health Promotion, 7*, 37–44.

Quick, J. C., Barab, J., Hurrell, J. J., Ivancevich, J. M., Mangelsdorff, A. D., Pelletier, K. R., et al. (1992). Health promotion, education, and treatment. In G. P. Keita & S. L. Sauter (Eds.), *Work and well-being: An agenda for the 1990s* (pp. 43–67). Washington, DC: American Psychological Association.

Quick, J. C., Camara, W. J., Hurrell, J. J., Jr., Johnson, J. V., Piotrokowski, C. S., Sauter, S. L., & Spielberger, C. D. (1997). Introduction and historical overview. *Journal of Occupational Health Psychology, 2,* 3–6.

Quick, J. C., Joplin, J. R., Nelson, D. L., Mangelsdorff, A. D., & Fiedler, E. (1996). Self-reliance and military service training outcomes. *Military Psychology, 8,* 279–293.

Quick, J. C., & Tetrick, L. E. (Ed.). (2003). *Handbook of occupational health psychology.* Washington, DC: American Psychological Association.

Radloff, R., & Helmreich, R. (1968). *Groups under stress: Psychological research in SEALAB II.* New York: Appleton-Century-Crofts.

Raimy, V. C. (Ed.). (1950). *Training in clinical psychology.* New York: Prentice-Hall.

Ralph, J. A., & Sammons, M. T. (in press). Future directions of military psychology. In C. Kennedy (Ed.), *Military psychology: Clinical and operational applications.* New York: Guilford Press.

Ramsberger, P. F. (2001). *HumRRO: The first 50 years.* Alexandria, VA: Human Resources Research Organization.

Rand Corporation. (2004). *History and mission.* Retrieved September 1, 2004, from http://www.rand.org/about/history/

Ransom, M. M. (2001). The boy's club: How "don't ask, don't tell" creates a double-bind for military women. *Law & Psychology Review, 25,* 161–177.

Raphael, B. (1986). *When disaster strikes.* New York: Basic Books.

Raphael, B., & Wilson, J. P. (2000). *Psychological debriefing: Theory, practice and evidence.* Cambridge, England: Cambridge University Press.

Rash, C. E. (Ed.). (1999). *Helmet mounted displays: Design issues for rotary-wing aircraft.* Fort Rucker, AL: U.S. Army Aeromedical Research Laboratory.

Rash, C. E., Verona, R. W., & Crowley, J. S. (1990). Human factors and safety considerations of night vision systems flight using thermal imaging systems. In R. J. Lewandowski (Ed.), *Proceedings of SPIE: Vol. 1290. Helmet-mounted displays II* (pp. 142–164). Bellingham, WA: International Society for Optical Engineering.

Rasmussen, J. E. (Ed.). (1973). *Man in isolation and confinement.* Chicago: Aldine Press.

Ratnaparki, M. V., Ratnaparki, M. M., & Robinette, K. M. (1992). Size and shape analysis technologies for design. *Applied Ergonomics, 23,* 181–185.

Reasor, R. D. (1996, December 4). *Distributed mission training engineering development.* Mesa, AZ: Armstrong Laboratory, U.S. Air Force Scientific Advisory Board Human Centered Technology Panel Science and Technology Review.

Redmond, D. P., & Hegge, F. W. (1985). Observations on the design and specification of a wrist-worn human activity monitoring system. *Behavior Research Methods, Instruments, & Computers, 17,* 659–669.

Reeves, J. J. (2002). Perspectives on disaster mental health intervention from the USNS *Comfort. Military Medicine, 167*(9, Suppl.), 90–92.

Regian, J. W. (1994). *Training research for automated instruction (TRAIN)* (Brochure, U.S. Air Force Scientific Advisory Board). Brooks Air Force Base, TX: Armstrong Laboratory.

Regian, J. W. (1996, December 5). *Cognitive tutoring*. Brooks Air Force Base, TX: Armstrong Laboratory, U.S. Air Force Scientific Advisory Board Human Centered Technology Panel Scientific and Technology Review.

Reidel, S. (2003). Critical thinking training for army schoolhouse and distance learning. *ARI Newsletter, 13*(2), 16–17. (Available from U.S. Army Research Institute for the Behavioral and Social Sciences, 2511 Jefferson Davis Highway, Arlington, VA 22202-3926)

Reiser, R. A. (2001a). A history of instructional design and technology: Part I. A history of instructional media. *Educational Technology, Research and Development, 49*(1), 53–64.

Reiser, R. A. (2001b). A history of instructional design and technology: Part II. A history of instructional design. *Educational Technology, Research and Development, 49*(2), 57–67.

Reister, F. A. (Ed.). (1954). *Battle casualties and medical statistics: U.S. Army experience in the Korean War*. Washington, DC: U.S. Government Printing Office.

Remarks in a discussion on the benefits of health care information technology in Baltimore, Maryland. (2004, April 27). *Weekly Compilation of Presidential Documents, 40*(18), 697–699. Available at http://frwebgate.access.gpo.gov/cgi-bin/getdoc.cgi?dbname=2004_presidential_documents&docid=pd03my04_txt-9

Reskin, B. F., & Padavic, I. (1994). *Women and men at work*. Thousand Oaks, CA: Pine Forge Press.

Reskin, B. F., & Roos, P. A. (1990). *Job queues, gender queues: Explaining women's inroads into male occupations*. Philadelphia: Temple University Press.

Richards, R. L. (1910). Mental and nervous diseases in the Russo–Japanese war. *Military Surgeon, 26*, 177–193.

Ridenour, N. A. (1961). *Mental health in the United States: A fifty-year history*. Cambridge, MA: Harvard University Press.

Ritchie, E. C., Friedman, M., Watson, P., Ursano, R., Wessely, S., & Flynn, B. (2004). Mass violence and early mental health intervention: A proposed application of best practice guidelines to chemical, biological, and radiological attacks. *Military Medicine, 169*, 575–579.

Ritchie, E. C., & Hoge, C. W. (Eds.). (2002). The mental health response to the 9-11 attack on the Pentagon. *Military Medicine, 167*(9, Suppl.).

Robert T. Stafford Disaster Relief and Emergency Assistance Act, P.L. 93-288, 92nd Cong. (1974) (enacted); amended by P.L. 103-181, P.L. 103-337, and P.L. 106-390 (2000).

Roberts, B. J., & Barko, W. (1986). Organizational development in the United States Army: A conceptual case analysis. *Public Administration Quarterly, 10*, 325–335.

Robertson, D. (1994). *Sly and able: A political biography of James F. Byrne*. New York: Norton.

Robinette, K. M., & Whittestone, J. J. (1992). *Methods for characterizations of the human head for the design of helmets* (Publication No. AL-TR-1992-0061). Wright–

Patterson Air Force Base, OH: Crew Systems Directorate, Human Engineering Division, Armstrong Laboratory.

Rockway, M. R., & Yasutake, J. Y. (1974). The evolution of the Air Force advanced instructional system. *Journal of Educational Technology Systems, 2,* 217–239.

Rohall, D. E., Segal, M. W., & Segal, D. R. (1999). Examining the importance of organizational supports on family adjustment to army life in a period of increasing separation. *Journal of Political and Military Sociology, 27,* 49–65.

Romano, J. A., & King, J. M. (2002). Chemical warfare and chemical terrorism: Psychological and performance outcomes [Special issue]. *Military Psychology, 14*(2).

Root, M. J. (2002, October). Keller, Fred S. *American national biography online, October 2002 update.* Retrieved August 5, 2003, from http://www.anb.org/articles/14/14-01126.html

Rosen, G. (1975). Nostalgia: A "forgotten" psychological disorder. *Psychological Medicine, 5,* 340–354.

Rosen, L. N., & Durand, D. B. (2000). Marital adjustment during deployment. In J. A. Martin, L. N. Rosen, & L. R. Sparacino (Eds.), *The military family: A practice guide for human service providers* (pp. 153–165). Westport, CT: Praeger.

Rosen, L. N., & Martin, L. (1998). Incidence and perceptions of sexual harassment among male and female U.S. Army soldiers. *Military Psychology, 10,* 239–257.

Routh, D. K. (1994). *Clinical psychology since 1917: Science, practice, and organization.* New York: Plenum Press.

Routh, D. K. (1996). Lightner Witmer and the first 100 years of clinical psychology. *American Psychologist, 51,* 244–247.

Rowe, J. N. (1971). *Five years to freedom.* New York: Random House.

Russ, C. R., Fonseca, V. P., Peterson, A. L., Blackman, L. R., & Robbins, A. S. (2001). Weight gain as a barrier to smoking cessation among military personnel. *American Journal of Health Promotion, 16*(2), 79–84.

Russell, T. L. (1994). *Job analysis of Special Forces jobs.* Alexandria, VA: U.S. Army Research Institute for the Behavioral and Social Sciences.

Russo, M. B., Vo, A., LaButta, R., Black, I., Campbell, W., Greene, J., et al. (2005). Human biovibrations: Assessment of human life signs, motor activity, and cognitive performance using wrist-mounted actigraphy. *Aviation, Space, and Environmental Medicine, 76*(7, Section 2, Suppl.), C-64–C-74.

Rutte, C. G., Diekmann, K. A., Polzer, J. T., Crosby, F. J., & Messick, D. M. (1994). Organization of information and the detection of gender discrimination. *Psychological Science, 5,* 226–231.

Ryan, P. B. (1985). *The Iranian rescue mission: Why it failed.* Annapolis, MD: Naval Institute.

Saba, F. (2002, May 1). Robert Gagné dies at age 85. *Distance-Educator's Daily News.* Retrieved August 5, 2003, from http://www.distance-educator.com/dnews/Article6712.phtml

Salas, E., DeRouin, R. E., & Gade, P. A. (in press). The military's contribution to our science and practice: People, places, and findings. In L. Koppes, P. Thayer, A. Vinchur, & E. Salas (Eds.), *The science and practice of industrial and organizational psychology: Historical aspects from the first hundred years.* Mahwah, NJ: Erlbaum.

Salas, M., & Besetsny, L. (2000). Transition into parenthood for high-risk families: The new parent support program. In J. A. Martin, L. N. Rosen, & L. R. Sparacino (Eds.), *The military family: A practice guide for human service providers* (pp. 197–205). Westport, CT: Praeger.

Samelson, F. (1977). World War I intelligence testing and the development of psychology. *Journal of the History of the Behavioral Sciences, 13,* 274–282.

Sammons, M. T., & Brown, A. B. (1997). The Department of Defense psychopharmacology demonstration project: An evolving program for postdoctoral education in psychology. *Professional Psychology: Research and Practice, 28,* 107–112.

Sandler, S. (1992). *Segregated skies: All Black combat squadrons of World War II.* Washington, DC: Smithsonian Institution Press.

Sands, W. A., Gade, P. A., & Knapp, D. J. (1997). The computerized adaptive screening test. In W. A. Sands, B. K. Waters, & J. R. McBride (Eds.), *Computerized adaptive testing: From inquiry to operation* (pp. 69–80). Washington, DC: American Psychological Association.

Sands, W. A., Waters, B. K., & McBride, J. R. (Eds.). (1997). *Computerized adaptive testing: From inquiry to operation.* Washington, DC: American Psychological Association.

Sanford, E. C. (1888). Personal equation: I. *American Journal of Psychology, 2,* 1–38.

Sauter, S. L., Murphy, L. R., & Hurrell, J. J., Jr. (1990). Prevention of work-related psychological disorders: A national strategy proposed by the National Institute for Occupational Safety and Health (NIOSH). *American Psychologist, 45,* 1146–1158.

Scarborough, E. (1992). Women in the American Psychological Association. In R. B. Evans, V. S. Sexton, & T. C. Cadwallader (Eds.), *The American Psychological Association: A historical perspective* (pp. 303–325). Washington, DC: American Psychological Association.

Scarville, J. (1990). *Spouse employment in the Army: Research findings.* Alexandria, VA: U.S. Army Research Institute for the Behavioral and Social Sciences.

Schein, E. H. (1957). Reaction patterns to severe, chronic stress in American Army prisoners of war. *Journal of Social Issues, 13,* 21–30.

Schemmer, B. F. (1976). *The raid.* New York: Harper & Row.

Schmitt, N., Pulakos, E. D., Whitney, D. J., Tsacoumis, S., Keenan, P. A., Anderson, R. T., & Copeland, L. A. (1994). *Development of entry-level tests to select FBI special agents* (Publication No. FR-PRD-94-06). Alexandria, VA: Human Resources Research Organization.

Schoomaker, P. J., & Vassalo, A. W. (2004). The way ahead. *Military Review, 82*(2), 2–16.

Schrot, J., Thomas, J. R., & Shurtleff, D. (1996). *Administration of L-tyrosine prevents cold-induced memory deficits in naval special warfare personnel* (NMRI Report No. 96-11). Bethesda, MD: Naval Medical Research Institute.

Schumm, W. R., Bell, D. B., Milan, L. M., & Segal, M. W. (2000). *Family support group (FSG) leaders' handbook.* Alexandria, VA: U.S. Army Research Institute for the Behavioral and Social Sciences.

Schumm, W. R., Bell, D. B., Segal, M. W., & Rice, R. E. (1996). Changes in marital quality among MFO couples. In R. H. Phelps & B. J. Farr (Eds.), *Reserve component soldiers as peacekeepers* (pp. 395–408). Alexandria, VA: U.S. Army Research Institute for the Behavioral and Social Sciences.

Schumm, W. R., Bell, D. B., & Tran, G. (1993). *Family adaptation to the demands of Army life: A review of findings.* Alexandria, VA: U.S. Army Research Institute for the Behavioral and Social Sciences.

Scoville, S. L., Gardner, J. W., & Potter, R. N. (2004). Traumatic deaths during U.S. Armed Forces basic training, 1977–2001. *American Journal of Preventive Medicine, 26,* 194–204.

Seaquist, M. R. (1968, August 6). *Recommendations for the enhancement of careers for clinical psychologists in the Air Force* (Report addressed to Colonel Alvin F. Meyer, Jr., AFMSGGB, Headquarters U.S. Air Force). Washington, DC: U.S. Air Force.

Segal, D. R. (1986). Measuring the institutional/organizational change thesis. *Armed Forces & Society, 12,* 351–375.

Segal, D. R., & Segal, M. W. (2004). The demography of the American military. *Population Bulletin, 59*(4), 1–40.

Segal, D. R., Segal, M. W., Applewhite, L. W., Bartone, J. V., Bartone B. T., Eyre, D. P., et al. (1993). *Peacekeepers and their wives: American participation in the Multinational Force and Observers.* Westport, CT: Greenwood Press.

Segal, D. R., Segal, M. W., Holz, R. F., Norbo, G. J., Seeberg, R. S., & Wubbena, W. L., Jr. (1976). Trends in the structure of Army families. *Journal of Political and Military Sociology, 4,* 135–139.

Segal, M. W. (1986a). Enlisted family life in the U.S. Army: A portrait of a community. In D. R. Segal & H. W. Sinaiko (Eds.), *Life in the rank and file: Enlisted men and women in the armed forces of the United States, Australia, Canada, and the United Kingdom* (pp. 184–211). Washington, DC: Pergamon–Brassey.

Segal, M. W. (1986b). The military and the family as greedy institutions. *Armed Forces & Society, 13,* 9–38.

Segal, M. W. (1989). The nature of work and family linkages: A theoretical perspective. In G. L. Bowen & D. K. Orthner (Eds.), *The organization family: Work and family linkages in the U.S. military* (pp. 3–36). New York: Praeger.

Segal, M. W. (1999). Military culture and military families. In M. F. Katzenstein & J. Reppy (Eds.), *Beyond zero tolerance: Discrimination in military culture* (pp. 251–261). Lanham, MD: Rowman & Littlefield.

Segal, M. W., & Harris, J. J. (1993). *What we know about Army families.* Alexandria, VA: U.S. Army Research Institute for the Behavioral and Social Sciences.

Seglie, L. R., & Selby-Cole, A. (2000). The Army transformation—Learning while doing. *Military Review, 80*(5), 51–56.

Seidenfeld, M. A. (1966). Clinical psychology. In A. Glass & R. Bernucci (Eds.), *Neuropsychiatry in World War II: Zone of interior* (pp. 567–603). Washington, DC: U.S. Government Printing Office.

Seligman, M. E. P. (2002). *Authentic happiness: Using the new positive psychology to realize your potential for lasting fulfillment.* New York: Free Press.

Senge, P. (1990). *The fifth discipline.* New York: Doubleday.

Sergeant, R. L. (1966). *Voice communication problems in spacecraft and underwater operations* (Tech. Rep.). Groton, CT: Naval Submarine Medical Research Laboratory.

Servicemen's Readjustment Act (GI Bill of Rights), P.L. 78-346, 78th Cong. (1944) (enacted).

Shakow, D. (1978). Clinical psychology seen some fifty years later. *American Psychologist, 33,* 148–158.

Shalev, A. Y., Peri, T., Rogel-Fuchs, Y., Ursano, R. J., & Marlowe, D. (1998). Historical group debriefing after combat exposure. *Military Medicine, 163,* 494–498.

Shannon, R. H., & Carter, R. C. (1981). *Technical report: A comparison of tactical naval work stations within air and sea environments.* New Orleans, LA: Naval Biodynamics Laboratory.

Shepard, R. (2000). Carl Iver Hovland: Statesman of psychology, sterling human being. In G. A. Kimble & M. Wertheimer (Eds.), *Portraits of pioneers in psychology* (Vol. 4, pp. 285–301). Washington, DC: American Psychological Association.

Sheridan, C. L., & Radmacher, S. A. (1992). *Health psychology: Challenging the biomedical model.* New York: Wiley.

Sherrow, V. (1996). *Women in the military: An encyclopedia.* Denver, CO: ABC–CLIO.

Shields, J., Hanser, L. M., & Campbell, J. P. (2001). A paradigm shift. In J. P. Campbell & D. J. Knapp (Eds.), *Exploring the limits of personnel selection and classification* (pp. 21–29). Mahwah, NJ: Erlbaum.

Shobe, K., & Severinghaus, R. (2004, May). *Submarine operations—New technology insertion impacts on the practice of navigation & piloting.* Paper presented at the Undersea Human Systems Integration Symposium, Newport, RI.

Shoenberger, R. W., Wherry, R. J., Jr., & Berkshire, J. R. (1963). *Technical report: Predicting success in aviation training.* Pensacola, FL: Naval School of Aviation Medicine.

Shuckit, M. A., Squire, H. S., Hurtado, S. L., Tschinkel, S. A., Minagawa, R., & Shaffer, R. A. (2001). A measure of the intensity of response to alcohol in a military population. *American Journal of Drug and Alcohol Abuse, 27,* 749–757.

Shurtleff, D., Thomas, J. R., Schrot, J., Kowalski, K., & Harford, R. (1994). Tyrosine reverses a cold-induced working memory deficit in humans. *Pharmacology, Biochemistry, and Behavior, 47,* 935–941.

Shute, V. J., & Regian, J. W. (1993). Principles for evaluating intelligent tutoring systems. *Journal of Artificial Intelligence and Education, 4,* 245–271.

Simmons, R. R., Caldwell, J. A., Stephens, R. L., Stone, L. W., Carter, D. J., Behar, I., et al. (1989). *Effects of the chemical defense antidote atropine sulfate on helicopter pilot performance: A simulator study* (USAARL Tech. Rep. No. 89-17). Fort Rucker, AL: U.S. Army Aeromedical Research Laboratory.

Simon, R., & Teperman, S. (2001). The World Trade Center attack: Lessons for disaster management. *Critical Care, 5,* 318–320.

Simpson, C. M. (1983). *Inside the Green Berets: The first thirty years.* Novato, CA: Presidio Press.

Simpson, D. (1998). *Women in the Armed Forces.* Retrieved August 1, 2002, from http://www.au.af.mil/au/aul/bibs/women/womtoc.htm

Smalter, D. J., & Ruggles, R. L., Jr. (1966). Six business lessons from the Pentagon. *Harvard Business Review, 44,* 64–68.

Smelser, J. J., & Mitchell, F. (Eds.). (2002). *Terrorism: Perspectives from the behavioral and social sciences.* Washington, DC: National Academy Press.

Smith, R. H. (1972). *OSS: The secret history of America's first central intelligence agency.* New York: Dell.

Sobel, D. (1995). *Longitude.* New York: Penguin.

Sokol, R. J. (1989). Early mental health intervention in combat situations: The USS *Stark. Military Medicine, 154,* 407–409.

Solomon, P., Kubzansky, P. E., Leiderman, P. H., Mendelson, J., Trumbull, R., & Wexler, D. (Eds.). (1961). *Sensory deprivation.* Cambridge, MA: Harvard University Press.

Sowder, B. J., & Lystad, M. (Eds.). (1986). *Disasters and mental health.* Washington, DC: American Psychiatric Press and Center for Mental Health Studies of Emergencies, National Institute of Mental Health.

Sparrow, J. C. (1951). *History of personnel demobilization in the United States Army.* Washington, DC: Center of Military History, U. S. Army.

Staal, M. A. (2004). A descriptive history of military psychology. *Air Force Psychologist, 23*(1), 6–12.

Staal, M. A., Cigrang, J. A., & Fielder, E. (2000). Disposition decisions in U.S. Air Force basic trainees assessed during mental health evaluations. *Military Medicine, 12,* 187–203.

Stager, P. (1987). Obituary: John L. Kennedy (1913–1984). *American Psychologist, 42,* 1127.

Stanley, J., Segal, M. W., & Laughton, C. J. (1990). Grass roots family action and military policy responses. *Marriage and Family Review, 15,* 207–223.

Stanton, S. L. (1985). *Green Berets at war: U.S. Army Special Forces in Southeast Asia 1956–1975.* Novato, CA: Presidio Press.

Stein, R. J., Haddock, C. K., Talcott, G. W., & Klesges, R. C. (1996). A collaborative effort to reduce smoking in the United States Air Force. *Health Psychologist, 18,* 4–5.

Sternberg, R. J., Forsythe, G. B., Hedlund, J., Horvath, J. A., Wagner, R. K., Williams, W. M., et al. (2000). *Practical intelligence in everyday life*. Cambridge, England: Cambridge University Press.

Stevens, G., & Gardner, S. (1987). But can she command a ship? Acceptance of women by peers at the Coast Guard Academy. *Sex Roles, 16*, 181–188.

Stiehm, J. (1981). *Bring me men and women*. Berkeley: University of California Press.

Stouffer, S. A., Guttman, L., Suchman, E. A., Lazarsfeld, P. F., Star, S. A., & Clausen, J. A. (1950). *Measurement and prediction* (*Studies in social psychology in World War II, Vol. 3*). Princeton, NJ: Princeton University Press.

Stouffer, S. A., Lumsdaine, A. A., Lumsdaine, M. H., Williams, R. M., Jr., Smith, M. B., Janis, I. L., et al. (1949). *The American soldier: Combat and its aftermath* (*Studies in Social Psychology in World War II, Vol. 2*). Princeton, NJ: Princeton University Press.

Stouffer, S. A., Suchman, E. A., DeVinney, L. C., Star, S. A., & Williams, R. M., Jr. (1949). *The American soldier: Adjustment during army life* (*Studies in Social Psychology in World War II, Vol. 1*). Princeton, NJ: Princeton University Press.

Street, L. L., Niederehe, G., & Lebowitz, B. D. (2000). Toward greater public health relevance for psychotherapeutic intervention research: An NIMH workshop report. *Clinical Psychology: Science and Practice, 7*, 127–137.

Street, W. R. (2001, 13 November). *February 10 in psychology*. Retrieved May 16, 2004, from http://www.cwu.edu/~warren/cal0210.html

Strosahl, K., Robinson, P., Heinrich, R. L., Dea, R. A., Del Toro, I., Kisch, J., & Radcliff, A. (1994). New dimensions in behavioral health/primary care integration. *HMO Practice/HMO Group, 8*, 176–179.

Stuart, J. A., & Halverson, R. R. (1997). The psychological status of U.S. Army soldiers during recent military operations. *Military Medicine, 162*(11), 37–43.

Suedfeld, P. (1968). Isolation, confinement, and sensory deprivation. *Journal of the British Interplanetary Society, 21*, 222–231.

Summers, G. M., & Cowan, M. L. (1991). Mental health issues related to the development of a national disaster response system. *Military Medicine, 156*(1), 30–32.

Summers, H. G. (1982). *On strategy: A critical analysis of the Vietnam War*. Novato, CA: Presidio Press.

Swank, R. L. (1949). Combat exhaustion. *Journal of Nervous and Mental Disease, 109*, 475–508.

Swets, J. A., & Bjork, R. A. (1990). Enhancing human performance: An evaluation of "new age" techniques considered by the U.S. Army. *Psychological Science, 1*, 85–96.

Taylor, A. J. W. (1982). The stress of post-disaster body handling and victim identification work. *Journal of Human Stress, 8*(4), 4–12.

Taylor, A. J. W. (1983). Hidden victims and the human side of disasters. *UNDRO News, March/April*, 6–9, 12.

Taylor, A. J. W. (1987). A taxonomy of disasters and their victims. *Journal of Psychosomatic Research, 31*, 535–544.

Taylor, F. V. (1947). Psychology at the Naval Research Laboratory. *American Psychologist, 2*, 87–92.

Taylor, H. L. (1985, September). *Training effectiveness of flight simulators as determined by transfer of training.* Keynote address presented at the North Atlantic Treaty Organization Defense Research Group Panel VIII Symposium, Brussels, Belgium.

Taylor, H. L., & Alluisi, E. A. (1994). Military psychology. In *Encyclopedia of human behavior* (Vol. 3, pp. 191–201). New York: Academic Press.

Taylor, H. L., & Stokes, A. F. (1986). Flight simulators and training devices. In J. Zeidner (Ed.), *Human productivity enhancement: Training and human factors in system design* (Vol. 1, pp. 81–129). New York: Praeger.

Taylor, J. E., & Fox, W. L. (1967). *Differential approaches to training* (Professional Paper 47-67). Alexandria, VA: Human Resources Research Organization.

Taylor, V., & Whittier, N. (1993). The new feminist movement. In L. Richardson & V. Taylor (Eds.), *Feminist frontiers III* (pp. 533–548). New York: McGraw-Hill.

Tedeschi, R. G., & Kilmer, R. P. (2005). Assessing strengths, resilience, and growth to guide clinical interventions. *Professional Psychology, 36*, 230–237.

Terman, L. M. (1916). *The measurement of intelligence: An explanation of and a complete guide for the use of the Stanford revision and extension of the Binet–Simon Intelligence Scale.* Boston: Houghton Mifflin.

Thomas, E. D., Yellen, T. M., & Polese, S. J. (1999, October). *Voices from the past—Command history post WWII to November 1999: An historical account of the Naval Personnel Research & Development Center (NRPDC) of San Diego, CA.* San Diego, CA: Naval Personnel Research and Development Center.

Thomas, J. A. (1988). *Race relations research in the U. S. Army in the 1970s: A collection of selected readings.* Alexandria, VA: U.S. Army Research Institute for the Behavioral and Social Sciences.

Thomas, J. R., Ahlers, S. T., Schrot, J., House, J. F., Van Orden, K. F., & Hesslink, R. H. (1990). Adrenergic response to cognitive activity in a cold environment. *Journal of Applied Physiology, 68*, 962–966.

Thomas, M. (1993). *Display for advanced research and training (DART).* Mesa, AZ: Armstrong Laboratory, Aircrew Training Research Division Brochure.

Thomas, M., Balkin, T., Sing, H., Wesensten, N., & Belenky, G. (1998). PET imaging studies of sleep deprivation and sleep implications for behavior and sleep function. *Journal of Sleep Research, 7*(Suppl. 2), 274.

Thomas, M., Sing, H., Belenky, G., Holcomb, H., Mayberg, H., Dannals, R., et al. (2000). Neural basis of alertness and cognitive performance impairments during sleepiness: I. Effects of 24 hours of sleep deprivation on waking human regional brain activity. *Journal of Sleep Research, 9*, 335–352.

Thomas, P. J., & Thomas, M. D. (1993). Mothers in uniform. In F. W. Kaslow (Ed.), *The military family in peace and war* (pp. 25–47). New York: Springer Publishing Company.

Thomas, P. J., & Thomas, M. D. (1994). Effects of sex, marital status, and parental status on absenteeism among navy enlisted personnel. *Military Psychology, 6*, 95–108.

Thomas, P. J., & Thomas, M. D. (1996). Integration of women in the military: Parallels to the progress of homosexuals? In G. M. Herek, J. B. Jobe, & R. M. Carney (Eds.), *Out in force: Sexual orientation in the military* (pp. 65–85). Chicago: University of Chicago Press.

Thompson, M. (2000, March 27). Aye, aye, ma'am. *Time*, pp. 30–34.

Thorne, D., Genser, S., Sing, H., & Hegge, F. (1985). The Walter Reed performance assessment battery. *Neurobehavioral Toxicology and Teratology, 7*, 415–418.

Tierney, K. J. (2000). Controversy and consensus in disaster mental health research. *Prehospital Disaster Medicine, 15*, 181–187.

Time magazine man of the year: American fighting man. (1951, January 1). *Time, 62*(1).

Time magazine person of the year: The American soldier. (2003, December 29).*Time*, 32–41.

Tolhurst, G. C. (1957). *The relationship of speaker intelligibility to the sound pressure level of continuous noise environments of various spectra and octave-band widths* (Tech. Rep.). Pensacola, FL: Naval School of Aviation Medicine.

Tolhurst, G. C. (1959). *Further approaches to microphone evaluation* (Tech. Rep.). Pensacola, FL: Naval School of Aviation Medicine.

Tormes, F. R., & Guedry, F. E. (1975). Disorientation phenomena in naval helicopter pilots. *Aviation Space & Environmental Medicine, 46*, 387–393.

Trent, L. K. (1992). Nutrition knowledge of active-duty Navy personnel. *Journal of the American Dietetic Association, 92*, 724–728.

Trent, L. K. (1998). Evaluation of a four- versus six-week length of stay in the Navy's alcohol treatment program. *Journal of Studies on Alcohol, 59*, 270–279.

Trent, L. K. (1999). *Para suicides in the Navy and Marine Corps: Hospital admissions, 1989–1995* (NHRC Tech. Rep. No. 99-4D). San Diego, CA: Naval Health Research Center.

Trent, L. K., & Conway, T. L. (1988). Dietary factors related to physical fitness among Navy shipboard men. *American Journal of Health Promotion, 3*(2), 12–25.

Trent, L. K., Hilton, S. M., & Melcer, T. (2004, March). *Pre-service tobacco use among graduating male recruits at MCRD, San Diego.* Paper presented at the 43rd Navy Occupational Health and Preventive Medicine Workshop, Chesapeake, VA.

Trent, L. K., & Stevens, L. T. (1995). Evaluation of the Navy's obesity treatment program. *Military Medicine, 160*, 326–330.

Tupes, E. C., & Christal, R. E. (1961). *Recurrent personality factors based on trait ratings* (USAF ASD Tech. Rep. No. 61-97). Lackland Air Force Base, TX: U.S. Air Force. (Reprinted 1992, *Journal of Personality, 60*, 225–252)

Turnage, J. J., & Kennedy, R. S. (1992). The development and use of a computerized human performance test battery for repeated-measures applications. *Human Performance, 5*, 265–301.

Tyler, M. P., & Gifford, R. K. (1990). Fatal training accidents: The military unit as a recovery context. *Journal of Traumatic Stress, 4*, 233–249.

Uhlaner, J. E. (1967a, September). *Chronology of military psychology in the Army.* Paper presented at the 75th Annual Convention of the American Psychological Association, Washington, DC.

Uhlaner, J. E. (Ed.). (1967b). *Psychological research in national defense today* (Report No. S-1). Washington, DC: U.S. Army Research Institute for the Behavioral and Social Sciences.

Uhlaner, J. E. (1968). *The research psychologist in the Army—1917 to 1967* (U.S. Army Behavioral Sciences Research Laboratory Tech. Rep. 1155). Washington, DC: Department of the Army.

Uhlaner, J. E. (1977). *The research psychologist in the Army—1917 to 1977* (Report No. 1155). Washington, DC: U.S. Army Research Institute for the Behavioral and Social Sciences.

U.S. Adjutant General's Office. (1919). *The personnel system of the United States Army: Vol. 1. History of the personnel system*. Washington, DC: Author.

U.S. Army. (1983). *Army Regulation 40-10: Health hazard assessment program in support of the Army materiel acquisition decision process*. Washington, DC: Headquarters, Department of the Army.

U.S. Army. (1994a). *Field Manual 8-51: Combat stress control in a theater of operations*. Washington, DC: Department of the Army.

U.S. Army. (1994b). *Field Manual 22-51: Leader's manual for combat stress control*. Washington, DC: Department of the Army.

U.S. Army Aeromedical Research Laboratory. (1991). *Annotated bibliography of USAARL technical and letter report publications: Vol. 1. June 1963–September 1987*. Fort Rucker, AL: Author.

U.S. Army Aeromedical Research Laboratory. (1996). *Annotated bibliography of USAARL technical and letter report publications: Vol. 2. October 1987–September 1995*. Fort Rucker, AL: Author.

U.S. Army Human Engineering Laboratory. (1990). *MILESTONES: An annotated bibliography of publications of the U.S. Army Human Engineering Laboratory from 1953–1990*. Aberdeen Proving Ground, MD: Author.

U.S. Army Medical Research and Materiel Command. (2002, October). *Automated Neuropsychological Assessment Metrics (ANAM ™): NEUROCOG & MODcog*. Fort Detrick, MD: U.S. Army Medical Research and Materiel Command. (CD-ROM available from L. Banderet, U.S. Army Research Institute of Environmental Medicine, Kansas Street, Natick, MA 01760)

U.S. Army Military History Institute. (2000, December 22). *Russo–Japanese War, 1904–05: A working bibliography of MHI sources*. Available at http://carlisle-www.army.mil/usamhi/bibliographies/referencebibliographies/russojapanesewar/genmisc.doc

U.S. Army Research Institute for the Behavioral and Social Sciences. (1977). *Women content in units forces development test (MAX WAC)* (Special Report No. 6). Alexandria, VA: Author.

U.S. Army Research Institute for the Behavioral and Social Sciences. (1995). *ARI support of tactical engagement simulation*. Arlington, VA: Author.

U.S. Army Research Institute for the Behavioral and Social Sciences. (2002). *Women in the U.S. Army: An annotated bibliography* (Special Report No. 48). Alexandria, VA: Author.

U.S. Census Bureau. (2004). *International data base*. Retrieved September 1, 2004, from IDB Data Access Options: http://www.census.gov/ipc/www/idbacc.html

U.S. Department of the Army, Office of the Surgeon General. (1949). The U.S. Army's Senior Psychology Student Program. *American Psychologist, 4*, 424–425.

U.S. Department of Health and Human Services. (1999). *Mental health: A report of the surgeon general*. Rockville, MD: Author.

U.S. Department of State. (n.d.). *National Security Act of 1947*. Retrieved December 9, 2005, from http://www.state.gov/r/pa/ho/time/cwr/17603.htm

U.S. Navy. (2005, August 12). *Manual of the Medical Department: Change 126* (NAVMED P-117, chap. 15). Available at http://www.vnh.org/admin/mmd/mmdchapter15.pdf

U.S. War Department. (1919). *Air service medical*. Washington, DC: U.S. Government Printing Office.

van Cott, H. P., & Kinkade, R. G. (1972). *Human engineering guide to equipment design*. Washington, DC: U.S. Government Printing Office.

VandenBos, G. R., & DeLeon, P. H. (1988). The use of psychotherapy to improve physical health. *Psychotherapy, 25*, 335–343.

Van Orden, K. F., Ahlers, S. T., House, J. F., Thomas, J. R., & Schrot, J. (1990). Non-hypothermic cold stress shortens evoked potential latencies in humans. *Aviation, Space, and Environmental Medicine, 61*, 636–639.

Van Orden, K. F., Jung, T. P., & Makeig, S. (2000). Combined eye activity measures accurately estimate changes in sustained visual task performance. *Biological Psychology, 52*, 221–240.

Veterans Administration/Department of Defense Clinical Practice Guideline Working Group. (2004). *Management of tobacco use* (Publication No. 10Q-CPG/TUC-04). Washington, DC: Author.

Veterans Benefits Improvement Act, P.L. 103-446, 103rd Cong., H.R. 5244 (1994) (enacted).

Vickers, R. R., Conway, T. L., & Hervig, L. K. (1990). Demonstration of replicable dimensions of health behaviors. *Preventive Medicine, 19*, 377–401.

Vineberg, R., Sticht, T. G., Taylor, E. N., & Caylor, J. S. (1971). *Effects of aptitude (AFQT), job experience, and literacy on job performance: Summary of HumRRO work units UTILITY and REALISTIC* (Tech. Rep. 71-1). Alexandria, VA: Human Resources Research Organization.

Vogel, J. A., & Gauger, A. K. (1993, May). *An annotated bibliography of research involving women, conducted at the U.S. Army Research Institute of Environmental Medicine* (USARIEM Tech. Note No. 93-5). Natick, MA: U.S. Army Research Institute of Environmental Medicine. (DTIC Document Reproduction Service No. ADA265497)

Wagg, W. L. (1981). *Training effectiveness of visual and motion simulation* (Publication No. AFHRL-TR-79-72). Williams Air Force Base, AZ: Air Force Human Resources Laboratory.

Wain, H. J., Grammer, G. G., Stasinos, J. J., & Miller, C. M. (2002). Meeting the patients where they are: Consultation-liaison response to trauma victims of the Pentagon attack. *Military Medicine, 167*(Suppl. 9), 19–21.

Waite, L. J., & Berryman, S. E. (1986). Job stability among young women: A comparison of traditional and nontraditional occupations. *American Journal of Sociology, 92,* 568–595.

Waits, W., & Waldrep, D. (2002). Application of Army combat stress control doctrine in work with Pentagon survivors. *Military Medicine, 167*(Suppl. 9), 39–43.

Waldrep, D., & Waits, W. (2002). Returning to the Pentagon: The use of mass desensitization following the September 11, 2001 attack. *Military Medicine, 167*(Suppl. 9), 58–59.

Walker, C. B., & Rumsey, M. G. (2001). Application of findings: ASVAB, new aptitude tests, and personnel classification. In J. P. Campbell & D. J. Knapp (Eds.), *Exploring the limits of personnel selection and classification* (pp. 559–576). Mahwah, NJ: Erlbaum.

Waller, D. C. (1994). *The commandos: The inside story of America's secret soldiers.* New York: Simon & Schuster.

Wanous, J. P. (1992). *Organizational entry: Recruitment, selection, orientation and socialization of newcomers* (2nd ed.) Reading, MA: Addison-Wesley.

Wanous, J. P., Poland, T. D., Premack, S. L., & Davis, K. S. (1992). The effects of met expectations on newcomer attitudes and behaviors: A review and meta-analysis. *Journal of Applied Psychology, 77,* 288–297.

Ward, K. D., Kovach, D. W., Klesges, R. C., DeBon, M. W., Haddock, C. K., Talcott, G. W., & Lando, H. A. (2002). Ethnic and gender differences in smoking and smoking cessation in a population of young adult Air Force recruits. *American Journal of Health Promotion, 16,* 259–266.

Weeks, J. L., Mullins, C. J., & Vitola, B. M. (1975). *Airman classification batteries from 1948 to 1975: A review and evaluation* (Publication No. AFHRL-TR-75-8). Brooks Air Force Base, TX: Air Force Human Resources Laboratory.

Weigley, R. F. (1973). *The American way of war: A history of United States military strategy and policy.* New York: Macmillan.

Weingarten, K., Hungerland, J., Brennan, M., & Alfred, B. (1971). *The APSTRAT instructional model* (Professional Paper 6-71). Alexandria, VA: Human Resources Research Organization.

Weitzman, D. O., Kinney, J. A., & Ryan, A. P. (1966). *Technical report: A longitudinal study of acuity and phoria among submariners.* Groton, CT: Naval Submarine Medical Research Laboratory.

Weitzman, E., & Bedell, R. C. (1944). The Central Examining Board for the training of naval air cadets. *Psychological Bulletin, 41,* 57–59.

Wesensten, N., Balkin, T., Thorne, D., Killgore, W., Reichardt, R., & Belenky, G. (2004). Caffeine, dextroamphetamine, and modafinil during 85 hours of sleep deprivation: I. Performance and alertness effects. *Aviation, Space, and Environmental Medicine, 75*(4, Suppl.), B108.

Wesensten, N., Belenky, G., Kautz, M. A., Thorne, D. R., Reichardt, R. M., & Balkin, T. J. (2002). Maintaining alertness and performance during sleep deprivation: Modafinil versus caffeine. *Psychopharmacology (Berlin)*, *159*, 238–247.

Wessely, S., Rose, S., & Bisson, J. A. (1999). A systematic review of brief psychological interventions ("debriefing") for the treatment of immediate trauma-related symptoms and the prevention of post traumatic stress disorder. In Cochrane Collaboration (Ed.), *Cochrane Library* (Issue 4). Oxford, England: Update Software.

Weybrew, B. B. (1959). Patterns of reaction to stress as revealed by a factor analysis of autonomic change measures and behavioral observations. *Journal of General Psychology*, *60*, 253–264.

Weybrew, B. B. (1979). *History of military psychology at the naval submarine medical research laboratory* (NSMRL Report No. 917). Groton, CT: U.S. Naval Submarine Medical Research Laboratory.

Weybrew, B. B., & Molish, H. B. (1979). *Technical report: Attitude change during and after long submarine missions*. Groton, CT: Naval Submarine Medical Research Laboratory.

Weybrew, B. B., & Noddin, E. M. (1978). *Technical report: Changes in psychological characteristics of enlisted submarine volunteers during the period 1950–1973*. Groton, CT: Naval Submarine Medical Research Laboratory.

Wherry, R. J., Jr., & Waters, L. K. (1960a). *Factor analysis of primary and basic stages of flight training: Advanced jet pipeline students* (Tech. Rep.). Pensacola, FL: Naval Aerospace Medical Research Laboratory.

Wherry, R. J., Jr., & Waters, L. K. (1960b). *Factor analysis of primary and basic stages of flight training: Advanced multiengine pipeline students* (Tech. Rep.). Pensacola, FL: Naval Aerospace Medical Research Laboratory.

Wigdor, A. K., & Green, B. F. (Eds.). (1991). *Performance assessment for the workplace*. Washington, DC: National Academy Press.

Wiggins, J. G. (2001). A history of the reimbursement of psychological services: The education of one psychologist in the real world. In R. H. Wright & N. A. Cummings (Eds.), *The practice of psychology: The battle for professionalism* (p. 231). Phoenix, AZ: Zeig, Tucker & Thiesen.

Wilcox, C. (1992). Race, gender, and support for women in the military. *Social Science Quarterly*, *73*, 310–323.

Williams, A. C., Jr., & Flexman, R. E. (1949). Evaluation of the School Link as an aid in primary flight instruction. *University of Illinois Bulletin*, *46*(71, Aeronautics Bulletin 5).

Williams, F. D. G. (1990). *SLAM: The influence of S. L. A. Marshall on the United States Army* (TRADOC Historical Monograph Series). Fort Monroe, VA: U. S. Army Training and Doctrine Command.

Williams, H. L., Lubin, A., & Goodnow, J. J. (1959). Impaired performance with acute sleep loss. *Psychological Monographs*, *73*(14, No. 484), 1–26.

Williams, J. H., Fitzgerald, L. F., & Drasgow, F. (1999). The effects of organizational practice on sexual harassment and individual outcomes in the military. *Military Psychology, 11,* 303–328.

Winkenwerder, W. (2003). *Force health protection briefing.* Retrieved September 1, 2004, from http://www.defenselink.mil/transcripts/2003/t03142003_t0313fhp.html

Wisher, R. A., & Olson, T. M. (2002). *The effectiveness of Web-based training* (Research Rep. No. 1802). Alexandria, VA: U.S. Army Research Institute.

Wisher, R. A., Sabol, M. A., Moses, F. L., & Ramsberger, P. F. (2002). *Distance learning: The soldier's perspective.* Alexandria, VA: HumRRO.

Wiskoff, M. F. (1997). Defense of the nation: Military psychologists. In R. J. Sternberg (Ed.), *Career paths in psychology: Where your degree can take you* (pp. 245–268). Washington, DC: American Psychological Association.

Wiskoff, M. F., & Rampton, G. M. (Eds.). (1989). *Military personnel measurement: Testing, assignment, evaluation.* New York: Praeger.

Wonderlic, E. F. (1992). *Wonderlic Personnel Test user's manual.* Libertyville, IL: Author.

Wood, D. P., Koffman, R. L., & Arita, A. A. (2003). Psychiatric medevacs during a six-month aircraft carrier battle group deployment to the Persian Gulf: A Navy force health protection preliminary report. *Military Medicine, 168*(1), 43–47.

Woods, N. F. (1995). Women and their health. In C. I. Fogel & N. F. Woods (Eds.), *Women's health care: A comprehensive handbook* (pp. 1–22). Thousand Oaks, CA: Sage.

Wright, K. M., Thomas, J. L., Adler, A. B., Ness, J. W., Hoge, C. W., & Castro, C. A. (2005). Psychological screening procedures for deploying U.S. Forces. *Military Medicine, 170,* 555–562.

Wright, K. M., Ursano, R. J., Bartone, P. T., & Ingraham, L. H. (1990). The shared experience of catastrophe: An expanded classification of the disaster community. *American Journal of Orthopsychiatry, 60,* 35–42.

Wright, K. M., Huffman, A. H., Adler, A. B., & Castro, C. A. (2002). Psychological screening program overview. *Military Medicine, 167,* 853–861.

Wu, S. K. (2001). Factors influencing sport participation among athletes with spinal cord injury. *Medical Science and Sports Exercise, 33,* 177–182.

Yerkes, R. M. (1917). The relation of psychology to military activities. *Mental Hygiene, 1,* 371–376.

Yerkes, R. M. (1918). Psychology in relation to the war. *Psychological Review, 25,* 85–115.

Yerkes, R. M. (Ed.). (1921). Psychological examining in the U.S. Army. *Memoirs of the National Academy of Sciences, 15,* 1–890.

Yoder, J. D. (2001). Military women. In J. Worell (Ed.), *Encyclopedia of women and gender* (Vol. 2, pp. 771–782). New York: Academic Press.

Yoder, J. D. (2002). 2001 Division 35 presidential address: Context matters: Understanding tokenism processes and their impact on women's work. *Psychology of Women Quarterly, 26,* 1–8.

Yoder, J. D., Adams, J., & Prince, H. (1983). The price of a token. *Journal of Political and Military Sociology, 11*, 325–337.

Zaccaro, S. J. (2001). *The nature of executive leadership: A conceptual and empirical analysis of success*. Washington, DC: American Psychological Association.

Zaccaro, S. J. (2002, June 12). *Testimony on fiscal year 2003 appropriations*. Retrieved May 15, 2004, from http://www.apa.org/ppo/issues/zaccarotest.html

Zaccaro, S. J., & Horn, Z. N. J. (2003). Leadership theory and practice: Fostering an effective symbiosis. *Leadership Quarterly, 14*, 769–806.

Zehner, G. F. (1994). Anthropometric accommodation in USAF cockpits. In K. Krishen (Ed.), *Seventh annual workday on space operations* (pp. 32–40). Washington, DC: National Aeronautics and Space Administration.

Zehner, G. F., Meindl, R. S., & Hudson, J. A. (1992). *A multivariate anthropometric method for crew station design: A bridge* (Report No. Al-7R-1992-0164). Wright–Patterson Air Force Base, OH: Armstrong Laboratory.

Zeidner, J. (1986). *Human productivity and enhancement: Vol. 1. Training and human factors in systems design*. New York: Praeger.

Zeidner, J. (1987). *Human productivity and enhancement: Vol. 2. Organization, personnel, and decision making*. New York: Praeger.

Zeidner, J., & Drucker, A. (1987). *Behavioral science in the Army: A corporate history of the Army Research Institute*. Alexandria, VA: U.S. Army Research Institute for the Behavioral and Social Sciences.

Zimbardo, P. G. (1973). On the ethics of intervention in human psychological research: With special reference to the Stanford Prison Experiment. *Cognition, 2*, 243–256.

Zimbardo, P. G. (2002). Psychology in the public service. *American Psychologist, 57*, 431–433.

Zimbardo, P. G. (2004). Does psychology make a significant difference in our lives? *American Psychologist, 59*, 339–351.

Zimbardo, P. G., Maslach, C., & Haney, C. (2000). Reflections on the Stanford Prison Experiment: Genesis, transformations, consequences. In T. Blass (Ed.), *Obedience to authority: Current perspectives on the Milgram paradigm* (pp. 193–237). Mahwah, NJ: Erlbaum.

Zook, L. M. (1996). *Soldier selection: Past, present, and future* (ARI Special Report 28). Alexandria, VA: U.S. Army Research Institute for the Behavioral and Social Sciences.

Zubek, J. P. (1969). *Sensory deprivation: Fifteen years of research*. New York: Appleton-Century-Crofts.

AUTHOR INDEX

Flanagan, J. C., 30
Fleming, J., 123, 124
Fletcher, J. D., 121, 122, 241
Flexman, R. E., 120
Foa, E. B., 26
Folen, R. A., 242
Foley, T., 171, 173, 176
Fonseca, V. P., 178
Ford, J. P., 108
Ford, L. A., 110
Fox, W. L., 107
Frank, L. H., 15
Franklin, J. H., 201–203
Franz, S. I., 17
Frelin, A. J., 22
Friedl, K. E., 74
Friedl, V. L., 212, 213, 215, 220
Friedman, M. J., 26, 249

Gade, P. A., 38, 189, 190, 193, 246
Gagné, R. M., 21, 34
Gaines, C., 229
Gal, R., 26, 31, 72, 74
Garamone, J., 221
Gardner, J. W., 23
Gardner, S., 217
Garland, F. C., 65
Garner, W. R., 31, 72
Gauger, A. K., 74
Gaydos, J. C., 22
Gayton, S. J., 243
Geldard, F. A., 53
General Accounting Office, 213, 214, 216–
218, 220, 221
Genser, S., 78
Gerber, P. D., 180
Gifford, R. K., 26, 245
Gillin, J. C., 67
Gimbel, C., 246
Ginn, R. V. N., 10, 17, 154, 155, 157, 158,
163
Glass, A. J., 14, 18, 19, 23, 24, 25
Glenn, J. F., 75
Glick, P., 218
Godbee, D. C., 11
Goffman, J. M., 57
Goldfrank, L. R., 248
Goldstein, J., 211
Goodman, J., 58
Goodnow, J. J., 78
Gorman, P. F., 191
Graham, P., 119

Graham, W. F., 60
Grammer, G. G., 26
Grau, L. W., 22
Gray, D. P., 10, 238
Gray, T. H., 119
Graybiel, A., 51
Green, B. F., 103, 104
Green, R. J., 116, 117
Greene, R. L., 88
Grefer, J. E., 148
Greig, J. L., 75
Griffith, J. D., 226
Grill, D. J., 143
Grinker, R. F., 23
Griswold, L. M., 69
Gropman, A., 200, 201, 203–205, 207, 209
Grubb, G. N., 190
Guedry, F. E., 51
Gum, D. R., 119
Gunderson, E. K. E., 15, 57, 58, 62, 63, 64
Guttmacher, M., 161

Haddock, C. K., 178, 181
Halverson, R. R., 23
Hamlin, E. R., II, 232
Hammes, J. A., 105
Hammond, J., 11, 248
Haney, C., 91
Hanna, T. E., 52
Hanser, L. M., 187
Harbeson, M. M., 48
Harford, R., 68
Harper, G. P., 238
Harrell, T. W., 32
Harris, D. M., 148
Harris, J. D., 52
Harris, J. J., 195, 226, 230, 233
Hartel, C., 193
Hartsough, D. M., 248
Hay, M. S., 219
Haythorn, W. W., 53
Headley, D. B., 79
Health Promotion Program, 19
*Healthy People: The Surgeon General's Report
on Health Promotion and Disease Pre-
vention*, 243
Hegge, F. W., 49, 78
Helba, C., 229
Helmkamp, J. C., 23
Helmreich, R., 53
Helms, R. F., 226
Hendrix, W. H., 33

Herbold, J. R., 64
Herek, G. M., 220, 221
Hervig, L. K., 60–62
Highfill-McRoy, R. M., 64
Hilgard, E. R., 31, 32, 35, 37, 38
Hilton, S. M., 58, 59, 63, 221
Hilton, T. T., 30, 31, 33
Hirsch, K., 59
Hodgdon, J. A., 61
Hoffman, H. S., 52
Hogan, D. W., 93
Hoge, C. W., 23, 26, 147, 150, 245, 249
Hoiberg, A., 37, 57, 60, 61, 212
Holland, V. M., 190
Hollander, J. R., 50
Holloway, J. D., 245
Holm, J., 212
Holsenbeck, L. S., 11
Horgen, K. E., 189
Horley, G. L., 73
Horn, Z. N. J., 192
Hosek, J., 230
Hourani, L. L., 19, 58–60
House, J. F., 68
Hovland, C. I., 35
Hudson, J. A., 118
Huffman, A. H., 20, 245
Hughes, R., 119
Huleatt, W. J., 26, 245
Human Resources Research Organization, 15
Hungerland, J., 107
Hunter, C. L., 180
Hunter, D. L., 170
Huntington, D. L., 14
Hupp, D. I., 51
Hurrell, J. J., Jr., 22
Hurrell, R. M., 214
Hurtado, S. L., 63, 64
Hutchins, C. W., Jr., 51
Hutt, M. L., 158

Iadeluca, R. B., 233
Iancu, J., 26
Incentives for the Use of Health Information Technology and Establishing the Position of the National Health Information Technology Coordinator, xiv
Ingraham, L. H., 77, 245
Inlow, M., 67
Institute of Medicine, xiv, 244, 245, 249
Ireland, M. W., 14

Iscoe, I., 248
Isler, W. C., 180
Ivie, R., 246

Jacobs, G. A., 248
Jacobs, T. O., 192
Jacobson, J. Z., 32
James, J. J., 22
James, L. C., 26, 242, 245
Janke, R., 178
Janowitz, M., 14
Jaques, E., 192
Jeffrey, R. J., 22
Jenkins, J. G., 49, 50
Jerome, L. W., 242
Jobes, D. A., 179
John, O. P., 114
Johnson, L. C., 49
Johnson, R. F., 75, 76
Johnson, W. B., 52, 238
Jones, A. P., 57
Jones, D., 59
Jones, D. E., 146
Jones, F. D., 14, 25, 26
Joplin, J. R., 241
Jorgensen, W. A., 22
Joy, R. J. T., 14
Jung, T. P., 67

Kamimori, G. H., 78
Kanter, R. M., 217, 231
Kaplan, J., 193
Kaplan, J. D., 190
Kaplan, Z., 26
Kasl, S. V., 60
Kaszas, S. L., 62
Katz, L. C., 190
Katz, P., 226
Keane, T. M., 26
Keenan, P. A., 105
Keene, J. D., 18
Keller, F. S., 34
Kellogg, R. S., 119
Kelly, H. O., 40
Kelly, T., 49
Kenardy, J., 150
Kendra, J. M., 11, 248
Kennedy, K., 59
Kennedy, R. D., 23
Kennedy, R. S., 48, 51
Kern, R. P., 102
Kilbourne, B., 58

Maslach, C., 91
Matarazzo, J. D., 244
Mattock, M., 230
Mavor, A. S., 104
McBride, J. R., 104, 132
McCallum, M., 79
McCaughey, B. G., 26
McCluskey, M. R., 108
McCormick, E. J., 31
McCrae, R. R., 114
McCullough, D. G., 204n4
McDaniel, J. W., 117
McDonough, J. H., 79
McEvoy, G. M., 14
McFann, H. H., 105
McGrath, J. E., 25
McGuire, F. L., 36, 141–143, 151
McLellan, T. M., 78
McNally, M. S., 60
McPherson, J., 201, 202n1
Meglino, B. M., 14, 189
Meier, N. J., 244
Meindel, R. S., 118
Melcer, T., 63
Meliza, L. L., 191, 242
Melton, A. W., 21, 30, 43, 45, 111, 112
Menninger, W. C., 154
Merill, L. L., 57
Messick, D. M., 218
Milan, L. M., 233
Military Family Resource Center, 227, 232
Miller, C. M., 26
Miller, G. A., xiii, 247
Miller, J., 178
Miller, J. G., 36
Miller, J. W., 51
Miller, M., 204n4
Milliken, C. S., 26
Misra, R., 214
Mitchell, B., 213
Mitchell, F., 248, 249
Mitchell, J. L., 115
Mitchell, J. T., 26
Mitler, M., 79
Moe, K. O., 178, 180
Moeller, G., 48, 51
Moes, G. S., 52
Moghaddam, F. M., 247
Moise, S., 78
Molish, H. B., 52
Moore, H. J., 51
Morgan, C. T., 31, 72

Morgan, W. J., 83, 85
Morrison, J. E., 191, 242
Morse, V. A., 52
Moses, G. R., 240
Moses, F. L., 241
Mosier, J., 179
Moskos, C., 14, 210
Motowidlo, S. J., 188
Mullins, C. J., 113
Murphy, L. R., 22
Murray, J. D., 173, 175
Myers, C. S., 25
Myers, D. G., 248
Myers, R. B., 11

Naitoh, P., 49, 52
Nalty, B., 201–203
National Academy of Sciences, 222
National Center for Post Traumatic Stress
    Disorder, 150, 245, 249
National Coalition on Health Care, 179
National Commission on Terrorist Attacks
    Upon the United States, 239
National Committee on Vital and Health
    Statistics, 179
National Institute for Occupational Safety
    and Health, 22
National Institute of Mental Health, 12, 15,
    26–27, 249
National Institutes of Health, 65
National Research Council, 247
National Security Strategy, 11, 239
Naval Research Laboratory, 15
Navy Fact File, 146
Neel, S., 14
Neff, W. M., 179
Nelson, C., 218
Nelson, D. L., 241
Nelson, P. D., 43, 50
Newhouse, P. A., 78, 79
Newman, R., 143, 166, 244, 247, 249
Nice, D. S., 221
Niederehe, G., 179
Noddin, E. M., 52

O'Connor, M. G., 21, 244
Odom, J. W., 11
O'Donnell, R. D., 78
O'Donnell, V. M., 78, 79
O'Donohue, W. T., 244
Office of Naval Research, 15, 21

Office of Strategic Services Assessment Staff, 83, 85
Ogilvy-Lee, D., 234
Older, H. J., 45
Olmstead, B., 43, 211, 213, 247
Olsen, J. R., 121, 241
Olson, G. M., 48
Olson, T. M., 241
Oordt, M., 180
Operation Iraqi Freedom Mental Health Advisory Team, 149, 249
Ormel, J., 180
Orthner, D. K., 226, 230
Osborn, W. C., 108, 109
Otis, G. A., 14
Oxman, T. E., 180

Padavic, I., 213
Page, R. C., 115
Palmer, V., 231
Panigrahi, B., 214
Panzer, A. M., 248
Paparone, C. R., 241
Parker, S. B., 69
Parsons, H. M., 72
Pastel, R., 80
Patriarca-Troyk, L. A., 64
Payne, D. M., 230
Pennell, R., 122
Penney, L. M., 189
Peri, T., 26
Peterson, A. L., 178
Peterson, K. A., 60
Pew, R. W., 112
Pflanz, S., 246
Phelps, R., 143, 166, 244
Phillabaum, R. A., 73
Phillips, B., 68, 202n1
Phillips, J. M., 14
Phillips, T., 201
Pickens, T., 119
Pickler, A. G., 52
Pickren, W. E., 12, 15
Pierce, P. F., 216, 221
Plag, J. A., 57
Planchon, L. A., 248
Poland, T. D., 14
Polzer, J. T., 218
Potter, R. N., 23
Povenmire, H. K., 120
Powell, J. B., 244
Prather, D. C., 119

Premack, S. L., 14
President's Commission on Mental Health, xiii
Prince, H., 217
Principi, A. J., 244, 245
Pryce, D. H., 234
Pryce, J. G., 234
Pryor, R., 244
Psychiatry Working Group, 173
Pugh, H. L., 14
Pugh, W. M., 57
Puig, J. A., 48
Pulakos, E. D., 105

Quarles, B., 201, 202
Quick, J. C., 22, 241

Radloff, R., 53
Radmacher, S. A., 179
Rahe, R. H., 62
Raimy, V. C., 36
Rakoff, S. H., 226
Ralph, J. A., 146, 147
Rampton, G. M., 18, 138
Ramsberger, P. F., 15, 106, 241
Rand Corporation, 15, 21
Ransom, M. M., 221
Raphael, B., 26
Rash, C. E., 75
Rasmussen, J. E., 53
Ratnaparki, M. M., 116
Ratnaparki, M. V., 116
Ravlin, E. C., 14, 189
Reasor, R. D., 121
Redmond, D. P., 78
Reeves, J. J., 26
Regan, J. J., 48
Regian, J. W., 122–124
Reidel, S., 190
Reiser, R. A., 33, 34, 241
Reisinger, C. W., 179
Reister, F. A., 14
Reskin, B. F., 213
Reynolds, D. H., 105
Reyst, H. E., 219
Rice, R. E., 229, 233
Richards, R. L., 14, 19, 23
Richardson, J. W., 58
Ridenour, N. A., 33
Ritchie, E. C., 26, 27, 245, 249
Roadman, C. H., 178
Robbins, A. S., 178

Van de Water, M., 247
Van de Wetering, B., 137
Van Orden, K. F., 67, 68
Vassalo, A. W., 243
Verona, R. W., 75
Veterans Administration/Department of
     Defense Clinical Practice Guideline
     Working Group, 181
Vickers, R. R., 60
Vickers, R. R., Jr., 61
Vineberg, R., 107
Vitola, B. M., 113
Vogel, J. A., 74

Wachtendorf, T., 11, 248
Wagg, W. L., 120
Wain, H. J., 26
Waite, L. J., 216
Waits, W., 26
Waldo, C. R., 219
Waldrep, D., 26
Walker, C. B., 110, 187
Waller, D. C., 87
Walsh, B. J., 69
Wanous, J. P., 14
Ward, K. D., 178
Warner, H. D., 119
Warner, J. T., 230
Warrack, A. G., 59
Waters, B. K., 102, 104, 132
Waters, L. K., 50
Watten, R. H., 61
Weeks, J. L., 113
Weigley, R. F., 14
Weinberg, E., 244
Weingarten, K., 107
Weissman, S., 51
Weitzman, D. O., 51
Weitzman, E., 49
Wesensten, N., 77–79
Wessely, S., 150
Weybrew, B. B., 15, 51, 52
Wherry, R. J., Jr., 50
Whicker, M. L., 248
White, R., 35

Whittestone, J. J., 116
Whittier, N., 213
Wicherski, M., 36
Wigdor, A. K., 103
Wiggins, J. G., 165
Wilcox, C., 214
Wilcox, V. L., 14
Williams, A. C., Jr., 120
Williams, F. D. G., 26
Williams, F. E., 14
Williams, H. L., 78
Williams, J. H., 220
Williams, R. M., Jr., 21, 35
Williford, R. C., 31
Wilson, J. P., 26
Wilson, M., 189
Wilson, P. G., 180
Winkenwerder, W., 19, 22
Wisher, R. A., 241, 242
Wiskoff, M. F., 18, 43, 138
Wonderlic, E. F., 88
Wood, D. P., 146
Wood, D. R., 244
Woods, N. F., 221
Woodward, J. J., 14
Worell, J., 211n
Wright, K. M., 20, 245
Wright, L. C., 229
Wright, V. R., 212
Wu, S. K., 148

Yagi, K., 102
Yasutake, J. Y., 121, 122
Yerkes, R. M., 17, 32
Yoder, J. D., 211n., 217
Yuan, H., 19, 60

Zaccaro, S. J., 40, 192
Zehner, G. F., 118
Zeidner, J., 15, 17, 21, 185, 187
Zimbardo, P. G., 91, 144, 247
Zion, C., 218
Zook, L. M., 103, 187
Zubek, J. P., 52

SUBJECT INDEX

AAAP. *See* American Association of Applied Psychology
AAF. *See* Army Air Force
AAF Aviation Psychology Program, 111–112
AAF Flying Training Command, 111
AAF Qualifying Examination, 111
AAR. *See* After action reviews
Aberdeen Proving Ground, 72–73
ACCT Battery. *See* Air-Crew Classification Test Battery
Acetylzolamide, 69
Actigraph, 68
Action research, 35
ACT–R theory, 194
Acute mountain sickness (AMS), 68–69
Acute stress, 150
Adams, Donald K., 85
Advanced Instruction System (AIS), 121–122
Advanced Personnel Test, 114
Advanced Research Projects Agency, 241n1
Advanced Simulator for Pilot Training (ASPT), 119–120
Advanced Simulator for Undergraduate Pilot Training (ASUPT), 119–120
Advancement Planning Model, 130
Aero-Medical Laboratory, 31
Afghanistan
    mental health assessment of troops in, 149–150
    Reserves deployed to, 243
AFHRL. *See* Air Force Human Research Laboratory
AFMS (Air Force Medical Service), 176
AFPTRC (Air Force Personnel and Training Research Center; Lackland Air Force Base, San Antonio, Texas), 112
AFQT. *See* Armed Forces Qualification Test
African Americans
    marriage rates of, 228
    military service history of, 201
African theater (World War II), 25
AFRP (Army Family Research Program), 195
After action reviews (AARs), 26, 191

AGCT (U.S. Army General Classification Test), 32
AIR (American Institutes for Research), 39
Air Combat Command, 121
Air Corps Physiological Research Laboratory, 31
Air-Crew Classification Test (ACCT) Battery, 18, 111, 115
Air Force. *See* U.S. Air Force
Air Force Basic Military Training program (Lackland Air Force Base, San Antonio, Texas), 170
Air Force clinical psychologists, 176
    internship programs for, 174–175
    national security needs met by, 173–174
    professional standards for, 174
Air Force Human Research Laboratory (AFHRL; Brooks Air Force Base, San Antonio, Texas), 21, 112
    and advanced simulator training for pilots, 120
    and SOPHIE, 123
Air Force Human Resources Research Center, 32–33
Air Force Medical Service (AFMS), 176
Air Force Office of Scientific Research, 120, 122
Air Force Office of Special Investigations, 170
Air Force Personnel and Training Research Center (AFPTRC; Lackland Air Force Base, San Antonio, Texas), 112
Air Force Personnel Laboratory, 113
Air Force Research Laboratory (Wright–Patterson Air Force Base, Dayton, Ohio), 112
Air Force Scientific Advisory Board, 112
Airman Classification Battery, 113
Airman Qualification Examination, 113
Air Service Medical Research Library, 30
AIS (Advanced Instruction System), 121–122
Alcohol abuse, 63–64. *See also* Substance abuse
Almond, Edwin M., 208

Al Qaeda, 13
Altitude-related cognitive deficits, 68–69
AMEDD. *See* Army Medical Department
American Academy of Sleep Medicine, 68
American Association of Applied Psychology (AAAP), 37, 38
American Board of Professional Psychology, 159
American Institutes for Research (AIR), 39
American Psychological Association (APA), xiii
    Council of Representatives, xiii
    counseling training programs developed by, 139
    history of, 37–39
    and National Naval Medical Center internships, 142
    and "Resilience in a Time of War" campaign, 5
    Resilience Initiative, 235
    response to national security challenges, 247–248
    response to September 11, 2001, 29
    Subcommittee on Psychology's Response to Terrorism, 247
American Psychological Association of Graduate Students (APAGS), xiii
American Red Cross, 11, 178, 248
American Soldier series (*Studies in Social Psychology*), 21, 35, 183
American War for Independence, 212
AMS (acute mountain sickness), 68–69
Andrews Air Force Base (Maryland). *See* Malcolm Grow Medical Center
Anthropometry, 116
Antisubmarine warfare training, 135–136
Anxiety disorders, 58–59
APA. *See* American Psychological Association
APAGS. *See* American Psychological Association of Graduate Students
APA Subcommittee on Psychology's Response to Terrorism, 247
APP (Aviation Psychology Program), 32
Applied social psychology, 34–35, 183
APSTRAT, 107
Aptitude, 31–33
ARI. *See* U.S. Army Research Institute for the Behavioral and Social Sciences
Armed forces. *See* U.S. armed forces
Armed forces color vision test, 51

Armed Forces Qualification Test (AFQT), 18, 187
Armed Services Vocational Aptitude Battery (ASVAB), 18, 56, 102–103, 187
    CAM vs., 114
    CAT–ASVAB, 104
    female disadvantage on, 217
    to replace individual service test batteries, 114
Armstrong, Harry G., 31
Armstrong Aerospace Medicine Research Laboratory, 116
Armstrong Laboratory, 112
    and cognition study, 114
    and Distributed Mission Training, 120–121
    Intelligent Computer Assisted Training Testbed, 123
    and TRAIN Project, 122
    whole-body scan installed by, 116
Army. *See* U.S. Army
Army Air Corps, 31, 201–203
Army Air Corps Aviation Psychology Program, 31
Army Air Force (AAF)
    aviation psychologists in, 111–112
    Aviation Psychology Program, 111–112
    Flying Training Command, 111
    Qualifying Examination, 111
Army Alcohol and Drug Abuse Prevention and Control Program, 19
Army Alpha test, 17, 32, 102
Army Beta test, 17, 32, 102
Army clinical psychologists. *See also* Clinical psychology (in U.S. Army)
    contributions of, to military psychology, 160–166
    duties of, 158–160
    educational standards for, 156
    obtaining, after World War II, 153–157
    organizational structure for, 156–157
    recruiting/training for, 160
    role of during World War II, 158
    training of, 163
Army Community Service, 225, 233
Army Europe Stress Management Team, 26
Army Experience Survey, 188
Army Family Research Program (AFRP), 195
Army General Classification Test (AGCT), 18, 32, 186
Army Knowledge Online portal, 242

Army Manpower and Personnel Integration (MANPRINT), 76–77, 193–194
Army Medical Department (AMEDD), 25, 154, 163
Army Medical Research Institute of Chemical Defense (USAMRICD), 79
Army Medical Research Institute of Infectious Diseases (USAMRIID), 80
Army Medical Service Corps, 156
Army–Navy Medical Services Corps Act, 10, 156
Army Nurse Corps, 212, 214
Army Rangers. *See* U.S. Army Rangers
Army Recruiter Course, 189
Army Recruiter School, 189
Army Recruiting Command. *See* U.S. Army Recruiting Command
Army Research Institute for the Behavioral and Social Sciences. *See* U.S. Army Research Institute for Environmental Medicine
Army Research Institute for Environmental Medicine (USARIEM; Natick, Massachusetts), 73–76
Army Research Institute for the Behavioral and Social Sciences. *See* U.S. Army Research Institute for the Behavioral and Social Sciences
Army Center for Health Promotion and Preventive Medicine, 76
Army Signal Corps, 34
Army Training Program, 20
Army War College (Carlisle Barracks, Pennsylvania), 158, 202
ARPANET, 241n1
Articles of War, 220
ASPT (Advanced Simulator for Pilot Training), 119–120
ASUPT (Advanced Simulator for Undergraduate Pilot Training), 119–120
ASVAB. *See* Armed Services Vocational Aptitude Battery
Atkins, Tommy, 246
Attrition, 58
    of U.S. Air Force, 171
    of U.S. Army Special Forces, 87
    of women, during Persian Gulf Wars, 216
Auditory signal detection, 51–52
Automated instructional design systems, 135
Automated Neuropsychological Assessment Metrics, 78

Aviation Psychology Program (APP), 32
Aviation Research Laboratory (University of Illinois), 120

Bafa Bafa, 133
Baker, Newton D., 32
Bank, Aaron, 86
Barbanel, Laura, 247
Base Realignment and Closure Commission (BRAC), 127
Basic Electricity and Electronics School (San Diego, California), 134
Battlefield training, 190–191
Battle of the Bulge, 206
Behavioral Analysis Service, 170
Behavioral Health Integration Project, 151n1
Behavioral Research Center, 35
Bell Telephone Laboratories, 35
Bennett, Rawson, II, 45
Berlin Wall, fall of, 16
Bethesda, Maryland. *See* National Naval Medical Center
Bevan, Bill, xiii–xiv
Binet–Simon test, 17
Bingham, Walter Van Dyke, 32
Bin Laden, Osama, 13
Biological weapons research, 80
Body fat ratios, 222
Boulder Conference, 36
Boys in the Barracks, 77
BRAC (Base Realignment and Closure Commission), 127
Bridgeport, California (Marine Corps Mountain Warfare Training Center), 69
Brief Vestibular Disorientation Test, 51
Broad Area Announcement, 3
Bronfenbrenner, Urie, 85
Brooks Air Force Base (San Antonio, Texas), 56
    Air Force Human Research Laboratory at. *See* Air Force Human Research Laboratory
    Armstrong Laboratory at. *See* Armstrong Laboratory
Bryan, Alice, 37
Bryan, Glen, 45
Building Healthy Communities initiatives, 181
Bureau of Medicine and Surgery (U.S. Navy), 151n2
Bureau of Naval Personnel, 130–131

Cold War, 44, 127, 153
Collins, James Lawton, Jr., 206, 207
Color vision test, 51
Combat exclusion policy, 212
Combat exhaustion, 23
Combat fatigue, 162
Combat operational stress response (COSR), 149–150
Combat Stress Control Detachments, 162, 164
COMBIMAN, 117
Community-oriented internships, 164
Comprehensive Occupational Data Analysis Program (CODAP), 115
Computer adaptive testing (CAT), 128
Computer-based training, 121–124, 190
    Advanced Instruction System, 121–122
    intelligent tutoring systems for, 122–124
Computerized Adaptive Screening Test (CAST), 189
Computerized adaptive testing (CAT), 104, 132
Computerized Adaptive Testing: From Inquiry to Operation, 104–105
Computerized training, 134
Computer simulations, 135
Contract research, 46
COSR (combat operational stress response), 149
Counseling psychology. See Clinical and counseling psychology
Counseling training programs, 139
Counterintelligence investigation, 84
Crawford, Meredith, 39, 100
CREW CHIEF, 117
Critical Thinking Skills Training, 190
Cyberterrorism, 249

DART (Display for Advanced Research and Training), 120
Darwin, Charles, 202
Davis, Benjamin O., Jr., 207
Davis, Benjamin O., Sr., 207
Dayton, Ohio
    Air Force Research Laboratory at, 112
    Environmental Health Effects Laboratory at, 56
    Wright Patterson Air Force Base. See Wright Patterson Air Force Base
Debriefing, 26–27
Defense Manpower Commission, 126
Defense Science Board, 12–13, 245

DeLeon, Pat, 159, 247
Demographics. See also specific groups, i.e.: Women
    military, 16
    of military casualties, 23
    of military families, 226–228
    national, 15–16
    and national security, 15–17
Department of Defense, 10. See also specific branches, e.g.: U.S. Navy
    directive on fitness, 19
    Force Health Protection program, 19, 22
    HIV/AIDS Prevention Program, 65
    Human Systems Integration Program, 77
    Psychopharmacology Demonstration Project, 159, 160, 165–166, 180, 244
    Risk Rule, 212
    role of, during disasters/emergencies, 11
    and sexual harassment policies, 219
    Worldwide Casualty System, 23
    Youth Attitude Tracking Survey, 189
Department of Homeland Security (DHS), 3, 12
    and FEMA, 11
    National Disaster Medical System, 12
Department of Human Behavioral Biology, 78
Department of Military Psychophysiology, 78
Department of the Navy Suicide Incident Report (DONSIR), 59
Department of Veteran Affairs (VA), 245
    and combat stress reports, 150
    and PTSD, 149
Deployment
    effect on military families, 228–230
    mental health issues affecting, 245
Desegregation (term), 200
Desert Shield. See Operation Desert Shield
Desert Storm. See Operation Desert Storm
Dewey, Thomas, 204
DHS. See Department of Homeland Security
Diamox, 69
Digitization, 241–242
Dimensional analysis, 114
Direct action, 86
Directed Energy Bioeffects Laboratory (San Antonio, Texas), 56
Disaster relief, 11
Disaster Relief Act, 11

Disasters, 11

Disease prevention, 181

Display for Advanced Research and Training (DART), 120

Disruption (term), 80

Disruption, weapons of mass, 79–80

Distance learning, 190

Distance training, 135

Distributed Mission Training, 120–121

Distributed Mission Training Centers (MTCs), 121

Distribution modeling, 130

Donovan, William, 84

DONSIR (Department of the Navy Suicide Incident Report), 59

"Don't ask, don't tell" policy, 221

Dunlap, Knight, 30

Durant, Michael, 90

Earthquakes, 11

Edgewood, Maryland, chemical threats research at, 79

Edison, Thomas, 47

Education. See Training

Edward, Idwa, 205

Efficiency ratio, 120

Eglin Air Force Base (Florida), 121

Ehime Maru, 66

Eisenhower, Dwight, 105, 206

Elmendorf Air Force Base (Alaska)
AWAS MTCs at, 121
F-15C MTCs at, 121

Emergency Committee on Psychology, 37, 38

Employment, of military spouses, 230–231

Engineering design, for systems environments, 47–48

Engineering psychology. See Human factors engineering

Enlisted personnel
classification of, 103–105, 132–133
selection of, 100–103, 132
training of, 105–108

Environmental change, 162–163

Environmental Health Effects Laboratory (Dayton, Ohio), 56

Ergonomics. See Human factors engineering

Exploring the Limits in Personnel Selection and Classification, 103–104

Extreme environments, 68–69

Families. See Military families; related topics, e.g.: Marriage

Family advocacy, 232

Family support services, 233

Family violence, 232

Fatigue, 66–68

Federal Bureau of Investigation (FBI), 84

Federal Emergency Management Agency (FEMA), 11–12

Fighters (behavioral classification), 101–102

1st Special Service Force, 83

First International Symposium on Military Psychology, 43

First Marine Division, 149

Fiske, Donald W., 85

Fit to Win—Substance Abuse Prevention, 19

Fitts, Paul M., 111–112
at the Aero-Medical Laboratory, 31
and German intuitive assessment, 33

Five-factor model, 114

5307th Composite Unit (Merrill's Marauders), 83

Flanagan, John C., 32, 34, 39

Fleet Surgical Teams, 146

Flight crews, human performance in, 49–51

Flight simulators, 119–120

Force Health Protection, 13–14

Foreign internal defense, 86

Fort Bragg (North Carolina)
Psychological Warfare Center at, 86
Special Forces candidates at, 87, 88
Special Warfare Center. See Special Warfare Center
Special Warfare School at, 87

Fort Leavenworth (Kansas)
Center for Army Lessons Learned, 242
military prison at, 157

Fort Ord, California, 164

Fort Rucker (Alabama; U.S. Army Aeromedical Research Laboratory), 74–75

Fort Sam Houston, San Antonio, Texas, 139

44th Medical Brigade, 11

Gagné, Robert, 34

Gardener, John W., 85

Gender integration, 216–217

Gengeralli, Joseph, 85

Geographic mobility, and military families, 230

George Washington University, 39, 99, 100

Gersoni, Charles S., 154, 156

GI Bill of Rights (Veterans Benefits Improvement Act), 14, 244–245
Gillem, Alvan, 206, 207
Gillem Board, 206, 207
Global war on terrorism. *See* Terrorism, war on
Goddard, H. H., 32
Goldwater–Nichols Department of Defense Reorganization Act of 1986, 11, 239
Groton, Connecticut. *See* Naval Submarine Medical Research Laboratory
Group debriefing, 26
g-seat, 119
Guedry, Fred, 51
Guerilla warfare, 86
Gulf Wars. *See* Persian Gulf Wars

Haislip, Wade, 208
Handbook for the Development of Skill Qualification Tests, 109
Hanfmann, Eugenia, 85
Hardware / Manpower Integration (HARDMAN), 193–194
Harlow, Harry, 39, 100
Harris, J. Donald, 52
Harry G. Armstrong Aerospace Medical Research Laboratory, 31
Hazardous material incidents, 12
Health behaviors, 60
Health care
    managing, by U.S. Air Force, 179
    for military women, 221–222
Health Evaluation Assessment Review, 19
Health Professions Scholarship Program (HPSP), 172
Health promotion, 19, 59–65, 181
    alcohol abuse, 63–64
    HIV/AIDS, 64–65
    nutrition, 60–62
    obesity, 59–62
    smoking, 62–63
Health research, in U.S. Navy, 55–70
    in adjustment to military life/culture, 56–58
    and health promotion, 59–65
    and human factors engineering, 65–69
Health Risk Appraisal, 19
Healthy People, 19, 243
HEL (Human Engineering Laboratory; Aberdeen Proving Ground, Maryland), 72–73
Helicopter pilot workload issues, 74–75

Historical debriefing, 26
HIV/AIDS, 64–65
HIV/AIDS Prevention Program, 65
Holley, Helen, 145
Homeland Security, Department of, 3
Homeland Security Act of 2002, 12, 248
Homosexuality, 220–221
Honolulu, Hawaii (Tripler Army Medical Center), 143
Hovland, Carl, 35
Hoyt, Gary, 149
HPSP (Health Professions Scholarship Program), 172
Human engineering. *See* Human factors engineering
Human Engineering Laboratory (HEL; Aberdeen Proving Ground, Maryland), 72–73
Human factors engineering, 20–21, 43–54, 71–81
    and AAF Aviation Psychology Program, 111–112
    and ARI, 193–194
    and Army Research Institute for the Behavioral and Social Sciences, 18
    and contract research, 46
    and coping with extreme stress, 77–79
    and health research, in U.S. Navy, 65–69
    and human factors engineering, 72–73
    and human performance research, in U.S. Army, 72–73
    laboratories conducting, 73–77
    and military performance, 18
    and National Academy of Sciences, 18
    NPRDC research into, 136
    and ONR, 43–47
    and system environments research, 47–53
    in U.S. Air Force, 116–118, 170
    and weapons of mass disruption, 79–80
    and World War II, 31
Human Goals Program, 126
Human performance
    in naval aviation and flight crews, 49–51
    in submariners and Navy divers, 51–53
Human Research and Engineering Directorate, 73n1
Human resource management, 13
Human Resource Management program (Navy), 133

Human Resources Directorate of the Air Force Research Laboratory, 33
Human Resources Research Office (HumRRO), 39, 99–110
   and classification of enlisted personnel, 103–105
   development of, 99–100
   Fighter/non-Fighter study in Korea, 101–102
   Project A, 103–104
   and selection of troops, 100–103
   and SQTs, 108–109
   and training of enlisted personnel, 105–108
Human Systems Department, 46
Human systems integration. *See* Human factors engineering
Human Systems Integration Program, 77
HumRRO. *See* Human Resources Research Office
Hurricanes, 11

ICAI (intelligent computer-assisted instruction) system, 122
ICS (Incident Command System), 12
Improved Performance Research Integration Tool (IMPRINT), 194
Incident Command System (ICS), 12
Individual differences, 30–33, 41
   in intelligence and aptitude, 31–33
   in performance abilities, 30–31
   in personality, 33
Inouye, Daniel, 159
Instruction technology/design, 33–34. *See also* Training
Integration, in U.S. armed forces
   desegregation (term) vs., 200
   gender, 216–217
   racial, 199–210
Intelligence, 31–33
Intelligence testing, 31
Intelligent computer-assisted instruction (ICAI) system, 122
Intelligent Computer Assisted Training Testbed, 123
Intelligent tutoring systems (ITSs), 122–124
Internal defense, foreign, 86
International Council of Psychologists, 37
Internet, 242–243
Intersociety Constitutional Convention of Psychologists, 38
Iran hostage crisis, 84

Iraq
   mental health assessment of troops in, 149–150
   psychiatrists deployed into, 149
   Reserves deployed to, 243
Irritable heart syndrome, 19
ITSs (intelligent tutoring systems), 122–123

Jacobs, Owen, 191–192
Jaques, Elliot, 192
Jenkins, John G., 38
Jenkins, William, 31
Job-Oriented Basic Skills (JOBS) training, 134
Job performance, 188
JOBS (Job-Oriented Basic Skills) training, 134
Job task analysis (U.S. Air Force), 115
John F. Kennedy Special Warfare Center and School (SWCS), 87, 90
Joint Optical Information Network, 189
Joint Personnel Recovery Agency, 92
Joint-Service Job Performance Measurement Project, 104
Jones, David, 145–146

Kennedy, John F., 86
Kinney, Jo Ann S., 51
Korean War
   Fighter/non-Fighter study in, 101–102
   racial integration of U.S. armed forces during, 207–209
   and U.S. Army Senior Psychology Student Program, 155, 157

Lackland Air Force Base (San Antonio, Texas)
   AFPTRC at, 112
   Airman Classification Battery administered at, 113
   Basic Military Training program at, 170
   Personnel Research Laboratory at, 112
   Wilford Hall Medical Center. *See* Wilford Hall Medical Center
LAMP. *See* Learning Abilities Measurement Program
Langley Air Force Base (Virginia), 121
Leadership and Motivation Technical Area, 192
Leadership development, 191–192
Learning Abilities Measurement Program (LAMP), 113, 114

Learning theories, 34
Lehman, John, 127
Lesbian discrimination, 220–221
Letterman Army Medical Center (San Francisco, California), 156–157
Levant, Ronald F., 247
Levy-Leboyer, C., 53
Lewin, Kurt, 35, 53
Life vests, color redesign of, 141
Link, Ed, 119
Link GAT-1, 120
Link instrument flight trainer, 119, 120
Lippitt, Ronald, 35
Lockbourne Air Force Base (Ohio), 205
Lodge Act, 86
Loma Prieta earthquake, 11
Louttit, Chauncey M., 38

MacArthur, Douglas, 208
MacKinnon, Donald W., 85
Macmillan, John, 45
Malcolm Grow Medical Center (Andrews Air Force Base, Maryland), 174, 180
Manpower, NPRDC research into, 130–131
Manpower and Personnel Resource Laboratory, 187
Manpower modeling, 130
MANPRINT. See Army Manpower and Personnel Integration
Manual of the Medical Department of the U.S. Navy, 57
Marburger, John, 29
Marine Corps. See U.S. Marine Corps
Marine Corps Mountain Warfare Training Center (Bridgeport, California), 69
Marr, Jack, 205
Marriage, 226–227
    and employment opportunities, 231
    military rates vs. civilian, 227
    rates of American Blacks, 228
    and women, 214–215
Marsh, John, 188
Marshall, George, 206
Mass disruption, weapons of, 79–80
McCloy Committee, 205
McNamara, Robert, 106, 186
Mechanical Aptitude Test, 186
Medicaid, 165
Medical Neuropsychiatric Research Unit (U.S. Navy), 55–56
Medical Service Corps, 142
Medicare, 165

Melton, Arthur W., 43, 112
Mental environment of soldiers, 161
Mental health, xiii, 58–59, 243–246
Mental Health Advisory Team, 149–150
Mental health care, 180
Mental Hygiene Consultation Service (MHCS), 161–162
Mental illness, 18–20
Merrill's Marauders (5307th Composite Unit), 83
MHCS (Mental Hygiene Consultation Service), 161–162
Microcomputer Intelligence for Technical Training (MITT) Writer, 124
MILES (Multiple Integrated Laser Engagement System), 191
Military. See U.S. armed forces
Military Academy. See U.S. Military Academy
Military demographics, 16
Military families, 225–234
    Army clinician's care of, 161
    demography of, 226–228
    effect of geographic mobility on, 230
    family violence in, 232
    separations/reunions in, 228–230
    and spousal employment, 230–231
    support services for, 233
Military Health System, 244
Military Language Tutor (MILT), 190
Military psychologists, 97
Military psychology
    changes after Vietnam War, 157
    contributions of Army clinical psychologists to, 160–166
Military treatment facilities (MTFs), 142
Miller, George A., xiii–xiv, 247
Miller, J. G., 36
Miller, James G., 85
Miller, Merle, 203–204n2
MILT (Military Language Tutor), 190
Minnesota Multiphasic Personality Inventory (MMPI), 88
Minorities
    and racial integration of U.S. armed forces, 199–210
    research on, in U.S. Army, 194–195
MITT (Microcomputer Intelligence for Technical Training) Writer, 124
MMPI (Minnesota Multiphasic Personality Inventory), 88
Modeling Cost and Performance for Military Enlistment, 104

Moe, Karl (Skip), 151n1
Montague, Ernest K., 154
Morse code, 34
Motherhood, 215, 226
Mountain Home Air Force Base (Idaho), 121
Mowrer, O. H., 85
MTCs (Distributed Mission Training Centers), 121
MTFs (military treatment facilities), 142
Multiple Integrated Laser Engagement System (MILES), 191
Murray, Henry, 85
Myers, Charles S., 24

Napier, Helen, 145
Natick, Massachusetts (U.S. Army Research Institute for Environmental Medicine), 73–76
National Academy of Sciences, 12, 18, 35
National Center for Post Traumatic Stress Disorder, 245, 246
National Center on Disaster Psychology and Terrorism, 248
National Commission on Terrorist Attacks Upon the United States (9/11 Commission), 239
National Council of Women Psychologists, 37
National Dairy Council's Nutrition Achievement Test, 61
National Defense Research Council, 47
National Defense University (Washington, DC), 158
National demographics, 15–16
National Disaster Medical System (NDMS), 12
National Fire Academy, 12
National Guard, 10
National Health Information Technology Coordinator, xiv
National Institute for Occupational Safety and Health (NIOSH), 22
National Institute of Mental Health (NIMH), 12, 36
    counseling training programs developed by, 139
    and intervention guidelines, 26–27
National Institutes of Health (NIH), 12, 44, 65
National military strategy (NMS), 11
National Naval Medical Center (Bethesda, Maryland), 142, 144–145

National Naval Medical Center internships, 142
National Research Council Committee on Techniques for the Enhancement of Human Performance, 193
National Response Plan, 12
National Science Foundation, 44
National security, 3, 7, 9–27
    APA's response to challenges of, 247–248
    challenges in, 13–15, 239–247
    and demographics, 15–17
    organizations dedicated to, 10–13
    and personnel needs of the U.S. armed forces, 17–21
    and psychology's contribution to, 22–27
    role of U.S. Air Force clinical psychology in, 176
National Security Act of 1947, 10
National Security Council, 10
National security psychology, 29–40
    and applied social psychology, 34–35
    and clinical and counseling psychology, 35–37
    and individual differences, 30–33
    and instruction technology/design, 33–34
    and professional organizations, 37–39
National Security Strategy, 11, 239
National Training Centers, 21, 191
National Vietnam Veterans Readjustment Study, 246
NATO. See North Atlantic Treaty Organization
Natural disasters, 11
Naval Academy, 132
Naval Aerospace Medical Research Laboratory (Pensacola, Florida), 56
Naval Air Warfare Center (Orlando, Florida), 127
Naval aviation, 49–51
Naval Health Research Center (NHRC), 55–56
Naval Health Research Center Laboratory (San Diego, California), 56
Naval Medical Research Institute (NMRI), 52–53
Naval Ocean Systems Command (San Diego, California), 127
Naval Research Laboratory (NRL), 47
Naval Reserve Officers Training Corps, 132

Naval Submarine Medical Research Laboratory (Groton, Connecticut), 52, 56, 57
Naval Training Devices Center (NTDC), 48
Navy. *See* U.S. Navy
Navy Basic Test Battery, 132
Navy Clinical Psychology, 151n2
Navy Department in-house laboratories, 46–47
Navy divers, 51–53
Navy Medical Neuropsychiatric Research Unit, 55–56
Navy Nurse Corps, 212
Navy Personnel and Training Research Laboratory (San Diego, California), 126
Navy Personnel Research, Studies and Technology Division, 127–128
Navy Personnel Research and Development Center (NPRDC), 125–138
  and CAT, 128
  history/development of, 125–128
  and human factors engineering, 136
  and manpower research, 130–131
  and personnel research, 131–134
  R & D projects of, 128–129
  and training, 134–136
Navy Personnel Research Field Academy, 125–126
Navy Science Assistance Program (NSAP), 129
Navy Sea–Air–Land (SEAL) teams, 144
Navy Special Psychiatric Rapid Intervention Teams (SPRINT), 26
Nazis, 33
NDMS (National Disaster Medical System), 12
Nelson, Paul D., 43n
Neuropsychiatric casualties, 25
Neuropsychiatric disorders, 36
Neuropsychiatric hospitals, 24
New age technologies, 192–193
New Recruit Survey, 188
NHRC. *See* Naval Health Research Center
NIH. *See* National Institutes of Health
NIMH. *See* National Institute of Mental Health
9/11 Commission (see National Commission on Terrorist Attacks Upon the United States), 239
99th Fighter Group, 203
NIOSH (National Institute for Occupational Safety and Health), 22

NMRI (Naval Medical Research Institute), 52–53
NMS (national military strategy), 11
Non-Fighters, 101–102
Nonmilitary psychologists, 249–250
North African theater (World War II), 25
North Atlantic Treaty Organization (NATO), 53
Nostalgia syndrome, 19
NPRDC. *See* Navy Personnel Research and Development Center
NRL (Naval Research Laboratory), 47
NSAP (Navy Science Assistance Program), 129
NTDC (Naval Training Devices Center), 48
Nutrition, 60–62, 222

Obesity, 59–62
Obsessive–compulsive disorders, 58, 59
Occupational analyses, 125
Occupational health, 22
Occupational Health and Safety Act (OSHA), 22
Occupational health psychology, 22
Occupational medicine research, 74
Office of Naval Research (ONR), 21, 43–47
Officer selection, 131–132
"Oh-So-Social," 85
Okinawa, Japan
  alcohol abuse study conducted in, 64
  Third Marine Division at, 146
160th Special Operations Aviation Regiment (SOAR), 84, 90–91
One-on-one tutoring, 122
ONR. *See* Office of Naval Research
ONR Psychological Sciences Division, 46
Operational behavioral health, 144–145
Operational psychology, 144–145
Operational Requirements Document, 66
Operational Stress Control and Readiness (OSCAR), 146–149
Operational stress reactions, 23–25
Operation Desert Shield
  preventative medical measures taken at, 26
  stress control during, 26
Operation Desert Storm, 20, 26
Orlando, Florida (Naval Air Warfare Center), 127
OSCAR (Operational Stress Control and Readiness), 146–149

during Korean War, 207–209
study of, in 1970s, 195
in U.S. Air Force, 203–205
in U.S. Army, 208
in U.S. Marine Corps, 201–203, 208
during World War I, 203
during World War II, 203–207
Racism, in military, 202
Rand Corporation, 21
Ranger Orientation Program, 93
Ranger Regiment, 93–94
Rangers. *See* U.S. Army Rangers
Rapid ITS Development Shell (RIDES), 124
Rating Scale for Selecting Captains, 32
Rating Scale for Selecting Salesmen, 32
RDC (Recruiter Development Center), 189
Reaction time, 30
Reagan, Ronald, 188
Realistic job preview, 189
Realistic Training (REALTRAIN), 191
Recruiter Development Center (RDC), 189
Recruiting, 14, 188–190
　　Air Force clinical psychology for, 171–173
　　for Army clinical psychologists, 160
　　market analysis for, 188–189
　　psychology clinicians for U.S. Air Force, 171–176
　　recruiter selection, training, and utilization for, 189
　　research by ARI, 188–190
　　sales aids for, 189–190
Recruiting systems (Navy), 131
Red Cross. *See* American Red Cross
Red Cross Service Centers, 178
Reeves, James, 149
Research Center for Group Dynamics, 35
"Resilience in a Time of War," 5, 235
Resilience Initiative, 5, 235
RIDES (Rapid ITS Development Shell), 124
Ridgway, Mathew, 208
Rifle shooting, 105–106
Risk Rule, 212
Robert T. Stafford Disaster Relief and Emergency Assistance Act, 11
Roosevelt, Franklin D.
　　and Black American appeal, 203–204n2
　　support for racial integration, during World War II, 203
Royal Army Medical Corps, 24
Russo–Japanese War, 19, 23, 24

SA (situational awareness), 66
Sailors Health Inventory Program questionnaire, 57
Salmon, Thomas, 24–25
San Antonio, Texas
　　AFHRL at, 112
　　AFPTRC at, 112
　　Armstrong Laboratory at. *See* Armstrong Laboratory
　　Army medical training center in, 157
　　Brooks Air Force Base at. *See* Brooks Air Force Base
　　Directed Energy Bioeffects Laboratory at, 56
　　Fort Sam Houston at, 139
　　Lackland Air Force Base at. *See* Lackland Air Force Base
　　School of Aerospace Medicine at, 170
San Diego, California
　　Basic Electricity and Electronics School at, 134
　　Naval Health Research Center Laboratory in, 56
　　Naval Ocean Systems Command at, 127
　　Navy Personnel Research and Development Center at, 125–138
　　Navy Personnel and Training Research Laboratory at, 126
　　and occupational analyses, 125
　　training programs established at, 142
San Francisco, California (Letterman Army Medical Center), 156–157
Sanitary Corps, 17–18, 238
Sarvis, Howard C., 105–106
Schizophrenia, 58
Schmidt-Nielson, Astrid, 47
School of Aerospace Medicine, 170
Schoomaker, Peter J., 238
SCOPES (Squad Combat Operations Exercise Simulation), 191
Scott, Walter Dill, 32
SEALAB, 53
SEAL (Navy Sea–Air–Land) teams, 144
SEAPRINT (Systems Engineering Personnel Integration), 77
Second Marine Division, 148
Selection. *See also* Classification
　　by HumRRO, 100–103
　　psychological adjustments due to, 56–58
　　into Special Forces, 86–90
　　in U.S. Air Force, 113–115

U.S. Air Force. *See also under* Air Force
    development/history of, 111–112
    health care of, 177–181
    and human engineering, 116–118
    job task analysis in, 115
    racial integration into, 203–205
    selection/classification of enlisted personnel in, 113–115
    smoking prevention programs of, 178, 181
    and training, 118–124
U.S. Air Force School of Aerospace Medicine, 30
U.S. armed forces. *See also specific branches, e.g.:* U.S. Air Force
    adjustment to life/culture of, 56–58
    family concerns surrounding, 225–234
    history of, 10–11
    mental health of personnel in, 243–246
    psychologists in, 238–239
    public opinion of, 246–247
    racial integration of, 199–210
    reasons for joining, 14
    roles of, 14
    training in, 241–243
U.S. Army, 22. *See also under* Army; *specific topics, e.g.:* Clinical psychology (in U.S. Army)
    Army Alcohol and Drug Abuse Prevention and Control Program, 19
    assignment of enlisted personnel in, 187–188
    clinical psychologists in. *See* Army clinical psychologists
    and Fit to Win—Substance Abuse Prevention, 19
    health protection for, 22
    human performance research in, 71–81
    and HumRRO, 99–110
    racial integration into, 208
    selection/classification of enlisted personnel in, 186–187
    Special Forces. *See* U.S. Army Special Forces
    and special operations psychology, 83–94
    and stress control, 26
    training systems, 34
    U.S. Special Operations Forces, 84
    War Department. *See* War Department
    women in, 213
    during World War I, 16

U.S. Army Aeromedical Research Laboratory (USAARL; Fort Rucker, Alabama), 74–75
U.S. Army Air Corps, 31, 201–203
U.S. Army Center for Health Promotion and Preventive Health, 76
U.S. Army General Classification Test (AGCT), 32
U.S. Army Medical Research Institute of Chemical Defense (USAMRICD), 79
U.S. Army Medical Research Institute of Infectious Diseases (USAMRIID), 80
U.S. Army Rangers
    Merrill's Marauders (5307th Composite Unit), 83
U.S. Army Recruiter School, 189
U.S. Army Recruiting Command (USAREC), 187, 188
U.S. Army Research Institute for Environmental Medicine (USARIEM; Natick, Massachusetts), 73–76
U.S. Army Research Institute for the Behavioral and Social Sciences (ARI), 21, 32, 183, 185–197
    and Army Family Research Program, 195
    and human factors and ergonomics, 193–194
    and human factors engineering, 18
    and leadership/leader development, 191–192
    new age technologies used by, 192–193
    recruiting research for, 188–190
    and research into minorities in U.S. Army, 194–195
    and research into women in U.S. Army, 195
    and soldier assignment, 187–188
    and soldier selection, 186–187
    and training research, 190–191
    and U.S. Army Special Forces research, 195–196
U.S. Army Senior Psychology Student Program (SPSP), 154–155, 157
U.S. Army Signal Corps Aviation Section, 30
U.S. Army Special Forces, 86. *See also under* Special Forces; Special Operations
    acute mountain sickness among, 68–69
    assessment/selection of, 86–90

attrition of, 87
candidates at Fort Bragg, 87, 88
research on, by ARI, 195–196
role of, during federally declared disasters, 11
Special Warfare School, 87
U.S. Navy psychology and, 144–145
in Vietnam, 83n1, 87
in Vietnam War, 86
U.S. Army Special Operations Command (USASOC), 94
U.S. Census, 15
U.S. Constitution, 9
U.S. Marine Corps, 147
behavioral health services for, 146
exclusion of women in some units of, 217
and health services, 147, 148
lesbian discharges in, 220–221
Marine Corps Mountain Warfare Training Center, 69
racial integration into, 201–203, 208
women in, 213
U.S. military. *See* U.S. armed forces
U.S. Military Academy (West Point, New York), 157, 217
U.S. National Guard, 10
U.S. Navy. *See also under Naval; specific topics, e.g.:* Clinical psychology, in U.S. Navy
clinical psychology in, 141–151
health research conducted by. *See* Health research, in U.S. Navy
Office of Naval Research, 21
personnel policies of, 126
Personnel Research and Development Center of. *See* Navy Personnel Research and Development Center
suicide rates in, 147
U.S. Navy Medical Neuropsychiatric Research Unit, 55–56
U.S. Navy psychology, and U.S. Army Special Forces, 144–145
U.S. Public Health Service (USPHS), 36, 139
U.S. Special Operations Forces (SOF), 84
USAARL (U.S. Army Aeromedical Research Laboratory; Fort Rucker, Alabama), 74–75
USAMRICD (U.S. Army Medical Research Institute of Chemical Defense), 79
USAMRIID (U.S. Army Medical Research Institute of Infectious Diseases), 80

USAREC. *See* U.S. Army Recruiting Command
USARIEM (U.S. Army Research Institute for Environmental Medicine; Natick, Massachusetts), 73–76
"Uses of Negro Manpower in War," 202
USNS *Comfort*, 26
USS *Eisenhower*, 212
USS *Enterprise*, 145–146
USS *Greenville*, 66
USS *John Kennedy*, 146
USS *Kittyhawk*, 145
USS *Stark*, 26
USS *Virginia*, 66

VA. *See* Department of Veteran Affairs
VA hospitals. *See* Veterans Administration hospitals
VCASS (Visually Coupled Airborne System Simulatory) project, 118
Veterans Administration (VA) hospitals, 36, 139
Veterans Benefits Improvement Act (GI Bill of Rights), 14, 244–245
Video teletraining (VTT) research laboratory, 135
Vietnam War
changes to military psychology following, 157
military personnel issues after, 16
and posttraumatic stress disorder, 149
U.S. Army Special Forces in, 83n1, 86, 87
Violence, in military families, 232
Visually Coupled Airborne System Simulatory (VCASS) project, 118
VTT (video teletraining) research laboratory, 135

Walter Reed Army Institute of Research (WRAIR; Washington, DC), 77–79
and combat stress reports, 150
Department of Military Psychophysiology, 78
Division of Neuropsychiatry, 77
Performance Assessment Battery, 78
War Department, 10, 35
and Army Alpha and Beta tests, 32
health protection for employees of, 22
Warfare, 20, 86
War neuroses, 24, 25

War on terrorism. *See* Terrorism, war on
Warrior Knowledge Network, 242
Washington, DC
  National Defense University at, 158
  Personnel Research and Development
    Laboratory at, 126
  WRAIR at. *See* Walter Reed Army In-
    stitute of Research
Watson, John B., 30
Weapons of mass disruption, 79–80
Webb, James H., 127
Weight control, 60–62, 222
West Point, New York (U.S. Military Acad-
    emy), 157, 217
White, Ralph, 35
Whole-body scan, 116
Wickham, John, Jr., 195
Wilford Hall Medical Center (Lackland Air
    Force Base, San Antonio, Texas),
    173, 174
Wilkin, Wendell R., 154
Witmer, Lightner, 141
Women (military), 37, 211–223
  ASVAB disadvantage of, 217
  attitudes surrounding, 145, 214
  exclusion of, in Project 100,000, 106–
    107
  exclusion of, in Special Forces, 87
  and gender integration, 216–217
  and health care, 221–222
  history of, 212
  and lesbian discrimination, 220–221
  in Navy clinical psychology, 143–144
  parasuicide attempts of, 59
  reasons for leaving (Army), 218–219
  role in national security, 212–213
  and sexual harassment, 219–220
  stresses of, 222
  and "Test of Women Content in Units,"
    195
  in U.S. Army, 195
  work–family concerns of, 214–216
  workplace discrimination of, 217–221
Wonderlic Personnel Test, 88
Woodworth, Robert S., 33
Work–family concerns, 214–216
Workplace accommodation, 116–118
Workplace discrimination, of women in mili-
    tary, 217–221

World Trade Center, 29. *See also* September
    11, 2001 terrorist attacks
World War I, 3, 16
  military personnel issues during and fol-
    lowing, 16, 17
  military training during, 20
  national security researchers before, 41
  personality predictors for, 33
  psychiatric casualties of, 23–25
  racial integration of U.S. armed forces
    during, 203
  *Studies in Social Psychology in World War
    II*, 35
World War II, 3
  applied social psychological research
    following, 183
  military personnel issues during and fol-
    lowing, 16
  military training during, 20
  national security researchers before, 41
  obtaining Army clinical psychologists
    after, 153–157
  personality predictors for, 33
  psychiatric casualties of, 25
  racial integration of U.S. armed forces
    during, 203–207
  role of Army clinical psychologists dur-
    ing, 158
  *Studies in Social Psychology in World War
    II* (book series), 21, 35, 183
  U.S. Army tests during, 18
Worldwide Casualty System, 23
WRAIR. *See* Walter Reed Army Institute of
    Research
WRAIR Division of Neuropsychiatry, 77
WRAIR Performance Assessment Battery, 78
Wright Patterson Air Force Base (Dayton,
    Ohio), 56. *See also* Armstrong Labo-
    ratory
Wright–Patterson Medical Center (Wright–
    Patterson Air Force Base, Ohio), 174

Yerkes, Robert M., 17, 31–32, 235, 247
Young, Charles, 202
Youth Attitude Tracking Survey, 189

Zimbardo, P. G., 247
Zuckert, Eugene, 205
Zumwalt, Elmo, 126

ABOUT THE EDITOR

A. David Mangelsdorff, PhD, MPH, has been a professor and civilian health psychologist with the U.S. Army–Baylor University Graduate Program in Health and Business Administration at the U.S. Army Academy of Health Sciences at Fort Sam Houston in San Antonio, Texas, since 1972. He attended Dartmouth College, in Hanover, New Hampshire; conducted research with W. Lawrence Gulick; and after graduation was commissioned an ensign in the U.S. Naval Reserve, served clinical research clerkships, and conducted psychophysiological research at the Submarine Medical Research Laboratory in Groton, Connecticut, with Benjamin Weybrew. He also studied English literature at Oxford University, Oxford, England.

He earned master's and doctoral degrees in psychology from the University of Delaware in Newark under the direction of Marvin Zuckerman. He has taught drug and alcohol education at the Medical Field Service School, San Antonio, Texas, as a uniformed psychologist and later served as technical director for the Health Care Studies and Clinical Investigation Activity at Fort Sam Houston, San Antonio, Texas. He also earned a master's degree in English from St. Mary's University, San Antonio, Texas, and in public health from the University of Texas School of Public Health, San Antonio.

Dr. Mangelsdorff retired as a colonel from the Army Reserves in 2000. He is active in the leadership of the Society for Military Psychology of the American Psychological Association (APA) and is an energetic writer on military psychology, operational stress, terrorism, patient attitudes, professional retention factors, psychological support, and curriculum outcomes. He has served as adjunct editor for the journal *Military Psychology* and on the editorial review board of numerous other journals. Among his more than 500 papers, presentations, and reports, his edited works include the *Handbook of Military Psychology* (which includes an edition in Chinese), *Military Cohesion*, the U.S. Army–Baylor University Health Care Administration's 50-

year history, the military psychology section of the *Encyclopedia of Psychology*, and numerous conference and stress workshop proceedings. The present volume evolved from his interest in studying military history, stress, and occupational health; continuing his teaching and research interests toward strengthening homeland defense and national security; and enhancing performance, healthy behaviors, and resilience.

Dr. Mangelsdorff is a fellow of the APA, the American Association for the Advancement of Science, the American Psychological Society, and Sigma Xi. He won a Fulbright scholarship in 2003 to study demographics in Germany. He has taught as an adjunct instructor at San Antonio College, Texas, and as an adjunct professor at the University of Texas School of Public Health and at Texas A&M University, College Station. He served as an adviser to the Target 90 Goals for San Antonio, cochairperson of the disaster mental health response committee of the San Antonio American Red Cross, board member of the 24th Street Theatre Experiment, member of the Alamo Theatre Arts Council (ATAC) board of directors, and ATAC judge for local theater productions. He is a consultant to the Veterans Administration and is an adjunct professor in the Department of Family Medicine at the Uniformed Services University of the Health Sciences in Bethesda, Maryland.

He organized numerous combat and operational stress workshops between 1981 and 1994. He chaired the North Atlantic Treaty Organization (NATO) Research Study Group on Psychological Support, represented the United States as a consultant to the embassy of Suriname, chaired U.S. research groups for NATO and The Technical Cooperation Program, and served as program director of the German–American Officer Exchange Program. His professional recognition has included an APA Presidential Citation, Secretary of the Army Decoration for Exceptional Civilian Service, Decoration for Meritorious Civilian Service (two awards), Federal Executive Association Employee of the Year, Legion of Merit, Order of Military Medical Merit, and numerous U.S. Army–Baylor Health Care Administration Faculty Researcher of the Year awards.

His philanthropic interests have included endowing professorships that continue his research and teaching in homeland defense and national security at Dartmouth College (psychology and brain sciences), St. Mary's University (international relations), the University of Texas Health Science Center at San Antonio (Center for Public Health Preparedness and Biomedical Research), the University of Delaware, and Baylor University.

Naming Evil
Judging Evil

Edited by Ruth W. Grant

With a Foreword by Alasdair MacIntyre

The University of Chicago Press | *Chicago & London*

Ruth W. Grant is professor of political science and philosophy at Duke University and a senior fellow at the Kenan Institute for Ethics. She is the author of *John Locke's Liberalism* and *Hypocrisy and Integrity,* both published by the University of Chicago Press.

The University of Chicago Press, Chicago 60637
The University of Chicago Press, Ltd., London
© 2006 by The University of Chicago
All rights reserved. Published 2006
Printed in the United States of America

15 14 13 12 11 10 09 08 07 06   1 2 3 4 5

ISBN-13: 978-0-226-30673-5   (cloth)
ISBN-10: 0-226-30673-9   (cloth)

Library of Congress Cataloging-in-Publication Data

Naming evil, judging evil / edited by Ruth W. Grant ; with a foreword by Alasdair MacIntyre.
    p. cm.
  Revisions of papers presented at a conference held Jan. 27–29, 2005 at Duke University. Includes bibliographical references and index.
    ISBN 0-226-30673-9 (cloth : alk. paper)   1. Good and evil—Congresses.
2. Judgment (Ethics)—Congresses.   I. Grant, Ruth Weissbourd, 1951–
    BJ1401.N36 2006
    170—dc22   2006010697

∞ The paper used in this publication meets the minimum requirements of the American National Standard for Information Sciences—Permanence of Paper for Printed Library Materials, ANSI Z39.48–1992.